First International Conference

RENEWABLE ENERGY – SMALL HYDRO

SELECT PAPERS

Editors
C.V.J. Varma
A.R.G. Rao

Central Board of Irrigation and Power
New Delhi

International Association for Small Hydro
New Delhi

A.A. BALKEMA/ROTTERDAM
1997

SPONSORED BY

UNDP-GEF Hilly Hydro Project
Ministry of Non-Conventional Energy Sources
Government of India

Ministry of Science & Technology, Govt. of India

Indian Renewable Energy Development Agency Limited

Kirloskar Brothers Limited

Renewable Energy Project Support Office (REPSO)
Winrock International

Sulzer Flovel Hydro Limited

CO-SPONSORED BY

Orissa Hydro Power Corporation Ltd.

Subhash Projects and Marketing Limited

ISBN 90 5410 749 9

Printed in India

FOREWORD

Hydro Power is probably the oldest and yet the most reliable source of all renewable energy, with bulk of its potential yet to be harnessed in many countries. Though Hydro Power Development started with small units of 100 kw and more in the beginning, attention was diverted to harnessing medium and major hydro because of their comparative economics. However, the oil crunch in early seventies forced the attention of all concerned for harnessing the small hydro, at the toe of existing dams or canal drops besides exploiting hill streams, to supplement the energy generation.

Being environmentally benign and having small gestation period, harnessing of Small Hydro Resources is receiving worldwide attention both in the developed and developing countries to supplement energy generation.

Financial assistance is forthcoming from funding agencies like the World Bank, and Asian Development Bank etc. for development of small hydro especially by private developers, who would like to harness these resources for meeting mostly their own requirements.

Considering the rapid advances in technology for harnessing these resources, it is considered opportune to provide a forum for exchange of experiences to facilitate flow of technology from one country to another. It is in this context, that the Central Board of Irrigation and Power organised this conference, as a part of Hydro Centenary-1997 to commemorate 100 years of hydro generation in the country.

The Board is thankful to all the authors for their valuable contributions. The Board owes a debt of gratitude to the Chairmen, Co-Chairman and Members of the Organising Committee and the Technical Committee for their invaluable guidance in organisation of this Conference.

(C.V.J. Varma)
Member Secretary
Central Board of Irrigation and Power

CONTENTS

INTERNATIONAL SCENARIO AND COUNTRY REPORTS

RESOURCE ASSESSMENT AND PLANNING

DESIGN AND CONSTRUCTION

PLANT AND EQUIPMENT

HIMALAYAN HYDRO (UNDER THE AUSPICIOUS OF UNDP-GEF HILLY HYDRO PROJECT)

INTERNATIONAL SCENARIO AND COUNTRY
REPORTS

SMALL HYDROPOWER: A GLOBAL PERSPECTIVE

Carl Vansant

HRW-Hydro Review Worldwide Magazine, HCI Publications Inc.
410 Archibald Street, Kansas City, MO 64111, USA

ABSTRACT

The development of small hydropower around the world is on the increase. Much of the world has huge potential to further develop this resource. Small hydro offers a wide range of bene-fits -- especially for rural areas and developing countries. The resource is environmentally responsible and has substantial eco-nomic advantages. Governments, financiers, and developers are finding new ways to fund and promote small hydro development. Efforts also are being made to improve the exchange of ideas and technology related to small hydro. By addressing these issues, small hydro can improve its value worldwide and assure itself of a role in the electricity supply marketplace.

INTRODUCTION

In the hydropower industry in many parts of the world, developers, financiers, and government leaders are all discover-ing that small is beautiful. Small hydropower stations offer a wide range of benefits -- especially for rural areas and develop-ing countries. Small hydro has always been considered an en-vironmentally responsible choice for electricity generation. Today, thanks to initiatives from governments and lenders, small hydropower is also being recognized as a smart economic choice.

Worldwide, the contributions of small hydropower have grown substantially in the last ten to 15 years, and the potential for future growth is even stronger. In a recent World Energy Council (WEC) report, the agency reported that small hydro stations throughout the world contribute more than 34,000 MW, representing about 5 percent of the installed hydro capacity worldwide. In addition, small hydro plants that are under construction or planned could contribute another 23,000 MW.

The reasons for the increase in small hydropower are varied. In developing countries, small hydro offers perhaps the greatest benefits. Small hydro plants often offer the fastest way to increase electrification, improve living standards, stimulate industrial development, and enhance agriculture, without the time and expense required to expand central transmission grids into rural areas. In some countries, the development of such projects also creates a local industry of its own, to manufacture the equipment required for the plants, and to construct the schemes.

While the benefits are tremendous, small hydro historically has faced challenges -- most notably in obtaining the financing that is required to develop a project. Over the past couple of decades, developers and governments have found ways to address these challenges. Because of that, small hydro is finding its place in today's electricity generation market.

This paper will address the current state of small hydropower, projections for future development, and some of the factors that are influencing the current development of small hydro schemes, and are likely to do so in the future.

THE DEVELOPMENT OF SMALL HYDRO

A discussion of small hydropower would not be complete without a definition of "small." Countries have widely differing definitions of what constitutes small hydropower. The most common definition for small hydro is 10 MW or less, however, there are significant deviations from this. For example, in China, small hydro is considered to be anything smaller than 25 MW; the Philippines classifies small hydro as smaller than 50 MW. Because of the variations, it is difficult to develop accurate statistics on installed capacity and the potential for additional development. Because the 10-MW break is the most common, for the purposes of statistics in this paper, small hydro will be defined as 10 MW or smaller. Exceptions from that will be noted where appropriate.

While the outlook for small hydro development appears strong in many parts of the world, it is perhaps the strongest in Asia. Thanks in part to programs to encourage small hydro in China, Japan, and India, installed generating capacity from small hydro plants in Asia accounts for more than 19,000 MW -- more than half of the worldwide installed capacity. In China alone, small hydro contributes more than 15,000 MW to the grid, which equals about a fourth of the total hydroelectric generating capacity in the country.

Many countries in Europe, also, have successfully exploited their small hydro capabilities. In Europe, small hydro contributes nearly 9,000 MW to the grid. France and Spain each have more than 1,000 MW of installed capacity from small hydro; in Italy, nearly 2,000 MW can be attributed to small hydro plants. Table 1 shows a breakdown of installed capacity by country and region of the world.

Table 1: Small hydro installed capacity worldwide (in MW)

Country	MW	Country	MW	Country	MW
ASIA		**AFRICA**		**EUROPE**	
Bhutan	5.7	Angola (<2)	11.9	Austria (<5)	1,930.0
China (<25)	15,055.0	Burundi (<1)	4.0	Belarus	7.0
India (<15)	243.0	Cameroon	2.0	Belgium (<12)	26.0
Japan	3,329.0	Comoros	2.0	Bosnia-Herzegovina	20.0
Korea	38.7	Cote d'Ivoire	5.0	Bulgaria	15⁵.0
Mongolia	0.5	Equatorial Guinea	1.0	Croatia	26.6
Nepal	15.0	Ethiopia	6.1	Czech Republic	197.0
Sri Lanka	22.5	Gabon	160.0	Denmark	9.5
Taiwan	75.0	Guinea	14.2	Faroe Islands (<8)	17.5
Total: Asia	*18,784.4*	Kenya	13.6	Finland	420.0
		Lesotho	2.2	France	3,000.0
LATIN AMERICA		Liberia	0.1	Germany	327.0
Argentina	84.0	Madagascar	26.1	Greece	42.7
Bolivia	104.0	Malawi	2.0	Hungary	12.4
Brazil	928.0	Mali	6.0	Iceland	51.0
Chile	11.0	Mauritius	24.0	Ireland	15.8
Colombia	100.0	Morocco	19.8	Italy	2,344.0
Costa Rica	83.9	Mozambique	2.8	Latvia	0.8
Dominica	14.0	Namibia	0.1	Lithuania	5.3
Dominican Republic	2.5	Nigeria (<5)	32.0	Luxembourg	17.0
Ecuador	26.0	Rwanda	6.0	Netherlands	4.1
El Salvador	18.0	Sao Tome & Principe	2.0	Norway	1,048.0
Falkland Islands	2.0	Sierra Leone	4.0	Poland (<5)	250.0
Guadeloupe	10.0	Somalia	4.8	Portugal	69.0
Guatemala	4.9	South Africa	21.2	Romania (<4)	401.0
Guyana	2.0	Swaziland	9.0	Russia	35.1
Haiti	9.2	Tanzania	14.9	Serbia	443.0
Honduras	1.2	Togo	4.0	Slovakia	36.3
Jamaica	26.7	Tunisia	15.8	Slovenia	88.0
Mexico	76.0	Uganda	5.0	Spain (<5)	1,200.0
Nicaragua	5.3	Zaire	69.8	Sweden (<1.5)	320.0
Panama	7.8	Zambia (<1)	14.8	Switzerland	1,187.0
Peru	50.0	*Total Africa*	*506.1*	Turkey	152.0
St. Vincent	6.0			United Kingdom	102.0
Venezuela	11.3	**OCEANIA**		*Total: Europe*	*13,959.9*
Total: Latin America	*1,583.8*	Australia	45.0		
		Fiji (<2)	0.9	**MIDDLE EAST/SOUTHWEST ASIA**	
SOUTHEAST ASIA		French Polynesia	10.0	Afghanistan	7.0
Cambodia	10.0	New Zealand	116.7	Cyprus	0.7
Indonesia	67.2	Pacific Islands	10.0	Iran	5.0
Laos	5.0	Western Samoa	11.6	Israel	7.1
Malaysia (<2)	12.4	*Total Oceania*	*194.2*	Jordan	8.0
Papua New Guinea (<5)	22.3			Kyrgyzstan	40.0
Philippines (<50)	95.5	**NORTH AMERICA**		Pakistan	6.1
Thailand	48.0	Canada	897.0	Uzbekistan	195.0
Vietnam	3.7	United States	4,198.0	*Total Middle East*	*268.9*
Total: Southeast Asia	*264.0*	*Total: North America*	*5,095.0*	**WORLD TOTALS:**	**40,656.3**

Note: If small hydro is defined as something other than 10 MW or smaller, that number is indicated following the country name.

Source: Data comes from the World Energy Council and from on-going tracking of project development by HCI Publications.

5

In addition to leading the world in the amount of installed capacity, Asia is also the region that could see the most small hydro development during the coming years. If all of the schemes that are planned or under construction are developed, Asia could add some 18,000 MW -- nearly double the currently installed small hydro capacity. Again, activities in China, Japan, and India are largely responsible for the future growth in the region.

Africa also shows great potential for additional capacity from small hydro, especially when compared to current installed capacity. Currently, Africa has about 500 MW of capacity that can be attributed to small hydropower schemes. In comparison, more than 1,300 MW is either under construction or planned. Europe also has the potential to add a significant amount of hydropower. Russia alone has nearly 800 MW planned and under construction -- nearly a fourth of the 3,100 MW that could be added in Europe. Table 2 shows a breakdown of capacity that is under construction and planned by country and region of the world.

As is the case with any type of development, the projects that are planned will be phased in over time. As various factors fall into place, new plants will be developed. Among those factors that influence development, perhaps the most significant are the demand or need for the electricity, and the required funding to finance the power plants. In mountainous regions especially, huge potential for small hydro development has been identified. However, without a realistic demand and the funding to build the plants, such sites will not be developed.

Using a formula that measures demand and potential, the International Energy Agency has predicted that by 2005, an additional 8,000 MW of small hydro capacity could be in service throughout the world; by the year 2010, small hydro capacity could total more than 42,000 MW; and by 2020, that number could reach 65,000 MW.

In order for that capacity to be developed, several factors must fall into place. A couple of those factors have already been identified: demand and financing. In addition, other factors that are influencing the development of small hydro include government efforts to enable development through new laws and regulations that provide protections to private investors, and a push to share technologies, ideas, and expertise.

THE FACTORS INFLUENCING SMALL HYDROPOWER DEVELOPMENT

The first challenge that small hydro developers must overcome is finding a suitable site for a small hydro plant, which supports a financable project. When financiers look at a proposed project, they will study several areas:

• Is the project technically, environmentally, and legally sound? Are engineering studies complete and valid? Have environmental studies been completed? Are environmental and other government permits all in place?

Table 2: Small hydro under construction or planned worldwide (in MW)

Country	MW	Country	MW	Country	MW
ASIA		**AFRICA**		**EUROPE**	
China (<25)	10,000.0	Angola (<2)	16.7	Austria (<5)	210.0
India (<15)	800.0	Burkina Faso	14.6	Bosnia-Herzegovina	4.0
Japan	6,842.0	Cote d'Ivoire	31.9	Bulgaria	24.0
Korea	30.8	Ethiopia	11.1	Croatia	50.8
Mongolia	2.0	Gabon	5.9	Finland	3.0
Nepal	14.5	Guinea	4.6	Germany	2.6
Total: Asia	*17,689.3*	Lesotho	0.2	Greece	36.0
		Madagascar	11.1	Ireland	23.1
LATIN AMERICA		Morocco	23.0	Italy	220.0
Argentina	9.0	Mozambique	1.0	Latvia	3.0
Bolivia	57.0	Nigeria (<5)	1,100.0	Luxembourg	5.0
Brazil	643.0	Sierra Leone	2.0	Norway	50.0
Colombia	35.0	Swaziland	0.3	Poland (<5)	218.0
Costa Rica	73.9	Tanzania	72.0	Portugal	10.0
Dominica	1.5	Tunisia	18.0	Romania (<4)	392.0
Dominican Republic	4.8	Zaire	4.2	Russia	786.0
Ecuador	7.5	Zimbabwe	0.7	Slovakia	210.0
El Salvador	13.8	*Total Africa*	*1,317.4*	Slovenia	44.0
Grenada	3.3			Spain (<5)	310.0
Guyana	50.0	**OCEANIA**		Switzerland	497.5
Haiti	3.9	Australia	51.0	Turkey	11.6
Jamaica	15.3	Fiji (<2)	1.0	United Kingdom	24.3
Nicaragua	7.0	New Zealand	11.0	*Total: Europe*	*3,134.9*
Panama	4.1	Solomon Islands	8.2		
Peru	35.0	Western Samoa	4.0	**MIDDLE EAST/SOUTHWEST ASIA**	
Uruguay	17.3	*Total: Oceania*	*75.2*	Iran	24.0
Venezuela	5.7			Israel	22.5
Total: Latin America	*987.1*	**NORTH AMERICA**		Jordan	22.0
		Canada	550.0	Kyrgyzstan	21.0
SOUTHEAST ASIA		United States	422.0	Pakistan	6.0
Cambodia	11.0	*Total: North America*	*972.0*	*Total: Middle East*	*95.5*
Indonesia	52.9			**WORLD TOTAL:**	**24,541.3**
Malaysia (<2)	17.7				
Philippines (<50)	169.0				
Thailand	9.4				
Vietnam	10.0				
Total: Southeast Asia	*270.0*				

Note: If small hydro is defined as something other than 10 MW or smaller, that number is indicated following the country name.

Source: Data comes from the World Energy Council and from on-going tracking of project development by HCI Publications.

• Is there a verifiable demand for the power that will be generated and is the market prepared to pay to receive the service?

• If a power purchase agreement is required, is that agreement in place?

Determining the amount of capital required for a hydro project is not a simple task. Many of the sites that are most suitable for small hydro development are remote, rural sites. Those rural sites, in turn, can lead to high costs in transporting equipment and materials to the site. Another possible added expense at rural sites is locating and training workers to construct the projects. On the opposite side, however, many small hydro projects can be developed as run-of-river plants, alleviating the need for an expensive dam and other civil works. And, frequently, construction costs can be considered an investment in the local community, as local workers are hired and in turn spend their wages in the community.

A general rule-of-thumb that developers use when figuring the expense of building new projects is US$1,000 per kilowatt (kW) of installed capacity. In Nepal, schemes have been built in the remote, mountainous region at costs ranging from US$1,200 to US$2,000 per installed kW. Using numbers such as these, and the assumption presented earlier that another 17,000 MW of small hydro capacity could be added worldwide by 2010, a capital investment exceeding US$20 billion could be required for small hydro development over the next 15 years.

Where this funding will come from is an important issue. Historically, the development of electricity generating stations most often has been a function of the national government. However, in recent years, governments have found that their funds available for infrastructure developments such as power stations are dwindling. This has led generally to two development paths: development is being opened to the private sector, and local governments are taking on a larger share of the responsibility for the development of infrastructure projects.

This in turn has led to two basic types of financing packages that are proving to be available most readily to small hydro developers: private financing, including equity provided by the developer and his partners; and local government funding for rural electrification.

Equity Financing

In areas where the government has revised laws to allow the private development of hydropower projects, it many times falls upon the shoulders of the developer to find a market for his power. Some countries require that a privately developed project sell power to the national utility, which in turn will distribute the power to its customers. However, some laws allow the development of projects for a dedicated customer -- such as a metals or mining firm that has a large electricity load.

The private sector has had success in developing small hydro projects in locations where they have a dedicated customer who is willing to purchase all or most of the electricity generated by the plant. Among the deals that have been announced in recent months are several in India, Uganda, and Brazil, with -- in each case -- an industrial firm serving as the primary customer for the projects. In instances such as these, many times the industrial firm is also the lead developer behind the project.

Examples of this include two in India: the 30-MW Tunga hydroelectric power plant being developed by Dandeli Steel & Ferro Alloys Ltd., and the 21-MW Kuthungal plant, which will be owned by metals firm Indsil Electrosmelts Ltd.; a 10-MW plant being developed near Kilembe, Uganda, by a joint venture that includes cobalt processor Kasese Cobalt Company; and the 10-MW Mello project, owned in part by state minerals conglomerate Companhia Vale do Rio Doce in Brazil.

In addition, several private developers have made substantial equity investments in projects they are developing to sell power to local utilities, rather than to an industrial customer. Again, these developers were willing to commit funding because of a guaranteed market in which they could sell resulting electricity. An example of this type of arrangement is the 22-MW Lake Mainit project being developed in the Philippines by HEI Power Corporation of the United States, Agusan Power Corporation of the Philippines, and Cumming Cockburn Ltd. of Canada. The project will sell power to Agusan del Norte Electric Cooperative, with national utility National Power Corporation (Napocor) buying any excess power.

Rural Electrification Financing

Small hydro also is benefitting from a shift in overall infrastructure finance from the national level to state and local responsibility. As national governments have decreased the amount of funding they provide for infrastructure development, local governments many times have picked up the slack, finding ways to bring electricity to their communities.

China provides perhaps the best example of this type of system at work. A national policy calls on local governments and communities to take an active role in developing small hydropower stations to benefit their towns and villages. Because of small hydro, rural areas of China have been able to lessen their dependence on agriculture, as electricity has encouraged the development of other industries in rural areas. Fujian Province's Yongchun County provides an example of this policy at work. Through local efforts, about 40 MW of small hydro projects have been developed in the county. Every town in the county now has its own power stations and every village has a grain-processing factory. Agriculture and by-products processing has been mechanized and the development of other agri-related industries has accelerated. In the town of Yidu, more than a dozen small hydro stations were built during a two-year period, all of them funded locally.

Other Financing

Although equity financing and local government funding have become common in small hydro project financing, those methods aren't the only methods. Traditional lenders, including federal governments, and multi-lateral lending organizations continue to provide some funding for small hydropower development. Many of the projects that receive funding through these efforts are programs that go beyond the development of a single project. Instead, these projects typically involve a series of generating plants throughout a river basin, or a program that also develops technology and skills in the process of developing the generating plants.

Such a program is being developed in India. Hilly Hydro is sponsored by India's Ministry of Non-Conventional Sources, along with support from the United Nations Development Program. Hilly Hydro is designed to foster the development of small hydro projects in the Himalayan and Sub-Himalayan regions of the country.

Through the Hilly Hydro project, the government is planning to encourage the development of 20 demonstration projects at various locations. The development of these 20 projects though, is secondary. The main thrust of the program is to put together a master development plan, develop the required technology, and develop the expertise required to create a sustainable small hydro industry in the region.

The master plan, which will include detailed investment proposals, will include a framework to most effectively utilize the small hydro potential of the region. In addition, Hilly Hydro also will produce a commercially viable and environmentally sound technology that can be used throughout the country. Finally, through the development of these small hydro sites, the organizers hope to develop the institutional and human resource capabilities, from local to national levels, that are needed to implement the project and to provide for the sustainable development of the small hydro sector in the affected regions.

In mid-1996, project organizers received bids for the development of the first 14 of the demonstration sites. To encourage participation, the organizers offered several incentives, including technical assistance, an interest rebate on loans, grants for some components for non-grid connected projects, and assistance in obtaining government clearances.

A Government Push

During the past couple of decades, more governments have stepped forward to encourage the development of small-scale hydro. Governments from the Americas to Asia have introduced new policies, tax incentives, and financing opportunities that allow and encourage the development of small hydropower projects.

The reasons for the government initiatives is varied. When the United States approved legislation that encouraged small hydro development nearly two decades ago, the goal was to reduce the consumption of imported fossil fuels. In China, small hydro is considered to be one method of alleviating poverty. In many developing countries, small hydro is viewed as the best alternative to bring electricity to remote, rural areas, while also providing opportunities to add or improve irrigation schemes for agriculture.

And, the forms of assistance are as varied as the governments' reasons for promoting small hydro. In Romania, for instance, the government approved new policies early this decade, which resulted in foreign firms being allowed to take part in the rehabilitation of existing small hydro plants. This, in turn, enabled the government utility, the Romanian Electricity Authority (RENEL), to acquire some expertise that allowed them to further develop the small hydro potential in Romania. In other countries, the initiatives have taken other forms: policies that allow and encourage the private development and ownership of small hydro facilities; programs that offer tax incentives for using renewable fuels, such as hydropower, or disincentives for using fossil fuels; and procedures that open up financing opportunities.

Three key models in place today are in India, the United States, and China. India's program provides a good example of a government that is trying to reduce government impediments, and in turn stimulate project development. The U.S. initiative can be described as an economic regulatory approach -- offering developers guaranteed markets for their power under certain regulations. And, in China the government's initiatives are a result of the country's recognition that it needs to increase private development, while at the same time, it attempts to maintain central control over the electric utility industry. Following is a closer examination of these programs.

The India Initiative

India has struggled with establishing programs to encourage the development of power projects of any type. Legislation is in place that allows the private development of electricity generating projects. However, developers have been required to request and receive a myriad of permits and clearances from the federal, state, and local governments before projects could proceed. Because of delays in approval processes, developers many times would lose patience and/or funding before permits were in place.

In recent months, the government has put forth new rules, in an effort to remove impediments to private development. The number of clearances required of hydropower projects has been reduced significantly for both large and small schemes. This effort is expected to stimulate the private development of hydropower.

11

In addition, the federal government has effectively removed itself from the approval process for power projects costing less than 10 billion rupees (US$285 million). Under the new procedures, these projects will not have to get any clearances from the federal government. Instead, those responsibilities were shifted to state governments, which were already responsible for negotiating with private project developers for the right to build a project within the state. This change is expected to have the most benefit for small- and medium-sized hydropower schemes, which now will be able to skip one level of project clearance.

Another initiative that is in place in India is a funding effort to promote mini-hydropower. At the suggestion of the Parliamentary Standing Committee on Energy, the Ministry of Non-Conventional Energy Sources has boosted investment in mini-hydropower -- to the tune of 200 MW by 1997. By 1995, more than 100 MW of that total had been sanctioned, and about half of that had been put into service.

The United States Experience

It has been nearly two decades since the United States put forth the Public Utility Regulatory Policies Act (PURPA), which effectively promoted the development of small hydro through the country.

The premise of PURPA was fairly simple. Utilities are required to purchase all of the output from certain types of generating plants, at pre-set, standardized costs. Utilities and state regulators work hand-in-hand to set the rates that are paid to qualifying facilities. These "avoided costs" represent the cost that the utility would have to pay to add generating capacity to its system or purchase the power from another source. While state regulators oversee the setting of avoided costs, federal regulators are responsible for granting qualifying facility (QF) status to those generators that qualify for the program.

The program offers additional benefits for QFs: the utilities also are required to provide power for backup, maintenance, and other services to the QFs at nondiscriminatory rates; the utility must provide an interconnection for the QF; and the QFs are exempt from certain state and federal regulations to which utilities are subject.

Between 1980 and 1995, the Federal Energy Regulatory Commission (FERC) certified nearly 900 small hydropower projects as qualifying facilities. Those projects represent more than 3,900 MW of generating capacity. Many new small hydro plants were developed at sites that had been developed early during the eighteenth-century industrial revolution, but had been abandoned along the way.

Such sites offered multiple benefits. Developers were able to minimize costs by using existing infrastructure and providing updated or rehabilitated equipment. Many times, obtaining the required permits from state and federal agencies was simple, owing to the fact that a project -- although inactive -- was already existing at the site. Developers also found that obtaining project financing for such projects was fairly simple, in that long-term power purchase contracts provided a guaranteed return on investment.

While small hydro is still being developed in the United States, the development environment is not as good as it was a decade ago. Utilities' avoided-costs have dropped drastically compared to the early days of PURPA. Generally, the rates are too low to attract financing to allow such projects to be built.

Small Hydropower in China

A discussion of successful small hydro development programs would not be complete without mention of China's initiatives. In recent years, China has witnessed rapid small hydro development. By the end of 1995, small hydro accounted for more than 16,650 MW, including some 950 MW that was put into service during 1995. Because of small hydropower, in China, more than 96 percent of the towns, 93 percent of the villages, and nearly 90 percent of the households in the country have access to electricity. Currently, about half of the territory and 300 million people are supplied with electricity from rural medium- and small-sized hydro stations. Rural electricity consumption has increased at an annual rate of 14.4 percent in the past five years.

Several policies have been put forth from the government level, which have encouraged this rapid development of small hydropower. One is a long-standing policy "he who constructs is he who manages, owns, and benefits." One example of this policy is in Guangdong Province, where, until the 1960s, the province was responsible for all electricity development. From 1949 to 1969, only 30 MW of small hydro was installed in the province. Since 1970, small hydro has been developed more effectively. Today, the province boasts more than 2,340 MW of small hydro capacity.

A second policy adopted by the Chinese government in the mid-1980s its "1-2-3" poverty alleviation program. Under this program, small hydro sites will be developed in 600 counties during a 15-year period. During the first five-year program (1986-1990), small hydro was added in 100 counties; over the second five-year period (1991-1995), projects were developed in 200 additional counties. During the final phase (1996-2000), projects will be developed in another 300 counties.

Other Programs

While India, China, and the United States offer some of the larger small hydro programs, they are not the only programs in place today. Some other programs in place include:

13

• The United Kingdom has put forth the Non-Fossil Fuel Obligation (NFFO) program. NFFO requires regional electricity companies to purchase a portion of their electricity from non-fossil fuel resources. Government support for this program allows the electricity companies to pay a premium price for a limited amount of time to those developers who win NFFO contracts. By 1995, 77 small hydro projects, with an installed capacity of 57 MW had won contracts under NFFO.

• In the Philippines, the government has offered a group of hydropower projects, including several small hydro sites, for development by the private sector. Development of the sites will take place under a build-operate-transfer (BOT) mechanism. National utility National Power Corporation is expected to announce the winners of the first set of projects in March 1997; it will receive bids on the second group of projects in July 1997. To encourage participation, the government has revised pricing rules to make them more attractive to potential developers.

• Nepal's hydropower potential is estimated to be 83,000 MW, of which about half is considered to be realistically exploitable. To date, only 240 MW has been developed, and the country was dealt a large setback in 1995, when the World Bank decided it would not provide US$175 million in funding for the 400-MW Arun III project. In response, the government is looking to the private sector to help develop the small hydroelectric potential. Initiatives include the passage of legislation that will allow BOT development of hydropower projects by independent power producers.

Technology Transfer, Training, Other Efforts

With all of the varied efforts underway to promote the development of small hydropower, a need has developed to share information on what has worked, and what hasn't. To address that need, a number of organizations have been formed to assist project developers, lenders, and government representatives in learning about experiences on the other side of the country -- or the other side of the globe.

While we have recognized China for the policies and initiatives it has put forth to promote small hydro development internally, we also must recognize the Chinese for their efforts in promoting small hydro development worldwide. One of the better known programs is headquartered at the Hangzhou Regional Center for Small Hydropower (HRC). HRC was established by the People's Republic of China, the United Nations Development Program, the United Nations Industrial Development Organization, and the Economic and Social Commission for Asia and the Pacific.

Among its tasks, the HRC has trained people to develop small hydro projects. More than 300 people from 45 countries outside China have received training from HRC. In 1996, HRC hosted two training workshops, both sponsored by the Chinese government. As sponsor, China covered costs for boarding, lodging, and training fees for the participants.

An outgrowth of the HRC was the development in 1994 of the Hangzhou International Center on Small Hydropower (HIC). The HIC's goals are to promote small hydro as a means for social and economic development of rural areas, especially in developing countries. This will be achieved through exchanging technical know-how, experience, information, and by sharing facilities for training and demonstration purposes.

An outgrowth of HIC is the International Network for Small Hydro Power (IN-SHP), which promotes small hydropower development through policy and technical cooperation. A relatively new organization, IN-SHP is still setting up its institutional framework and is seeking funding for its work. Once fully established, the network plans to promote small hydro through exchanges of information and advice, technical assistance, promotion of private investment in specific projects, and by promoting the worldwide market for small hydropower technology.

Another relatively new organization is the International Association for Small Hydro (IASH). IASH was formed in 1995 to promote the exchange of information and technical expertise, which would allow for accelerated development of small hydro. IASH accomplishes its mission through courses, workshops, and meetings, as well as through the publication of a quarterly newsletter to keep its members informed of advances.

Several organizations operate in Europe, including numerous country-specific groups and the European Small Hydro Association (ESHA). ESHA's focus is on promoting the independent development of small hydro inside and outside of the European Union. Among its tasks are education of government leaders and the general public; and lobbying efforts to remove impediments to project development, and to standardize regulations. In addition to ESHA, many European countries have country-level organizations dedicated to fostering communication, technology exchange, and the promotion of small hydropower. Similar organizations can be found on every continent.

MEASURING THE BENEFITS OF SMALL HYDROPOWER

Some benefits of small hydropower projects are obvious: like its larger siblings -- the medium- and large-sized hydro-power plants -- small projects offer a clean, renewable source of electricity. Additionally, many small hydropower plants can be developed as run-of-river projects, which limits any harmful effects that would develop from altered flows or from reservoir impoundment.

However, in addition to the environmental benefits, small hydro contributes many economic benefits as well. Especially in remote areas with limited or no electricity, economic development and living standards have improved as a result of small hydropower. In many developing countries, rural residents are able to switch from using firewood for cooking to electricity. This cuts emissions of carbon dioxide and also limits deforestation in those areas. On the economic development end of the spectrum, rural communities are able to attract new industries many of them agriculture-related -- owing to their ability to draw electric power from the small hydropower stations.

The development of small hydropower has also led to the birth of small, local manufacturers to support the hydropower plants. South Africa and China are among the countries that have witnessed the formation of companies to manufacture components for small hydro facilities.

In South Africa, the state electricity company, Eskom, supported an effort to develop standardized equipment and a standardized project design that could be easily and inexpensively installed in remote areas. Designers came up with a Francis-type design that met two primary criteria: keeping costs to a minimum, and ease of construction in the remote, rural areas. The design utilizes local, mainly unskilled labor, for installation. The turbines for the projects are manufactured by South African firms and the castings for runners and other components are supplied by South African foundries.

Meanwhile, in China, the vast development of small hydropower has also developed a number of manufacturing companies that focus on that market. The General Machinery Works, in Yongchun County, is an example of such a firm. The company started more than a decade ago with three employees and one lathe. Over the years, the company has produced more than 800 turbines for small hydropower plants. The machines not only meet the demand within the county, but are also sold to other places. Today, the firm also manufactures generators and sends teams to rural areas to help farmers install and repair both the hydraulic equipment and agricultural machinery. Such teams are called "iron doctors" by the farmers.

CONCLUSION

While the development of small hydropower schemes is increasing in momentum, obstacles still stand in the way. Governments, lenders, and developers are finding ways to address these impediments: new ways of funding projects, new legislation to encourage development, and efforts to exchange ideas and technology. By continuing to address these issues, small hydropower can improve its value worldwide and assure itself of a role in the electricity supply marketplace.

REFERENCES

Beggs, Sandy L., 1995. "Fulfilling Small Hydro's Worldwide Potential," *HRW-Hydro Review Worldwide*, Winter 1995, Volume 3, No. 5, pp. 18-24, Kansas City, Missouri, United States.

ETSU, 1996. "Small-scale Hydro Schemes," ETSU, Department of Trade and Industry, National Association of Water Power Users, March 1996, Harwell, Oxon, United Kingdom.

HydroWorld Alert, 1995-1996. Volumes 1 and 2, various issues, HCI Publications, Kansas City, Missouri, United States.

Tong Jiandong, 1996. "Strategy of Global SHP Development and the Role of IN-SHP;" Presentation at HydroVision '96, August 20-23, 1996, Orlando, Florida, United States.

World Energy Council, 1995. Survey of Energy Resources Report, pp. 143-174, 1995, London, United Kingdom.

SMALL HYDRO ORGANIZATIONS

Address and contact information follows for small-hydro organizations mentioned in this paper (from the *HRW* magazine 1996-1997 Worldwide Hydro Directory).

> European Small Hydropower Association
> Rue Joseph II, 36 bte 7
> 1040 Brussels, Belgium
> (32) 2-2195815; fax (32) 2-2192134
>
> International Association for Small Hydro Power
> c/o Central Board of Irrigation and Power
> Malcha Marg, Chanakyapuri
> New Delhi 110021, India
> (91) 11-3015984; fax (91) 11-3016347
>
> International Network for Small Hydropower
> Hangzhou International Center
> P.O. Box 1206
> Hangzhou 310012, China
> (86) 571-8074424; fax (86) 571-8062934

ACKNOWLEDGMENT

The author acknowledges the contributions of Sandy Beggs and Ward Byers of HCI Publications to this paper. Much of the information for this report was drawn from HCI Publication's International Hydro Project Database, extensive surveying and interviews with market participants, and on-going tracking of project development.

SCENARIO OF 'SMALL HYDRO' IN INDIA

B.S.K. Naidu

Indian Renewable Energy Development Agency Ltd.
Core 4A, First Floor, India Habitat Centre, Lodi Road
New Delhi 110 003, India

Synopsis

Of all the non-conventional renewable energy sources, small hydro represents 'highest density' resource and stands in the first place in generation of electricity from such sources throughout the world. Global installed capacity of Small Hydro is around 47,000 MW against the estimated potential of 180,000 MW.

India has a history of 100 years in Small Hydro. However the country switched on early to larger hydro and pursued the same reaching an installed capacity of 21,000 MW with only 500 MW of Small Hydro. Of late, environment driven awareness seems to have reminded us that what India needs is not the mass production but production by masses in a decentralised manner respecting the 'carrying capacity' of modular eco-systems.

The paper highlights the benefits of Small Hydro, giving an account of its potential in hilly as well as plain regions of India. Mini Hydel potential at thermal power plant cooling water tail ends is also indicated as a fine example of energy retrieval, efficiency and conservation. Tracing the history of Small Hydro development in India, the paper details the recent Government support to the private sector participation in the national programme.

Fiscal and financial incentives by Central Government and policy framework provided by State Governments are attracting private investments in Small Hydro. A quality package of 100 MW World Bank initiative (to be followed by another line of credit for 200 MW) on Small Hydro operated in an entrepreneur friendly environment is likely to spin off a promising techno-commercial scenario paving the way for attractive business opportunity in Small Hydro. UNDP-GEF initiative on hilly hydels in Himalayas is also likely to enhance interest in Small Hydro.

Technology variations are necessary to suit different 'head-discharge' combinations and to optimize the designs in Small Hydro schemes. While the paper presents a technology matrix of World Bank sponsored projects, it needs to be pointed out that technology adaptation for Indian conditions of geology and silt loads is going to be an important feature of the National programme, besides appropriate technology for socially oriented programmes like portable micro hydel scheme for remote and isolated communities.

Environmental aspects of Small Hydro with emphasis on mitigation of already low levels of negative impacts and enhancement of positive impacts typically associated with water resource projects, are discussed in the backdrop of Indian perceptions of hydro development. The paper visualises the emerging scenario and concludes with strategies for future.

INTRODUCTION

Hydro represents non-consumptive, non-radioactive, non-polluting use of water resources towards inflation free energy development with most mature technology characterised by highest prime moving efficiency and spectacular operational flexibility. It contributes to 18% of World Electricity Supply today, Exhibit-1. Out of non-conventional renewable energy sources, Small Hydro represents 'highest density' resource. Even excluding large hydro, small hydro stands in first place in the generation of electricity from renewable sources throughout the world. World installed capacity of Small Hydro today is around 47,000 MW against an estimated potential of 180,000 MW. A general scenario is as under:

India	500 MW
Japan	3,300 MW
China	15,000 MW
Rest-of-Asia	200 MW
Europe	9,000 MW
Rest-of-world	19,000 MW
Total	47,000 MW

Capacity utilisation of 40% is expected from 'Small Hydro' on an average.

WHAT IS SMALL HYDRO ?

There is general tendency all over the world to define Small Hydro by Power output. Different countries are following different norms keeping the upper limit ranging from 5 to 50 MW e.g.

UK (NFFO)	≤ 5 MW
UNIDO	≤ 10 MW
India	≤ 15 MW
Sweden	≤ 15 MW
Colombia	≤ 20 MW
Australia	≤ 20 MW
China	≤ 25 MW
Philippines	≤ 50 MW
New Zealand	≤ 50 MW

In India, small hydro schemes are further classified by the Central Electricity Authority (CEA) as follows:

Classification of Micro, Mini & Small Hydel Schemes in India

Type	Station Capacity	Unit rating
Micro	Upto 100 kW	Upto 100 kW
Mini	101 kW to 2000 kW	101 kW to 1000 kW
Small	2001 kW to 15000 kW	1001 kW to 5000 kW

As a matter of fact, Small Hydro can not be defined in terms of Power Output or Physical Size of the power station, due to the influence of head on the scheme.

For example, a turbine with a runner diameter of 2m and a head of 10m would have an output of about 2 MW, and would be considered a small turbine. With a head of 300 m, however, a turbine of the same size would have an output of about 100 MW, and would be considered a large hydro ! Similarly, a 10 MW turbine with a head of 10 m would have a runner diameter of about 4.5 m, and would be considered a large one. With a head of 300 m, however, a turbine of the same power output would have a runner diameter of about 0.7 m, and would be considered small !

What therefore needed is a definition which combines both power output and physical size factors. A convenient parameter which can do this, is the cost of the project. In Indian context, small power projects costing less than or equal to Rs 1000 million do not require CEA clearance even with single sourcing. Keeping with the same spirit, 'Small Hydro' can be defined as those hydro projects which cost less than or equal to Rs 1000 million, equivalent to an installed capacity of approximately 25 MW. Such a definition is acceptable to the World Bank and incidentally matches with the Chinese definition.

The need for a definition arises so as to consider these projects for pormotional support, quicker development though de-regulation, standardisation and resource mobilisation from national and international sources keeping in view the environmental harmony of these schemes. The "smallness" in terms of size, capacity, cost, environmental and social impacts can be considered wholistically to arrive at a logical definition.

WHY 'SMALL HYDRO' ?

Now let us examine as to why we should develop 'Small Hydro'

a) if they are costlier than Big Hydro (Rs/MW, Rs/kWh) due to the denial of economy of scale.
b) if their contribution is going to be insignificant (a few thousand MW out of more than 100 thousand MW in Indian context)

Small Hydro merits development due to the following:

a) It needs limited investments affordable by Private houses
b) Short gestation, enabling quicker electricity and financial returns
c) Small hydro could be relevant at decent heads as-low-as 2m
d) Environmentally friendlier than conventional hydro,

      i)   Not involving setting up of large dams associated with problems of deforestation, submergence or rehabilitation.
     ii)   Least impact on flora & fauna (aquatic & terrestrial) and bio-diversity.
    iii)   No chances of impinging on "carrying capacity" of modular eco-systems unlike large scale development interventions.

e) One of the least 'CO$_2$ emission responsible' power sources, even considering full energy chain right from the impact of production of plant equipment etc. [Exhibit-2]

f) Small Hydro is significant for off-grid, rural, remote area applications in far flung isolated communities having no chances of grid extension for years to come.

Small Hydro is environmentally benign, operationally flexible, suitable for peaking support to the local grid as-well-as for stand alone applications in isolated remote areas. Even if we ignore the CO$_2$ abatement costs and 'acid rain' abatement costs etc., of conventional thermal route, small hydro is benevolent on the following known hard facts of economics:

1. Levelised cost of generation from Small Hydro would be less than half that of thermal.

2. Their real capital investment is less than that of thermal, considering infrastructural costs (small dams/diversion structure vis-a-vis large mines/transport of coal) and auxiliary factors (small hydro auxiliary consumption of 1/2 % vis-a-vis 10% in case of thermal).

On the basis of 'project life cycle cost' in real terms, inflation-free small hydro becomes several times cheaper than thermal option, due to the following weightage factors and cheaper operational costs with reference to maintenance and zero cost input.

Advantage	Weightage Factor
a) Double the life of small hydro vis-a-vis thermal	2
b) An escalation of 6% per annum in case of thermal over hydro having project life of 60 years and more	8.89
c) The actual generation in long term comes at least 20% higher than the 75% hydrological dependability	1.2

SMALL HYDRO VIS-A-VIS OTHER RENEWABLES

Now let us examine superiority of 'Small Hydro' even amongst renewables

1. It is 'highest-density' renewable energy source against widely spread and thinly distributed solar energy, biomass and wind resource, etc.

2. Its cost of generation is cheapest amongst renewables

SECTOR	RS/Kwh
Small Hydro	1.00 to 1.50
Cogeneration	1.25 to 1.50
Biomass Power	1.75 to 2.00
Wind Energy	2.00 to 2.75
Solar PV	10.00 to 12.00

3. Small Hydro efficiencies are highest amongst renewables

SECTOR	EFFICIENCY
Small Hydro	85 - 90%
Cogeneration	60%
Biomass Power	35%
Wind Energy	40%
Solar PV	15%

4. Small Hydro has longest project life

SECTOR	LIFE(YEARS)
Small Hydro	60
Cogeneration	30
Biomass Power	30
Wind Energy	20
Solar PV	20

5. Small Hydro resource is likely to have more consistency compared to other renewables like wind, solar, biomass in respect of availability for power generation. Relatively better capacity factors can be expected. Even run-of-the-river schemes can have small pondage to meet the daily peak requirements of power.

INDIAN POTENTIAL OF SMALL HYDRO

India has one of the world's largest irrigation canal networks with thousands of dams. It has monsoon fed, double monsoon fed as well as snow fed rivers and streams with perennial flows. An estimated potential of 10,000 MW of Small Hydro exists in India. However, nearly 5000 MW have been actually identified through more than 2000 sites in 13 states of India as follows:

Small Hydro Potential identified in India

State	No. of Sites	Capacity MW
1) Arunachal Pradesh	190	111
2) Bihar	110	134
3) Gujarat	197	157
4) Himachal Pradesh	298	1393
5) Jammu & Kashmir	131	859
6) Kerala	216	561
7) Madhya Pradesh	80	238
8) Maharashra	122	68
9) Meghalaya	36	8
10) Orrisa	48	28
11) Punjab	198	196
12) Uttar Pradesh	328	1128
13) West Bengal	69	69
TOTAL	**2023**	**4950**

Source : WP&DC, 1995

UNDP-GEF Hilly Hydel Project has further identified a potential of 1200 MW mini hydel sites of less than 3MW each, in the 13 Himalayan States as below:

23

State Wise Identified Sites by UNDP - GEF : Hilly Hydel Project
(Less than 3 MW)

Sl. No.	State	Site Identified		Additional Sites Identified	Total Sites	Remarks
		Nos.	MW			
1.	Jammu & Kashmir	94	112	45	139	
2.	Himachal Pradesh	141	152	22	163	
3.	Uttar Pradesh	180	106	80	260	
	North	*415*	*370*	*147*	*562*	
4.	Sikkim	47	48	--	47	
5.	West Bengal	49	47	--	49	Hilly area is rather small
6.	Bihar	68	80	--	68	Hilly area is rather small
	East	*164*	*175*	*--*	*164*	
7.	Assam	26	28	--	26	
8.	Arunachal Pradesh	364	346	--	364	
9.	Meghalaya	66	22	14	80	
10.	Manipur	66	52	13	79	
11.	Nagaland	10	9	42	52	
12.	Mizoram	17	7	21	38	
13.	Tripura	1	1	--	1	
	North East	*550*	*465*	*90*	*640*	
	Total	1129	1010	237	1366	Projected MW = 1200 MW

Mini hydel potential at thermal power plant cooling water tail ends is also indicated as a fine example of energy retrieval, efficiency and conservation. A large thermal - mini hydel hybrid system symbolically fulfils non-fossil fuel obligation by the generator of conventional power. A dozen important large thermal power plants of India have already proposed tail end mini hydels as follows:

Mini Hydel Projects proposed at the tail ends of Cooling Water systems of
Thermal Power Plants

Sl No	Power Station	State	Capacity Proposed (MW)	Plant Owner
1.	Singrauli	UP	3.00	NTPC
2.	Rihand	UP	1.50	-do-
3.	Unchahar	UP	0.25	-do-
4.	Farakka	WB	0.75	-do-
5.	Korba	MP	1.50	-do-
6.	Anta	Raj.	0.20	-do-
7.	Ramagundam	AP	0.50	-do-
8.	Korba - I	MP	0.80	MPEB
9.	Korba - II	MP	1.00	-do-
10.	Satpura	MP	1.00	-do-
11.	Parichha - I	UP	0.50	UPSEB
12.	Parichha - II	UP	0.50	-do-

SMALL HYDRO DEVELOPMENT IN INDIA

India has a century old history of Hydro Power and it is important to note that the beginning was from Small Hydro. The first Small Hydro Project of 130 kW commissioned in the Hills of Darjeeling in 1897 marked the development of Hydro Power in India. The Sivasamudram Project of 4500 kW was the next to come up in Mysore district of Karnataka in 1902, for supply of power to the Kolar Gold Mines at 25 Hz. The pace of power development including hydro was rather tardy. The planned development of Hydro Projects in India was taken up in the post independence era. This means that the 1362 MW capacity (including 508 MW hydro) installed in the country before independence was mainly coming from Small and Medium size projects.

Since focus was laid on large scale power generation through big hydro, thermal and nuclear routes, the Small Hydro Power (SHP) potential has remained largely untapped. During the last 10-15 years, however, the necessity for development of SHP has been felt globally due to its various benefits particularly concerning environment, and it has assumed importance in the power development programme.

In India, the initial enthusiasm shown and the pace of progress exhibited during the post-independence era to tap the enormous hydro potential in a big way, slowly started fading away. The success stories of building these "modern temples" could not be sustained for various reasons ranging from resource crunch to failures in optimizing the gestation.

Meanwhile the 'power starvation' still continues posing a lot of problems for country's ambitious developmental activities. While strategies are being formulated and tried for the accelerated hydro power development in India through large projects, pending significant results from that front, attempts are being made to develop this renewable source of energy through small hydros.

After independence in 1947, the need was felt for speedy development of infrastructure especially the power sector and the planners chose the big hydro electric projects to augment the capacity in big chunks. The establishment of over 20,500 MW of Hydel Power Stations was significant in 50 years compared to 500 MW of previous 50 years. However, pace of thermal power development was much higher than that of hydro. The growth period of hydro power generally gave less importance to small projects since the development was mainly in Central sector and the State Electricity Boards (SEB) were more or less tuned to the Central Planning System. This does not mean that there has been no development of Small Hydro Power. The experiments were continued by some enterprising SEB's and the installed small hydro capacity of about 500 MW in the country is mainly owned by these agencies, besides another 500 MW being under construction. The Indian Small Hydro Power Development Programme received a new dimension and tempo after the liberalization of economy and invitation to private sector for investment in Power. The private sector was attracted by these projects due to their small adaptable capacity matching with their captive requirements or even as affordable investment opportunities.

GOVERNMENT SUPPORT TO SMALL HYDRO

Government of India, through its full-fledged Ministry of Non-Conventional Energy Sources (MNES) formed in the year 1992 - the year of Rio Summit on Environment and Development, is extending multi-dimensional support to the development of mini hydels (upto 3 MW) as one of the environmentally benign renewable energy technologies, keeping in tune with the Government's overall thrust on liberalisation of economy and private sector participation in power development. Fiscal incentives available for 'Small Hydro' sector are as follows:

* Schemes involving capital upto Rs 1000 Million need no prior clearance from the Central Electricity Authority (CEA), even if they are single sourced.
* Schemes involving capital upto Rs 500 Million need no Environmental Clearance from Ministry of Environment & Forests (MOEF).
* Income Tax holiday for Power.
* Term loans through IREDA for schemes upto 25 MW.
* Concessional customs duty @ 20% / 10% for non-captive use.

For Schemes Upto 15 MW :

➤ No excise duty for turbines.
➤ Accelerated depreciation for income tax under consideration (Similar to other NRSE systems).

Additional Incentives under MNES - Small Hydro Programme

For Power Station Capacity ≤ 3 MW

* Incentives for Detailed Survey & Investigation
  100% Grant-in-Aid subject to certain ceilings depending upon the type of schemes.

* Incentives for Preparation of DPRs:
  Grant-in-Aid of 50% of the DPR costs subject to certain ceilings depending upon the type of schemes.

* Interest Subsidy Scheme through Financial Institutions.

  a) For Hilly Regions; North Eastern (NE) Region and Andaman & Nicobar (A&N) Islands
  Rs.11.20 million /MW,
  Applicable Project Cost: Maximum Rs 60 million /MW

  b) For Non-Hilly (Other) Regions:
  Rs. 3.83 million/MW
  Applicable Project Cost: Maximum Rs 40 million /MW

For Schemes Upto 100 kW:
(In Hilly Regions, NE and A & N)

＊ Capital Subsidy of Rs 15,000/kW

In addition to the above fiscal incentives, MNES has issued guidelines for off-take of power from renewables on concessional terms by the State Electricity Boards in respect of power wheeling, banking and buy-back. Following the model guidelines, 9 States of India have announced their Private Sector Policy Incentive package for Small Hydro Sector, Exhibit-3. These states have also identified Small Hydro sites for allotment to private sector for development as indicated in Exhibit-4.

The Issue of Water Royalty

In India, the producer is allowed to use the water for power generation. However, royalty on the water used for small hydro projects is charged at a rate around 10% of the prevailing electricity tariff for HT consumers. In this connection, following needs to be addressed:

1. In India royalty on water is charged in case of Central Sector Projects for the 'distress' caused to the home State like submergence, dislocation of population, loss of forests etc., and the other beneficiary States in the region share the burden of water royalty.

   The above analogy is irrelevant to private developers of 'Small Hydro' because of the fact that:

   ＊ The project is in one State only and there are no other beneficiary States to share the burden.

   ＊ Small Hydro projects do not cause environmental 'distress' of the kind mentioned above.

2. Royalty on water is not charged to State Sector Hydro Projects in India.

3. Royalty on water is not charged in case of irrigation projects even for the 'consumptive' use of water.

For non-consumptive use of water in Small Hydro projects (which are known for their ecological harmony with nature) therefore, water royalty may not be charged so as to improve their economic viability.

In United Kingdom also, where environmental concerns are at the peak, Small Hydros (below 5 MW) are exempted from the payment of 'water abstraction' charges.

Mini/Micro Hydro Projects (≤ 3 MW) - Achievements under MNES Programmes

Project Status

＊　Capacity Installed - 134 MW
＊　Plants under Construction - 247 MW

27

* Project Pipeline in the Private Sector - 550 MW
* Feasibility studies completed - 250 MW
* 13 Himalayan states covered under UNDP - GEF Hilly Hydel Project on Preparation of Master Plan and 20 Demonstration Schemes

Incentive Schemes/ Pilot Programmes

Programme	No. of Projects	Capacity (MW)
Projects Supported :		
Detailed Survey Investigation (DSI) Scheme	135	135
Detailed Project Report (DPR) incentive Scheme	38	59
Capital Subsidy Scheme	87	97
Pilot Programme on Portable Micro Hydel Sets	50	0.575
Projects Commissioned :		
Pilot Programme on Portable Micro Hydel Sets	14	0.16
Mini Hydel projects supported under capital subsidy scheme	23	10

PRIVATE SECTOR PARTICIPATION

As mentioned earlier, the States where small hydro potential is available have come out with attractive policies for private sector participation, in line with MNES guidelines. The pioneering efforts of the Karnataka State Government in establishing a Single Window Agency for clearance of SHP projects to private sector and the Karnataka Power Corporation(KPCL) as the nodal agency has sparked the growth of the sector and has become an example for other States to follow. Karnataka has already allotted about 39 sites aggregating to 273 MW to private sector , which is the highest capacity allocated by any State to private sector. Some of the success stories of private sector projects in the new era in the State are 18 MW Shivapur project of Bhoruka Power Corporation, Bangalore and 1 MW Gokak Falls Scheme of Gokak Mills, Belgaum. The neighbouring States like Kerala and Andhra Pradesh followed the example and many SHP sites were allotted to private sector companies mainly to meet their captive requirements. The first private sector project in Kerala namely, Maniyar 12 MW SHP, was completed in a record time of two years. The northern States of Uttar Pradesh and Himachal Pradesh, which have tremendous potential of hilly hydel schemes also announced their policies and their projects are under allotment for captive as well as power sale options. The real achievement of private sector involvement would become visible by 2000 A.D.

WORLD BANK INITIATIVE THROUGH IREDA

Realizing the potential and the importance of new and renewable sources of energy in the national development with particular reference to rural sector and to implement the Government's energy policy, Indian Renewable Energy Development Agency (IREDA) was incorporated as a public limited Government company in 1987. The mission of IREDA is to stimulate, promote, support and accelerate an efficient, environmentally sustainable infrastructure for effective utilization of New & Renewable Sources of Energy (NRSE) technologies for productive purposes.

IREDA operates a revolving fund to develop, promote and finance commercially viable NRSE technologies in the country. With the aids & grants from GEF, World Bank and other financial institutions, it has been extending soft-term loan assistance to the prospective developers in various NRSE sectors comprising Small Hydro. IREDA continues to be the main instrument for achieving the objectives of the Government in renewable energy.

IREDA started financing small Hydro Projects from middle of VIIth Five Year Plan [1985-90] and the projects sanctioned were for Government sector along with subsidy from MNES. In line with Govt. of India policy of privatisation of power sector, from 1992-93, IREDA started financing Private Sector small hydro projects from its own resources and a modest beginning was made through financing one private sector project of 12 MW and another renovation project of 1 MW. By the end of FY 1995-96 IREDA has financed a total of 33 Small Hydro projects with loan assistance amounting to Rs. 2255 million for installation of aggregate capacity of nearly 140 MW, which include 24 private sector projects.

World Bank Programme

A pre-investment study was carried out under the auspices of the Energy Sector Management Assistance Programme (ESMAP) jointly supported by UNDP and World Bank in the years 1989-90 in association with the Indian Govt. agencies. The principal objective of this study was to apply techno-economic criteria to improve the design and economic viability of irrigation based mini-hydro schemes in India, and to identify and prepare a medium term investment programme to develop a series of irrigation based hydro schemes in India. This study covered more than fifty prospective small hydro sites in five states. Detailed techno-economic analysis and cost-effective designs were made for some of these sites.

Consequent to the ESMAP study, World Bank (WB) offered a line of credit worth US $ 70 million to be utilized during 1993-97 with a target capacity sanction of 100 MW.

The objective of this World Bank Project was " *to promote, support and accelerate the development of the Small Hydro in India by tapping the hydro potential available with the irrigation dams and canals in order to provide energy for the rural, remote and far flung areas in a decentralized and environmentally benign manner by providing financial & technical assistance to the prospective developers"*.

The credit line has moved fast enough and IREDA could sanction 21 projects with an aggregate capacity of 116 MW, exceeding the target one year ahead of schedule. The enthusiasm shown by the private promoters as well as their understanding that the Small hydro is the most attractive long term option not only for their captive needs but also as a business opportunity by way of selling power at commercial rates have given a fillip to the sector and there is going to be a steady growth in the Small hydro sector. Out of the World Bank package, a capacity of 11 MW is likely to be commissioned by March 1997. In view of the success so far achieved, it has been agreed in principle by WB and IREDA that an extension of the present scheme for additional 200 MW be made.

UNDP-GEF INITIATIVE ON HIMALAYAN SMALL HYDRO

Against the background of depleting forest resources of Himalayas, the UNDP-GEF India Hilly Hydel Project was initiated in the year 1994 as the first Indian Project from GEF portfolio in order to develop a national strategy and master plan for optimum utilization of Small Hydro resources of Himalayan and sub-Himalayan regions with an outlay of US $ 15 million. The scheme also envisages implementation of 20 demonstration schemes, upgradation of 100 water mills for electricity generation and subsequent establishment of a revolving fund at the Indian Renewable Energy Development Agency (IREDA) for commercialization of hilly hydels.

The project is being implemented by the MNES through a Project Management Cell established at Tata Energy Research Institute (TERI). The scheme is also attempting to develop management and ownership models through community participation. The energy produced from Small Hydro projects will be used for cooking & heating etc., thus saving wood with consequent reduction in deforestation. Availability of electricity for water pumping for irrigation is also expected to change the type of cultivation from jhoom (shift & burn) to irrigation thereby reducing the denudation of forest. Enough institutional mechanisms are also envisaged to develop for testing, training of personnel for operation, maintenance and revenue collection, management etc. The project has replicability potential in the Himalayan territories of Afghanistan, Pakistan, Nepal, China and Bangladesh and indeed even other hilly areas of the world.

The project has succeeded in sensitizing the nodal agencies, manufacturers, Consultants, NGOs etc. The various role players in the 42 months target-oriented project are indicated below with allocation of their responsibilities:

BLOCK	SECTOR	INTERNATIONAL CONSULTANT	NATIONAL COUNTERPART
1	Water Mills	IT Power, U.K.	1. AHEC (N) 2. TERI (NE)
2	Low Wattage (LW) / Load Development(LD) / Rural Development (RD)	MHPG consortium, Europe	1. TIDE (LW) 2. CES (LD & RD)
3	Zonal Plan	Mead & Hunt, USA	AHEC
4	Master Plan	Mead & Hunt	CES

5	Training, Research & Consultancy	MHPG	AHEC
6	Environmental Assessment	Mead & Hunt	CES
7	Tech. Selection	1. Mead & Hunt (N) 2. MHPG (NE)	AHEC
8	Project Execution	1. Mead & Hunt (N) 2. MHPG (NE)	-
9	Publicity	MHPG	-

N - Northern States, NE - North East States

PORTABLE MICRO HYDEL PROGRAMME OF MNES

Another innovative concept that has been developed and is under implementation, is the installation of light weight **Portable Micro Hydel Sets** in the hilly areas of the country, particularly the Himalayan region. The main objective of this novel scheme is to provide power to non-grid connected remote areas by using 'stand alone systems' ranging from 5 to 15 kW, which would be easily installed and maintained by the local communities. MNES in the first instance has supplied 50 such sets free of cost and IREDA is the implementing agency for the same. The beneficiary will have to carry out all peripheral works essential for completion of the projects, including civil works, local distribution network etc. In the subsequent phases, based on the success of implementation and demonstration of the benefits of the first phase, the programme may be further expanded with phased reduction of subsidy. As of now, 14 projects have been commissioned and are generating power.

The above scheme addresses a socially relevant sector particularly for those who are virtually cut-off from rest of the country and forced to live without power with no chances of grid reaching them for many years.

TECHNOLOGY SCENARIO

Though the Hydro Turbine technology has achieved near perfection, the design of Small Hydro Turbines should not be considered as a miniaturization of large Turbines. The adaptability of technological innovations to suit the smaller versions and cost effective simplifications are being tried by some of the manufacturers, especially in smaller micro size sets. India has about 10 manufacturers of small hydro turbines, of which many are specialized in smaller capacities. Turbine application ranges of Indian manufacturers in general are given below:

Turbine Type	Head (m)	Output (kW)
Pelton	100 - 550	50 - 6,000
(Micro Pelton)	(30 - 70)	(5 - 15)
Turgo	30 - 200	15 - 3,000
(Micro Turgo)	(10 - 75)	(3 - 45)
Francis	15 - 250	15 - 7,000
Kaplan	2.5 - 30	500 - 7,500
Semi-Kaplan	2.5 - 25	10 - 6,000
Propeller	2.5 - 30	10 - 6,000
Bulb	2 - 12	100 - 2,600

The type of machines in low head schemes are generally 'S' type tubular turbines (Kaplan or Semi-Kaplan) with generator outside the water passage. Francis and Impulse turbines are preferred in medium and high head schemes. Micro pelton and turgo-impulse are seen in high head as well as micro hydel sets. Orientation of the turbine is chosen mainly based on the space availability, size, maintainability and civil works considerations.

A technology matrix of World Bank projects is presented at Exhibit - 5.

Some cost effective design concepts are found to be practiced in intake, penstocks, trash racks and power house construction. Heavy structures and costly equipments are avoided to improve the economics of the project. In the area of equipments, some innovative options like Automatic falling shutters, avoidance of Butterfly valve, Flow control in Bypass systems etc., are preferred.

In hilly hydel projects the design of turbine blades and other parts to withstand the silt conditions are given more importance. New materials and coatings are being developed to improve the metallurgy of the under water components. At the same time, silt chambers and silt control systems are being incorporated in the design to reduce the material damage due to silt in water. Appropriate technology is also being attempted for socially oriented programmes like portable micro hydel scheme for remote and isolated communities.

ENVIRONMENTAL ASPECTS

Till recently, in the hydro projects of India like in many other countries, the environmental aspects were not given the importance they deserved. Even now the so called 'awareness' or say 'thirst for preservation by confrontation' is confined to certain 'really effected' groups and the 'elite' lobbyists. In its backdrop, Small Hydro Projects are looked at as favourites and support is easily forthcoming from all. This advantage would have to be judiciously and honestly utilised for the speedier implementation of large number of Small Hydro Projects.

Small Hydro projects are generally environment friendly and non-polluting. They do not involve serious deforestation, rehabilitation and submergence. However depending on the site and the layout of the scheme, trees may have to be removed in marginal areas. This is invariably compensated by afforestation of equivalent area or double the area of degraded forest. These projects do not involve construction of dams and therefore generally no rehabilitation problems arise. However, in case of formation of small pondages, and also any removal of habitation along the diversion canals etc., suitable arrangements are made in consultation with the Government Authorities. Pollution and related negative effects are not expected in hydro projects. However the projects pass through the Pollution clearance mechanisms of the State Governments and also the forest clearance in case of involvement of forest land. The effect on the downstream water supply and drainage is one of the important concerns which is addressed during the SHP designs. The dry area of the stream of canal from the diversion structure till the tail race vis-a-vis the water needs of habitat for drinking and irrigation and effects on aquatic and fish life are to be studied in details. Necessary compensatory measures like provision of separate drinking water and irrigation lines and fish ladders etc. are to be incorporated in the design to mitigate such impacts, if any.

Due to the environmentally benign nature of these small schemes and in order to reduce the time involved in clearance procedures, Ministry of Environment & Forests have exempted hydel projects with an outlay of less than Rs 500 Million from Environmental clearance. However they have to obtain all necessary clearances from State Government agencies like Irrigation Department, Pollution Control Board, Forest Department etc.

We should remind ourselves, that the relaxations given by Government are only to shift the responsibilities to the concerned individuals and not to totally "do away" with such responsibilities. This is particularly relevant in the present context of Private Sector participation. Though it is well known that Small Hydro Projects are not contributing to any negative impacts on the environment and ecology, the strategies and thrusts shall be to carefully plan and implement them so as to multiply the "positive" impacts, normally associated with water resource projects such as tourism, employment, fisheries, social forestry, agro units, water supply and irrigation through electricity, community health and improved standard of living etc.

EMERGING SCENARIO & STRATEGIES FOR FUTURE

From the foregoing discussions, it is apparent that in India, it is an opportune time that Small Hydro should get a strategic thrust, as environment driven awareness has rediscovered 'Small Hydro' as a principal renewable energy source for sustainable development.

For a multi-dimensional strategic thrust, identification of weak areas and threat perceptions need to be visualised carefully and appropriate steps need to be taken. In this connection, following need to be addressed:

- Non-availability of hydrological data and pre-investment study reports of newly identified sites.
- Single window clearance facility not functional in all the States.
- Non-uniformity of Wheeling & Banking facility.
- Non-uniformity of buy-back and third party sale.
- Water royalty charged from Private Entrepreneurs.
- Economics depends on Government Policies which need to be consistent.

A country-wide scientific resource assessment needs to be done making full use of metered data of atleast 2 to 3 years as well as corroborative data from the existing irrigation and power projects. On policy front, uniformity of facilities from State Electricity Boards may be a desirable step besides withdrawing water royalty from Small Hydro Projects. State clearance mechanisms need to be further streamlined and fine-tuned to make single window clearance really effective. The various fiscal and financial incentives extended by the Government should continue in the forthcoming plans to ensure economic viability and attractive returns from the projects.

Target segment based approach would be desirable in a diverse country like India with complex, socio economic objectives. Following segments appear imminent for evolving strategies to achieve a balanced growth of Small Hydro:

1. Hilly Hydels
2. Existing irrigation canal falls
3. Run-of-the-river schemes
4. Socially relevant micro hydel schemes
5. Hybrid systems with other renewable sources as well as conventional sources of energy

6. Stand alone systems for off-grid applications
7. R&M of existing Small Hydro Schemes

In the overall scenario of Small Hydro, few trade-offs are necessary to be carefully perceived for a right kind of decision on strategic development of Small Hydro. These are, a) Detailed Vs. Adequate Investigation, b) Conventional Vs. Innovative designs, c) Standardisation Vs. Optimisation, d) Efficiency Vs. Effectiveness, e) Commercialisation desirability Vs. Socio-economic Objectives.

A. Detailed Vis-a-vis Adequate Investigation

In India the regulating agencies like CEA & MNES prefer hydrological data for at least 2 years for feasbility assessment. Although hydrological data over a larger period is considered desirable for establishing reliable flow duration curves, in view of accelerated and massive programme of development envisaged in the country, it may be alright to go ahead with 2 years data keeping in view the country's large network of irrigation canals and existance of hydro projects as much as 21,000 MW which provide enough data base for corroboration studies to reconfirm the 2 years data and its logical extrapolation, with an appropriate knowledge of catchment characteristics of the concerned water streams.

B. Conventional Vis-a-vis Innovative Designs

With enormous experience of conventional hydro in the country, the natural tendency of designers and consultants is to miniaturise the larger schemes, by scaling down the bigger machines, not even attempting to review the technology packages involved. Some of the requirements of the conventional hydro systems may be much simplified or eliminated for cost effectiveness. While losing on economy of scale, if enough care is not taken to gain on costs of simplified technologies, we would not be able to optimise the systems on techno-economics.

C. Standardisation Vis-a-vis Optimisation

The hydro turbine essentially is a tailor made equipment to suit a particular 'head-discharge' combination, although seveal standard design of machines are available in different bands of head-discharge ranges. Even within a band, if the manufacturer has a proper back up of well tested hydraulic designs, then picking up the most optimum alternative and tailoring it to the dimensions of hydrology and power requirements would hardly take any time compared to the time required for civil works and power evacuation systems. Therefore, the desirability of standardisation need not be stretched to 'off the shelf' sets of hydro turbines. However, standardisation could be a continuing effort in respect of auxiliary systems and control systems.

D. Efficiency Vis-a-vis Effectiveness

The hydraulic efficiency sensitivity in large size machines in terms of output is very high while in Small Hydro, it may not be so tangible. Moreover, the risks involved in the hydrological data of smaller time frames can have much more pronounced effect on efficiency compared to manufacturing compromises. Therefore, in a massive fast track programme, the planners, designers and manufacturers may give priority to the end result and effectiveness in lieu of fractional accuracies in efficiencies.

E. Commercialisation Vis-a-vis Socio-economic Objectives

While commercialisation of small hydro in a liberalising Indian economy could be a desirable objective to achieve productivity, profits and economic sustainability. In certain situations, socio-economic needs of communities such as rural energy needs of isolated villages may have to get priority over pure commercial objectives, giving due respect to the social cause. This calls for Govt. support for 'not-so-viable' projects also, keeping in view the social & environmental gains, which at times may not be tangible and quantifiable.

CONCLUDING REMARKS

Having been re-discovered as the most potent source of renewable energy for sustainable development, Small Hydro is passing through a most opportune phase of prospective development globally as well as in India. It meets the Indian perceptions so pragmatically defined by our Father of Nation Mahatma Gandhi in the following words:

- What India needs is not mass production but production by masses.
- India lives in villages.
- There is enough in nature for man's need but not enough for man's greed.

Acknowledgement: Author is thankful to MD, IREDA for permitting presentation of this paper.

Contribution of 'Hydro' in World Energy Supply

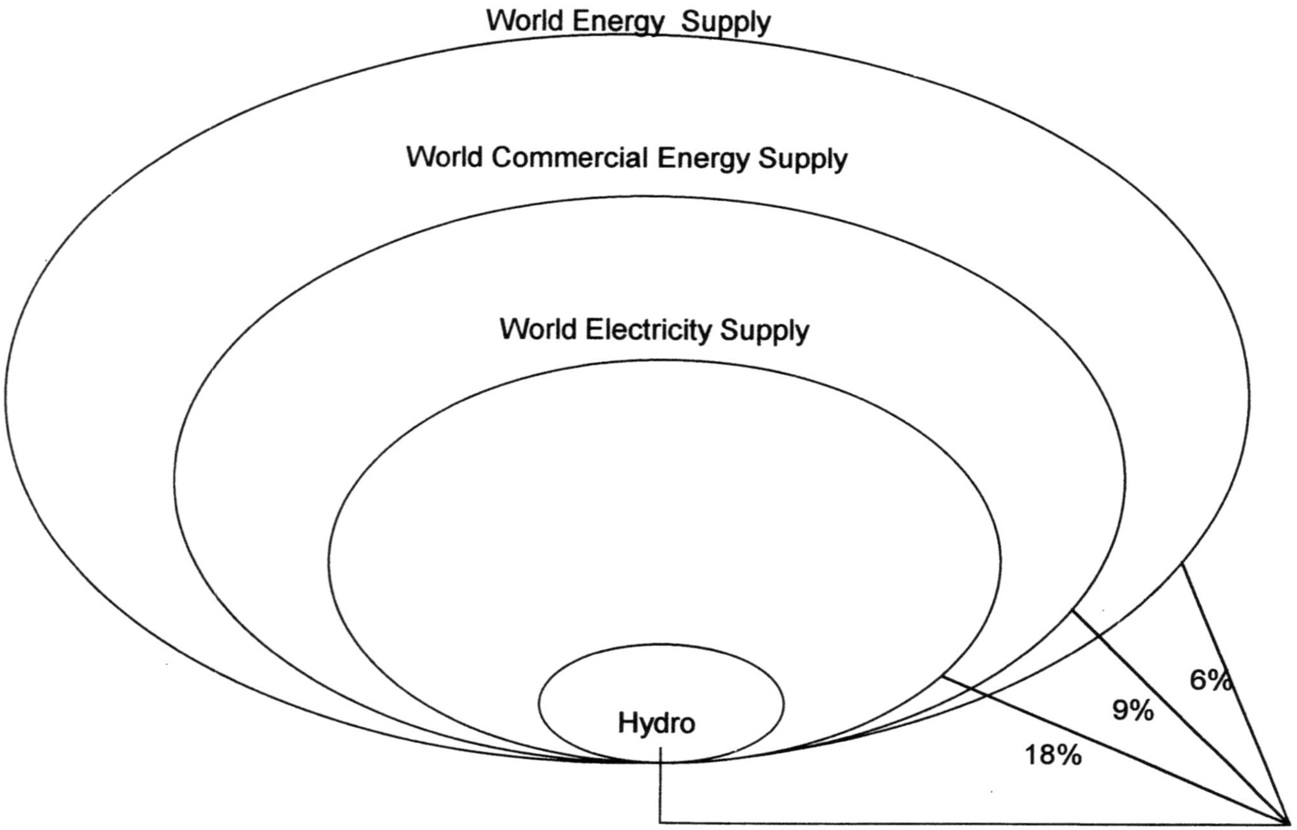

Exhibit - 1

World Energy Supply

World Commercial Energy Supply

World Electricity Supply

Hydro

6%

9%

18%

Carbon-di-Oxide Emission Factors *Exhibit - 2*
from
the full Energy Chain of Different Power Sources

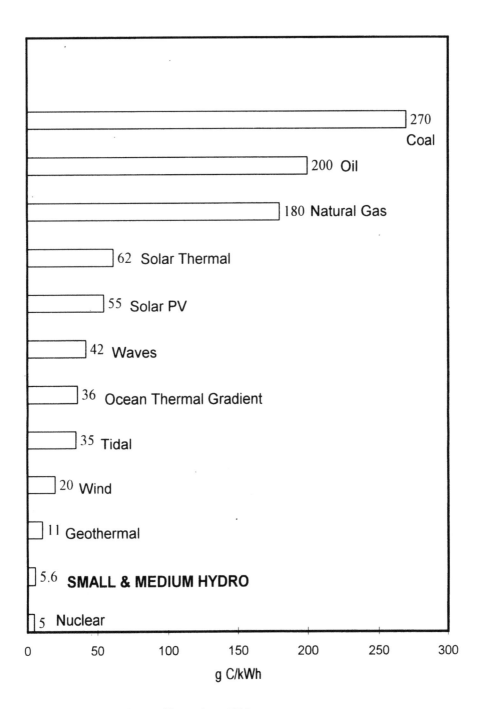

g C/kWh

Source :- Atoms in Japan, November, 1994

Exhibit - 3

POLICY / INCENTIVES FOR SMALL HYDRO SECTOR

ITEMS	MNES GUIDELINES	ANDHRA PRADESH	TAMIL NADU	KARNATAKA	ORISSA	KERALA	UTTAR PRADESH	MADHYA PRADESH	PUNJAB	HIMACHAL PRADESH
RATES/CHARGES										
a) Power Wheeling	2%	• 132 KV - 8% • 33 KV/11KV/LT * upto 50 km - 10% * 51 to 100 km -12% * beyond 100 km - 15%	15%	• 2% upto 1 MW • 5% upto 3 MW • 10% above 3MW	• 2% upto 3 MW • 8% upto 15 MW	12%	• 2% (Captive) • 2.5% (Third party)	2%	2%	2%
b) Power Banking	One year	Not Allowed	Allowed for captive	To be negotiated	At mutually agreed rate	At mutually agreed rate	Up to 1 year ; To be negotiated	Not Allowed	Not allowed	Allowed with additional charges
c) Buy back by SEB (Per kWh)	Rs 2.25	As per NTPC rate for Ramagundam	At mutually agreed rate	Rate not specified	At mutually agreed rate	At mutually agreed rate	Rs 2.25	Rs 2.25	Rs 2.25	Rs 2.25
d) Third party sale	Mutually agreed rate	Not Allowed	Not Allowed	Allowed	Allowed	---	Allowed	Allowed	Allowed	Allowed
e) Royalty on water	10% of elect. tarriff	As per rates fixed by concerned deptt. of govt.	Included in 1(a) above	10% of prevailing elect. tarrif	Included in 1(a) above	Included in 1(a) above	10% of electricity generated	-	-	• 1 - 3 MW-10 % • 3 - 15MW-12% Exempn for first 5 year upto 1MW
INCENTIVES										
a) Capital subsidy	---	-Up to 15 lacs per project -Up to 20 lacs in backward area	-10% of cost of equipment -Max 15 lacs	-As extended to other industries- Max 5 lacs -Additional subsidy: 5% of cost of equipment (Max 5 lacs)	-As given to other industries	---	---	---	---	---
b) Elecy. Duty Exemption	-Exemption of electricity duty							-Exemption for 5 years	Exemption for 5 years	-Exemption for 5 years
c) Sales tax Exemption	-Exemption from sales tax & other concessions applicable to industry/ backward area			-Exemption for 5 years for captive New Industry incentives				-Exemption from sales tax & other concession applicable to new industry	Exemption from sales tax&other concession applicable to industry/ba-ckward area	No Sales tax on power generation and transmission eqpt. And building material for power projects
d)Demand charge Exemption	Upto 30%							-Upto 30%	-Upto 30%	-Upto 30%
e) Promotional agency		APSEB	TNEB	KREDL	OPGCL	KSEB	UPLJVN/ UPSEB/ NEDA	MPUVN	PSEB/ PEDA	HPSEB/ HIMURJA

Exhibit - 4

TOTAL IDENTIFIED POTENTIAL OF SHP

(States where policies for Pvt. sector participation are in place)

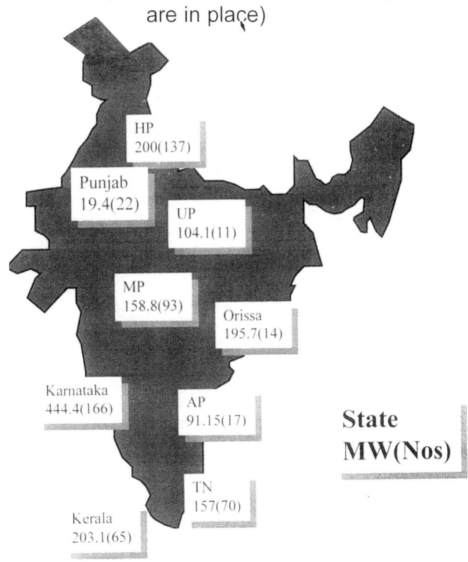

HP
200(137)

Punjab
19.4(22)

UP
104.1(11)

MP
158.8(93)

Orissa
195.7(14)

Karnataka
444.4(166)

AP
91.15(17)

TN
157(70)

Kerala
203.1(65)

State
MW(Nos)

Note: The national boundaries shown are only indicative and not actual

WORLD BANK PROJECTS FINANCED BY IREDA - TECHNOLOGY MATRIX

Exhibit - 5

State	Name of the Project	Capacity kW	Net Head (M)	Civil Works			Turbine	E & M Works		Generation Voltage KV	Transmission Details		
				Type of Scheme	Water Conductor System	Gate / Valve		Control	Power House		Poles	Cond	Meter
Andra Pradesh	Gundur BC- II	2 x 2150	8.5	Canal Drop	Open Channel	Gates	S Type Kaplan	Digital	RCC + S	11	LT	ACSR	DTM
	Lock in Sula	2 x 2000	12.2	- Do -	- Do -	- Do -	- Do -	- Do -	- Do -	11	- Do -	- Do -	- Do -
Karnataka	MalaPrabha	2 x 1200	10	Canal Drop	- Do -	- Do -	- Do -	- Do -	- Do -	3.3	- Do -	- Do -	- Do -
	Shiva	2 x 1500	8.3	- Do -	- Do -	B.V	SType Semi Kaplan	- Do -	- Do -	3.3	- Do -	- Do -	- Do -
	Dhupdal	2 x 1400	4.8	Dam Toe	Penstock	Gates+B.V	- Do -	- Do -	RCC + M	3.3	Con	- Do -	- Do -
	Shahpur - 1	1 x 1300	6.2	Canal Drop	- Do -	- Do -	S Type Kaplan	- Do -	RCC + S	3.3	LT	- Do -	- Do -
	Shahpur - 2	1 x 1300	6.2	- Do -	- Do -	- Do -	- Do -	- Do -	- Do -	3.3	- Do -	- Do -	- Do -
	Shahpur - 3	1 x 1300	6.2	- Do -	- Do -	- Do -	- Do -	- Do -	- Do -	3.3	- Do -	- Do -	- Do -
	Shahpur - 4	1 x 1300	6.2	- Do -	- Do -	- Do -	- Do -	- Do -	- Do -	3.3	- Do -	- Do -	- Do -
	Shahpur - 5	1 x 1400	9.8	- Do -	- Do -	- Do -	- Do -	- Do -	- Do -	3.3	- Do -	- Do -	- Do -
	Anveri	2 x 750	22	- Do -	- Do -	Gates	Francis	- Do -	- Do -	3.3	- Do -	- Do -	- Do -
	Hemavathy	4 x 4000	16	Dam Toe	Penstock	Gates+B.V	S Type Kaplan	- Do -	- Do -	11	- Do -	- Do -	- Do -
	Maddur	2 x 1000	13.2	Canal Drop	- Do -	Gates	S Type Semi Kaplan	- Do -	- Do -	3.3	- Do -	- Do -	- Do -
	Mudhol	1 x 1000	13.1	- Do -	- Do -	- Do -	- Do -	- Do -	- Do -	3.3	- Do -	- Do -	- Do -
	Deverebalekare	2 x 1000	10.9	Dam Toe	Penstock	- Do -	- Do -	- Do -	- Do -	3.3	- Do -	- Do -	- Do -
Kerala	Western Kallar	1 x 3000 + 2 x 1000	66	R-O-R	- Do -	Gates+B.V	Francis	- Do -	- Do -	6.6	- Do -	- Do -	- Do -
	Karikkayam	2 x 7500	20	- Do -	Penstock	- Do -	Vertical Kaplan	- Do -	- Do -	11	- Do -	- Do -	- Do -
	Ullungal	2 x 3500	10	- Do -	Penstock	- Do -	S Type Kaplan	- Do -	RCC + M	6.6	- Do -	- Do -	- Do -
	Boothathankettu	4 x 4000	9.5	Dam Toe	- Do -	Gates	S Type Kaplan	- Do -	- Do -	11	Con	- Do -	- Do -
	Kuthungal	3 x 7000	134	R-O-R	PressureTunnel	- Do -	Francis	- Do -	- Do -	11	- Do -	- Do -	- Do -
Tamil Nadu	Periyar Vaigai	3 x 2200	17	Canal Drop	Open Channel	- Do -	S Type Semi Kaplan	- Do -	- Do -	11	- Do -	- Do -	- Do -

Legend:
RCC + S : RCC Structure + Stone Masonary + Sheet Roof Top   B.V : Butterfly Valve   ACSR : Aluminium Conductor Steel Reinforced
RCC + M : RCC Structure + Stone Masonary   LT : Lattice Steel Structure   Con : Concrete Poles
DTM : Dual Tarrif Meter

Plate - 1

Return Canal Hydel Potential
at
Farakka Super Thermal Power Project of NTPC (West Bengal)
[750 kW can be retrieved from the condenser outlet]

Plate - 2

Ullunkal Hydroelectric Scheme [2 x 3.5 MW], Kerala
[Being executed under World Bank Programme]

Plate - 3

Typical Portable Micro Hydel Set standing on pedestals
[50 such sets have been provided by MNES free of cost
to hilly and isolated areas in India]

Plate - 4

Sarahan Portable Micro Hydel Scheme
Himachal Pradesh
[15 kW set working under 72 m head]

First International Conference on Renewable Energy–Small Hydro
3 – 7 February 1997, Hyderabad, India

MICRO-HYDROPOWER DEVELOPMENT IN INDONESIA — PAST AND FUTURE PERSPECTIVES — A CASE STUDY

Mark Hayton

Swiss Centre for Development Cooperation in Technology and Management (SKAT)
Vadianstr 42, CH - 9000, St.Gallen, Switzerland

ABSTRACT

Due to increasing global awareness of the negative impacts fossil fuels impose on the environment, an energy policy with a priority on the exploitation of available natural resources has obvious benefits for a developing country like Indonesia, especially for rural electrification
Indonesia's tropical mountainous terrain provides almost unlimited possibilities for small hydropower development. Stand alone *off-grid* schemes are an economically sound solution for the electrification of remote rural villages beyond the reach of utility grid lines. Spiraling energy demands primarily from the urban industrial centers meanwhile have forced the government to deregulate the power sector in order to meet demand, thus providing investment opportunities for the private power producer. As proof of the governments commitment to promote the exploitation of environmentally friendly energy sources wherever possible, non negotiable small power purchase tariffs (SPPT's) and standard power purchase contracts will be announced by the ministry of mines and energy shortly. GTZ has been active in the development and dissemination of micro-hydropower technology in Indonesia since 1991. After focusing on stand alone schemes supporting the governments rural electrification for the past five years, the project now enters a new phase hoping to pioneer the development of on-grid schemes selling power to the state owned utility. Past, present and future, the commitment to supply electricity throughout the archipelago determines that Indonesia's power sector is a dynamic ever changing environment.

BACKGROUND

The GTZ - MHP Project was initiated in 1991. Based in Bandung, West Java, the project works in close cooperation with the Directorate General of Electricity and Energy Development (DitjenLPE), government and non-government technology institutions and small private companies active in the field of small hydropower. After initial surveys were carried out to assess the existing micro-hydropower environment, it was clear that most efforts that had been made in the field of stand alone micro-hydropower had been unsuccessful. The majority of existing plants were either out of operation or running well below optimal performance. The reasons for this were the following :

⇒ poorly designed and constructed civil works
⇒ low quality E/M equipment not fulfilling the technical requirements
⇒ transmission and distribution grids far below PLN village electrification standards
⇒ weak organizational structures resulting in poor plant management

PILOT-PHASE - 1991-1996

Over the course of the pilot phase, emphasis has been placed on establishing a sound and comprehensive technology transfer facilitating the local manufacture and supply of electro-mechanical equipment for MHP plants. Working closely with international technology institutions and private companies specialized in small hydropower equipment design and manufacture, the project has developed a number of local manufacturing and assembly facilities. These are capable of supplying the complete electro-mechanical equipment requirement for MHP schemes up to 150 kW in size. The exploitation of local resources has resulted in competitive prices / kW installed capacity of the equipment being maintained. Moreover, equipment produced by the local manufacturers has been exported to projects in Uganda, Zaire and Papua New Guinea.

PROJECT OBJECTIVES

Overcome the shortfalls and initiate a 10-100kW MHP market by supporting an environment which boosts Indonesian promoted MHP development. Focus initially on off-grid schemes subsequently progressing to on-grid plants.
⇒ establishing MHP implementation models (financing, ownership...)
⇒ securing investment capital.
⇒ defining and transferring the appropriate MHP technology.
⇒ assisting partners in MHP project implementation starting at design stage through to commissioning
⇒ enhance successful management & operation capacity.

PROJECT ORGANISATION

⇒ The "Micro Hydro Power (MHP) Technology Dissemination Pilot Project" is
    - based on a Memorandum of Understanding between DitjenLPE & GTZ.
    - DitjenLPE's MHP program is backstopped by the GTZ project.
⇒ Implementing counterparts are :
    - various institutions (NGOs)/ companies/ consultants active in MHP development
⇒ GTZ technical assistance consists of approx. 1.5MioU$ over 4 years :
    - 1x MHP expert & 1 x backstopper
    - R&D for appropriate equipment and implementing procedures
    - training of design engineers, manufacturers, plant managers & operators
    - PR and MHP documentation
⇒ GTZ only supports Indonesian efforts to promote MHP by providing specific inputs in crucial areas of MHP development. Besides R&D components there was no contribution to the investment in power plants.

PROJECT ACTIVITIES

⇒ Technology transfer for MHP design, manufacturing & installation including:
    - cross flow and Pelton turbines
    - generating & control systems
    - civil construction
    - power distribution systems
⇒ Setting up MHP implementation, management & operation procedures :
    - surveying, feasibility studies & site selection
    - design & implementation supervision
    - project monitoring & evaluation
    - various training
⇒ Backstopping / supporting the small hydropower component of the DitjenLPE village electrification program:
    - potential site identification and selection

- commissioning of MHP plants
- Introduction of innovative institutional set-ups proven as applicable for decentralized MHP plant operation & management.
- regular monitoring of installed plants
- documentation of installed plants

MHP PLANT FINANCING, IMPLEMENTATION AND OPERATION

A total of 26 x MHP schemes throughout 6 provinces have been implemented with technical assistance provided via the GTZ - MHP Project. Numerous sources have been exploited to finance the installations. These consist primarily of central and provincial government budgets (APBN, APBD) allocated for infrastructure development related projects. This investment made by the Indonesian government has exceeded **Dm 1,000,000 (one million Dm)** since the project started. The involvement of local banks for the provision of soft loans to finance specific components of schemes is also project policy. Numerous ownership and management models have been applied amongst the different schemes implemented by the project. These range from pure commercially based projects to more community orientated set-ups such as village electricity cooperatives and electricity utilities managed as village owned enterprises.

⇒ MHP project financing :
  - DitjenLPE (Rural Electrification Program)
  - local government =>civil construction, power distribution
  - private investors
  - banks (bank Exim, BPD)
  - other donor organizations (foreign and local)

IMPACT AND ACHIEVEMENTS

Proportionally, the impact of the project on the nations power supply is negligible. When viewed in the context of the potential impact assuming a multiplication of the small hydropower development is attained in the future, the projects real impact takes on more relevance. The following achievements can be attributed directly to the presence of the project between 1991-95:

⇒ 26 MHP plants built (or under construction) with Technical Assistance from the DitjenLPE/GTZ MHP project

Table 1. Summary of projects implemented

	NGO	Govnt.	total	unit
Inst. Capacity	203	315	518	kW
Energy	1'684	1'052	2'737	MWh
Operation Time	180'000	41'000	221'000	hrs
Spec. Costs inst. capacity	1,600	2,800	2,200	US$/kW
Diesel avoided	505'268	315'733	821'000	liters
Fuel costs@UScents 22/l	187	117	180,000	US$
CO2 avoided	1'516	947	2'463	tons
On-grid @Uscents 7/kWh	211	132	155,500	US$

⇒ Standard MHP packages are available including:
  - technical engineering ((pre) feasibility, design...)
  - social preparation (awareness programs, impact studies...}
  - hardware manufacturing (electro mech. equipment, penstocks...)
  - construction of entire rural electrification schemes
  - training modules for operators & managers
⇒ several companies are now active in MHP development
⇒ 6'000 rural households electrified
⇒ More than US$ 180,000 savings in "avoided fuel costs"
⇒ 220'000 operating hours of locally manufactured generating equipment accumulated

⇒ 2'500 tons of CO_2 emission avoided

ECONOMIC VIABILITY

MHP is without question the most economically viable long term solution to meet this demand where there exists a hydraulic potential. When calculated over a period of >5 years, MHP plants are economically more attractive than fast track solutions such as diesel gensets, the most common solution for isolated grid rural electrification . Real financial viability, however, is only reached when repaid over periods exceeding 10 years.

Table 2: Typical financial viability of a stand alone MHP plant (11 kW)

Year	cash in	cash out	net in-out	Present worth of payment	increase of NPV
	Mio P/y	Mio P/y	Mio P/y	Mio P/y	Mio P/y
0		43.00	-43.00	-43.00	-43.00
1	9.80	1.00	8.80	8.32	-34.68
2	9.80	1.00	8.80	7.87	-26.80
3	9.80	1.00	8.80	7.45	-19.35
4	9.80	1.00	8.80	7.05	-12.31
5	9.80	3.15	6.65	5.04	-7.27
6	9.80	1.00	8.80	6.30	-0.96
7	9.80	1.00	8.80	5.96	5.00
25	9.80	1.00	8.80	2.19	68.11

Inflation Rate	5.00%	**Net Present Value**	68.11	
Discount Rate	11.00 %	**Internal Rate of Return**	20%	
real. inter. rate	5.71%			
O&M	1			
Investment	43			

The relatively high initial investment costs of Rp 43 Mio (Rp 4 Mio / kW installed capacity).is repaid within 6 years based on a revenue stream of Rp 8.8 Mio / year (using the net present value equation). Assuming an inflation rate of 5%, a discount rate of 11%, operation and maintenance costs of Rp 1 Mio / year, reinvestment of Rp 2.15 / 5 years and a lifetime of 25 years the *net present value* is positive and shows an amount of Rp 68.11 Mio. The *internal rate of return* is 20% which reflects the economic viability. For such conditions, a bank will offer a loan of up to 75% of the initial investment.

Despite its financial viability, effective dissemination of MHP as a solution for rural electrification is still hindered by the following :
⇒ There does not exist an Indonesian MHP information network;
⇒ There is a lack of awareness on behalf of local governments/villages with regard to the complexity of running a decentralized electricity utility;
⇒ No effective dissemination strategy exists;
⇒ Standard cost guidelines for decentralized MHP plants do not exist;
⇒ No technical or institutional guidelines exist for MHP project implementation;
⇒ Government budgets and regulations applied for MHP project implementation are unrealistic and discouraging for MHP contractors;

GOVERNMENT POLICY CHANGES IN THE POWER SECTOR WHICH WILL INFLUENCE THE FUTURE OF MHP DEVELOPMENT

Rural Electrification Is a National Program

In 1983 Rural Electrification became a national program. The objective of the program is to provide electricity to the entire population of the 67.000 Indonesia villages. The villages vary widely in size from an average of 732 households per village in Java to 265 in Sumatra and 144 in Kalimantan. Human resources also vary considerably influencing development strategies for programs such as rural electrification. The villages are categorized as Swasembada (modern), Swakarya (transitional) and Swadaya (traditional) which reflects their level of development.

Small Power Purchase Tariff and Standard Power Purchase Contract Announced

The ministerial decree announced at the beginning of 1996 is aimed at promoting and encouraging the development of renewable energies, not only for rural electrification but also to reduce the requirement of fossil fueled power plants. It is hoped that this legislation will attract investment from the private sector and cooperatives in the building of small grid connected power plants. Small Power Purchase Tariffs (SPPT) will be announced annually by PLN for each of the 11 PLN regions in Indonesia. Small hydropower is included in the first priority category. These rates are not negotiable and PLN is obliged to purchase from the small power producers up to limits announced annually by the government (<30MW for Java and <15MW for outside Java). These rates are calculated based on the "avoided costs" of PLN not having to produce the same amount of power. Particularly on the outer islands where generation costs are proportionally high; the purchase rates offer very lucrative investment opportunities for the private sector and cooperatives. Preference is given to priority one groups (i.e. hydropower). This is also reflected in the higher purchase rates offered to renewable energy producers. A standard power purchase contract will also be issued streamlining the negotiation process with PLN.

Table 3. Power purchase tariffs announced by the Indonesian government - January 1996

Region and interconnected system	Capacity charge in term of		Energy charge	
	Peak [UScent/ kWh]	Off Peak [UScent/ kWh]	Peak [UScent/ kWh]	Off Peak [UScent /kWh]
Zone 1 Java-Bali	7	0.5	5.5	4
Zone 2 Region II, Medan system	8	0.6	7	4
Zone 3 Region III, Padang system Region IV, Palembang system Region VI, Barito system Region VIII, Ujung Pandang system	9	0.7	8.5	6
Zone 4 Region V, Pontianak system Region VII, Manado system	8.5	0.6	5.5	5.5
Zone 5 Isolated diesel systems	11	1	6	5.7

Decentralization is Government Policy

The Indonesian government also prudently adopts a decentralization policy transferring more autonomy and responsibility to provincial and district governments. To reduce the financial dependence on central government budgets, provincial and district administrations are encouraged to develop, own and operate private, profitable companies. At present in each province one "autonomous district" (kabupaten) has been established to test decentralization legislation developed at central government level.

Cooperatives Are Encouraged to Participate In The Power Sector

To support the rural electrification process and relieve the burden on PLN, the Ministry of Cooperatives was given a mandate by the government outlined in the presidential decision (Keppres) No. 37/1992, to play a greater role in rural electrification. The POLA program implemented between the Ministry of Co-operatives and PLN provides alternatives with varying degrees of responsibility handled by the cooperatives from simply monthly tariff collections to the full operation and management of small grids via bulk supply from PLN. The implementation of the POLA program is coordinated by DitjenLPE on behalf of the Ministry of Energy and the Directorate Bina Koperasi Perindustrian dan Ketenagalistrikan on behalf of the Ministry of Co-operatives at central and provincial level. Reacting to the pressure for improved efficiency and a more commercial management approach ,

The Ministry of Cooperatives POLA scheme is summarized as follows ::

POLA 1

PLN assigns the task to village cooperatives (KUD) to undertake the following:
⇒ meter reading
⇒ tariff collection
⇒ house installation repair & maintenance
⇒ maintenance of distribution lines

POLA 2

PLN assigns the task to KUD to subcontract the following:
⇒ erection of low voltage lines
⇒ house wiring
⇒ electricity survey
⇒ staking out distribution line routes
⇒ purchase of elec. equipment

POLA 3

KUDs which have undertaken POLA1&2 and/or purchase bulk energy from PLN or private producers resell electricity to consumers within their grid.

POLA 4

KUD undertake the entire electrification (generation, installation, distribution)

POLA MSA Grid/ (supervisery office)

KUD manages/operates entire PLN assets within this PLN sub branch (office, grid...)

POLA MSA Diesel/ (sub ranting PLN)

Same for isolated Diesel grids (takes over all tasks of former PLN office)

For both this means:
⇒ meter reading
⇒ tariff collection

⇒ house installation repair & maintenance
⇒ maintenance of distribution lines
⇒ maintenance of generator & distribution
⇒ operation of generator & distribution
⇒ administration and reporting

EMPHASIS FOR FUTURE COOPERATION (POST 1995)

The rapid grid extension into rural areas has created a very different environment to that of a decade ago. In the forthcoming years the demands on the rural electrification program will be focused on electrifying clusters of houses located away from the PLN lines, not necessarily entire villages. It is stated in the five year development plan (REPELITA) that by 2003/04 all Indonesian villages will be electrified. Whilst this target may be somewhat optimistic, it is already apparent that over the next decade the majority of villages throughout the archipelago will be electrified.

In view of this fact it is necessary to support a strategy designed to address specific target areas within the target group such as isolated clusters of houses in villages already electrified as well as entire villages. To address the issue of rural electrification effectively, a diverse approach including not only micro-hydropower but also grid extension, diesel gensets, solar house systems etc. needs to be developed. Institutions need to be established at provincial level which can evaluate and select suitable solutions on a case to case basis. This approach has already been initiated by the government with the establishment of rural electrification cooperatives (Koperasi Listrik Perdesaan [KLP]).

There presently exists only two KLP's in Indonesia (Lampung and Lombok). These rely entirely on diesel generators for their supply capacity and as a result incur high operating costs and struggle to generate sufficient revenue to expand their operations into the more remote regions. New government legislation described earlier permitting the sale of electricity to PLN from grid connected micro-hydropower plants operated by village cooperatives presents an opportunity for KLP's to operate on a more profitable basis avoiding the high fuel costs incurred by diesel gensets. Once profitable, KLP's will thus be in a position to implement rural electrification programs in line with regional master plans at their own discretion.

In the past the implementation of micro-hydropower projects has been carried out largely independently of other activities contained in the national rural electrification program. The rapidly reducing numbers of villages without electricity, however, dictate that this approach is not longer acceptable. Investment in the building of micro-hydropower plants is only economically viable and can only be justified when calculated from a long term perspective. This limits the amount of feasible sites for stand alone schemes to sites located in extremely remote regions where PLN is unlikely ever to extend its grid lines. A more diverse policy for effective exploitation of the countries hydraulic resources must therefore be applied. The overall objective is :

Support the national rural electrification program providing access to electricity for rural people via the sustainable exploitation of the nations hydraulic resources

The deregulation of the power sector now allows for decentralized power stations on both a large and small scale to feed into the PLN grid network. This opens the door for small scale renewable energy producers to generate and sell their power to the PLN via the national grid network. With this procedural superstructure established, the development of the small hydropower sector takes on a new perspective. The opportunity to exploit the abundant hydraulic potential available in Indonesia with on-grid decentralized mini hydropower plants is presented. Whilst the demand for isolated grid MHP plants supporting the rural electrification efforts of the government still exists, the new frame conditions facilitating grid connected schemes dictate that a shift in project policy is necessitated in order to more effectively address the improved climate and the challenges of rural electrification. Future technical assistance in the field of SHP aims to exploit these favorable frame conditions and contribute to the overall development of the Small Power Project (SPP) sector.

Flow chart 1 in annex 1 illustrates a concept developed by GTZ for sustainable development of cooperative owned decentralized hydropower plants generating a constant revenue stream to finance local rural electrification activities.

PERSPECTIVES FOR MHP DEVELOPMENT:

⇒ the potential of $1m^3$/sec over a drop in elevation of 10m is 100kW
⇒ the production over a year would be 800'000 kWh
⇒ at 125Rp/kWh this amounts to 100 Mio Rp/year
⇒ with this a loan of 10Mio Rp/kW could be satisfied (7%,15yrs, inflation deducted)
⇒ the required investment varies between 3 and 8 Mio Rp/kW (see details in the Annex)
⇒ the IRR based on a 30% equity is 30% or higher!!!
⇒ a MHP plant normally outlives the loan by decades

MHP OPTIONS: WHERE TO INVEST?

⇒ MHP advantages:
      - producing high quality electricity locally
      - renewable energy resource with minimal environmental impacts
      - energy storage simple, efficient and often possible
⇒ Lucrative applications for "avoided cost" schemes :
      - selling electricity to a PLN grid
      - replace Diesel gensets
      - cut peak demand

MHP CLIENTEL : WHERE ARE THE POTENTIAL INVESTORS?

Besides developing new MHP schemes, there exists a large number of old MHP schemes which are either abandoned or run inefficiently due to the lack of reinvestment.
⇒ Decentralized industries (small)
      - ice factories...
⇒ Plantations
      - tea estates...

another large group of interested investors are all those who could replace electricity produced by inefficient isolated diesel grids (PLN)

⇒ Decentralized grids :
      - cooperatives
      - village enterprise
⇒ PU (Dept. of Public Works)
      - drop structures in irrigation channels

The economic viability calculation seen in Annex 2 was made for an existing scheme in West Sumatra (Zone 3). The scheme originally built in 1923 with a potential of approximately 1MW presently only produces 100kW supplying a nearby ice factory and a number of hamlets in the close vicinity of the plant.

WHICH PRIORITY AREAS NEED ADDRESSING?

A lucrative PLN purchase tariff (SPPT) based on a simple contract is just one factor in the Return on Investment (RoI) equation. Other factors are equally critical such as:

⇒ accurate site data
⇒ simple decentralized application procedures & MHP development permissions ("single window")
⇒ transparent legal requirements
⇒ suitable credit lines made available (loan duration, interest rates, grace periods, collateral required)
⇒ transparent minimum technical standards published
⇒ support & incentives for MHP plant owners / operators
⇒ local engineering & manufacturing competence

MHP DEVELOPMENT : WHAT ARE THE REQUIREMENTS?

⇒ easy access to relevant data :
⇒ hydrological yearbook (run off data)
  - meteorological (rainfall data)
  - topographical maps
  - master plans
  - existing feasibility studies and designs
⇒ simple transparent procedures:
⇒ legal requirements (water rights, liability...)
  - technical standards
  - SPPT, standard contract
  - taxes
⇒ Incentives :
⇒ tax/duty exemptions (import, VAT, income ...)
  - information dissemination
⇒ Technical / managerial support :
⇒ local engineering and manufacturing capacity
  - local maintenance support
  - training for all key players

SUMMARY

⇒ An MHP development program should address several potential areas (PLN grid connection, Diesel replacement...)
⇒ All over Indonesia a multitude of interested investors in lucrative MHP development areas can be found.
⇒ To launch the program, rehabilitating existing MHP schemes poses less difficulties and makes the "learning curve" steeper.
⇒ Besides favorable credit lines, tariffs and contracts, a number of incentives, support and services have to be provided.
⇒ Local engineering, manufacturing and management capacity is essential for a sustainable operation of MHP plants and for a developing MHP market.
⇒ A Technical Assistance (TA) component alongside an RED Small Hydro Power (SHP) Development Program is essential to facilitate the process.
⇒ The option to sell power to the grid creates an opportunity for municipalities and cooperatives to generate revenue to finance sustainable rural electrification programs.

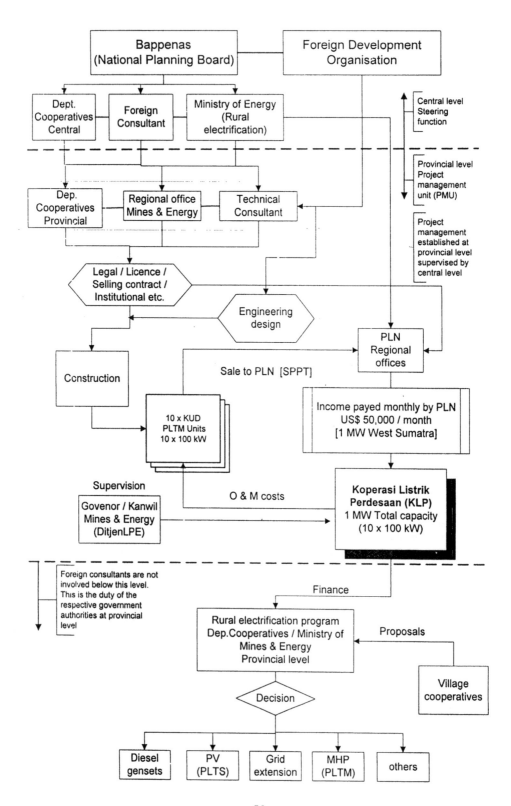

Bappenas
(National Planning Board)

Foreign Development
Organisation

Dept.
Cooperatives
Central

Foreign
Consultant

Ministry of Energy
(Rural
electrification)

Central level
Steering
function

Dep.
Cooperatives
Provincial

Regional office
Mines & Energy

Technical
Consultant

Provincial level
Project
management
unit (PMU)

Project
management
established at
provincial level
supervised by
central level

Legal / Licence /
Selling contract /
Institutional etc.

Engineering
design

PLN
Regional
offices

Construction

Sale to PLN [SPPT]

10 x KUD
PLTM Units
10 x 100 kW

Income payed monthly by PLN
US$ 50,000 / month
[1 MW West Sumatra]

Supervision

Govenor / Kanwil
Mines & Energy
(DitjenLPE)

O & M costs

Koperasi Listrik
Perdesaan (KLP)
1 MW Total capacity
(10 x 100 kW)

Foreign consultants are not
involved below this level.
This is the duty of the
respective government
authorities at provincial
level

Finance

Rural electrification program
Dep.Cooperatives / Ministry of
Mines & Energy
Provincial level

Proposals

Village
cooperatives

Decision

Diesel
gensets

PV
(PLTS)

Grid
extension

MHP
(PLTM)

others

MHP economic viability assessment :

Salido Kecil 1000 kW MHP supplying only the PLN grid (Zone 3) based on power purchase tariff 1996

	Year	Cash in US$/y	Cash out investment US$/y	Cash out O&M/other US$/y	Cash out total US$/y	cash flow Net (in-out) US$/y	discount factor	pres. worth of net in-out US$/y	increase of NPV US$/y	increase of IRR US$/y
construction	0		1'200'000		1'200'000	-1'200'000	1	-1'200'000	-879'545	
	1	414'545		30'000	30'000	384'545	0.833	320'455	-585795	-22%
	2	456'000		33'000	33'000	423'000	0.694	293'750	-316'525	3%
	3	501'600		36'300	36'300	465'300	0.579	269'271	-69693	17%
	4	551'760		39'930	39'930	511'830	0.482	246'832	146'861	25%
re-investment	5	606'936	24'158	43'923	68'081	538'855	0.402	216'554	354'268 ◄	30%
	6	667'630		48'315	48'315	619'314	0.335	207'407	544'391	34%
	7	734'393		53'147	53'147	681'246	0.279	190'123	718'671	36%
	8	807'832		58'462	58'462	749'370	0.233	174'280	878'427	38%
	9	888'615		64'308	64'308	824'307	0.194	159'756	1'018'587	39%
re-investment	10	977'476	38'906	70'738	109'645	867'832	0.162	140'160		

Parameter Box - enter values in the shaded boxes !!

Infl. factor	10.00%	Investment	1'200'000	US$
Interest rate	20.00%	Investment	1'200	US$ x1000
Lifespan	10 yrs.	O&M/Other		
Inst.capacity.	1'000 kW	O&M/Other	30.00	US$ x1000
Availability	0.60 fac.	Re-invest fac.	0.25	factor
Operation	8000 hrs/yr	Gross income	414'545	US$

Utilised energy production / year	4'800'000	kWh
Annual investment costs	121'599	US$
Present worth of reinvestment	15992	US$
Annual interest costs	120'000	US$

Viability indicators :

Net Present Value [NPV]	1'018'587
Internal rate of return [IRR]	39%

Investment costs :

Rs./kW installed capacity.	2'640'000
US$/kW installed capacity.	1'200

Energy production costs :

Prod.cost/kWh [UScents.]	5.033

Positive NPV is reached after only 5 years !!

First International Conference on Renewable Energy—Small Hydro
3 – 7 February 1997, Hyderabad, India

THE STUDY OF SMALL SCALE HYDROELECTRIC POWER SCHEMES IN THE UNITED KINGDOM

K.V. Rao[1] *and I.F. Kirby*[2]

[1] Director of Water Resources Engineering, City University
 Northhampton Square, London EC1V 0HB, UK
[2] Flynn and Rothwell Consulting Engineers
 Thomas Tredgold House, 231 London Road, Bishops Stortford
 Herts CM23 3LA, UK

INTRODUCTION

Hydropower energy is one of the most widely used methods of energy production in the world. Approximately one quarter of the world's energy demand is provided from using hydropower technology. From humble beginnings in the form of water wheel systems, hydropower has developed into a highly technical means of energy production. Modern hydroelectric schemes are extremely complex developments involving many disciplines and activities that require intensive study to ensure economic success. Such an examination is undertaken in the form of a feasibility study.

The objective of a feasibility study is to determine whether a particular project configuration, identified during a reconnaissance study, should be constructed. If the results of this study show that one of the project alternatives is economically feasible, a detailed economic appraisal and a financial feasibility analysis is then prepared. These studies require power marketing and power systems operating studies, as well as a more detailed concept design and cost estimate. In most cases, an environmental assessment is also required. A reconnaissance study, carried out before the main feasibility study, is used to determine whether a potential project is likely to be a success. If the results of this study suggest that the proposed scheme is viable, further time and effort can than be deployed on more detailed examinations through the feasibility study.

The objective of this paper is to give the reader an insight into the procedures involved in undertaking a modern feasibility study for a hydroelectric project. To achieve this the planning of a feasibility study is first considered as a whole. The individual elements or areas which go to form a feasibility study are then considered in detail. These areas include 1. Topographical Survey, 2. Feasibility Plan, 3. Hydrology, 4. Geology, 5. Power Potential Studies, 6. Storage Reservoirs, Dams and Weirs, 7. Environmental Aspects, 8. Legal Considerations and 9. Economics.

A discussion of the undeveloped hydroelectric generation potential of the United Kingdom is also included within this paper. This is intended to give the reader an idea of the scale of the potential hydroelectric development in the UK. The principles behind feasibility studies will be demonstrated using a particular case study. A catchment area in Mid Wales at Plynlimon has been selected for this purpose. The catchment is located on the River Severn between its source at OS grid reference 2822E 2899N and the town of Llanidloes at OS grid reference 2946E 2839N. The catchment is managed by the Institute of Hydrology, based in Wallingford, Oxfordshire. It has been used for many years as an experimental catchment for the study of hydrological processes and consequently a vast amount of data exists for the site. For this reason the site was considered ideal for use in this paper.

SMALL SCALE HYDROELECTRIC SCHEMES

Small Scale Hydroelectric Schemes can be designed to suit many different situations. Such schemes are usually associated with rural locations, though schemes in built up urban locations are not uncommon. Many small scale hydroelectric schemes are provided to supply electricity locally to remote industry or housing where connection to the grid network would be expensive. The type of scheme arrangement depends mainly upon the available head of water for power generation. Schemes may be divided into three categories, namely; High, Medium and Low Head Schemes. Figure 1 shows some typical site layouts for the different types of scheme.

For a small scale hydroelectric scheme to operate successfully, various electrical, mechanical and civil engineering structures must be constructed and installed. Each individual component plays an important role in the overall scheme. The typical components of small scale hydroelectric schemes are; Forebay, Intake and Weir, Spillway, Stilling Basin, Headpond, Desander, Headrace canal (Leat), Penstock, Powerhouse, Turbine, Switch Yard and Transmission Line.

Depending upon the specific site requirements and design criteria, particular combinations of these individual components may be used. Two of the more common scheme layouts are shown in Figure 2. One of the main advantages of a small scale hydroelectric scheme is the ability to adapt to suit unique situations. Often schemes may be designed to utilise and incorporate existing infrastructures. This helps to reduce the capital expenditure of proposed schemes, greatly increasing financial feasibility in many situations. The most important factor governing whether a proposed hydroelectric scheme will go ahead is cost. The principle aim, therefore, of any feasibility study is to project capital costs and revenue earning potential. An economic comparison can then be made between the capital cost and the potential revenue to establish the feasibility of the scheme. The capital cost of a scheme is determined during the feasibility study by assessing the individual components that constitute a hydroelectric scheme. The potential revenue of a scheme is determined during the feasibility study by assessing hydrological and technical characteristics of an optimum design scenario.

HYDROELECTRIC GENERATION POTENTIAL IN THE UK

The first comprehensive investigation into the potential for small scale hydroelectric development in the UK was commissioned by the Energy Technology Support Unit, under the Department of Energy in May 1987. The investigation was carried out by Salford University Civil Engineering Ltd. The aim of the study was to identify the potential for small scale development (25KW - 5MW) and factors governing future development. In particular, factors known to be barriers to development, such as Local Authority rating, water abstraction charges, land ownership, energy use and environmental issues, were investigated.

The results of the investigations showed that the amount of small scale hydroelectric generation potential in the United Kingdom is dependant upon the economic climate. A good indicator of the economic climate is a parameter referred to as the economic rate of return. Consequently, the report and study produced by Salford University Civil Engineering Department reflects the potential capacity of the UK against various rates of internal return. The results of these detailed investigations are summarised in Figure 3. The report also identified that an internal rate of return of 10% would be a typical figure that would attract potential developers. With this in mind figure 4 Summarries the distribution of capacity, by area, for schemes with an internal rate of return ≥ 10%. This figure shows that the United Kingdom currently has the capacity to develop some 322MW of electricity from small scale hydroelectric schemes. This equates to some 1312GWh of energy production annually. The figure also shows that 88.7% of this potential (286MW) may be found in Scotland.

In order to assess the number and type of potential developments, the study distributed the capacity of schemes into hydraulic head categories. Table 1 shows that most of the potential development may be associated with high head schemes located in Scotland. Such schemes account for some 81.1% (261.1MW) of the total capacity.

Table 1

Head (m)	Generating Potential (MW)		
	England, Wales & N.Ireland	Scotland	Total
2-3	3.6	0.0	3.6
3-5	6.8	8.4	15.2
5-10	1.6	2.6	4.2
10-30	4.6	2.7	7.3
30-50	4.4	11.2	15.6
>50	15.3	261.1	276.4
Total	36.3	286.0	322.3

DEVELOPMENT OF POTENTIAL

The study carried out was also used to identify factors preventing potential development. The report identified the following factors which are known to affect the implementation of potential schemes; 1. Local authority rating, 2. Water abstraction charges, 3. Land ownership issues, 4. Energy use and 5. Environmental issues. Both local authority rating and water abstraction charges were identified by the study as major barriers to the development of small scale hydroelectric schemes. In particular, water abstraction charges were noted as severe barriers to the development of low head / high flow schemes. The variability of such charges and the uncertainty of the actual charge, prior to development, discouraged many developers. However, in order to readdress these problems, recent legislation has excluded schemes under 5MW from water abstraction charges, and local authority ratings no longer apply to small scale hydroelectric schemes. Changes in legislation have also resulted in non-fossil fuel obligations being placed on the Regional Electricity Companies and this has recently encouraged development of small scale hydroelectric schemes. Land ownership issues, energy use and environmental issues were also identified as barriers to the development of new schemes. These issues still remain as barriers to development as there is little that can be done to address them generally. Every scheme must therefore be assessed on its specific requirements and critical issues.

FEASIBILITY PROJECT PLANNING

A feasibility study consists of a series of dynamic or iterative activities that are combined and manipulated to yield the most economically feasible scheme. The aim of a feasibility study is therefore to narrow the focus from the many possible solutions to the single most economically feasible solution. This is normally achieved by applying the principles of cost/benefit analysis on a number of the more favourable options. Ultimately it is the economics of the proposed scheme that will govern whether the development will go ahead and for this reason the economic potential of the scheme is considered at each stage of the project. Progress to the next stage only proceeds if the forecast is favourable. Planning of this process can be complex and difficult, but is essential if a feasibility study is to be a success. The project plan enables the engineer to identify critical issues and formulate the objectives to be worked towards.

The study plan is a document which describes the aims and objectives of the feasibility study and the method of approach to be adopted. The plan is used to describe the current state of the scheme and how the development will be assessed. The plan should therefore reflect the thoughts of the engineer at the early stages of the project and detail critical issues and relevant design criteria for the rejection or acceptance of particular scheme scenarios. Such a process can also help to identify areas of difficulty that may affect the time it will take to achieve a full feasibility study. Poor or inadequate planning of a feasibility study can have a marked effect on the success of the study. Every effort should therefore be made in the planning of a project to ensure that as many as possible of the critical issues and potential problems are identified at the earliest possible moment. To achieve this, many projects are broken down into a number of study stages.

A typical hydroelectric development study will normally consist of three stages, namely; 1. Reconnaissance Study, 2. Economic Appraisal and 3. Feasibility Study. The aim of a reconnaissance study is to determine whether a site has the potential to be successfully developed. Because of the limited resources available for the implementation of reconnaissance studies, technical evaluation of the development site is often kept to the minimum. The reconnaissance study plays an important role in the identification of design criteria and critical issues to be dealt with during the feasibility study. Critical issues that come to light during the course of the reconnaissance study usually originate from; 1. Power Demand, 2. Geological Aspects, 3. Hydrological Aspects, 4. Power Potential, 5. Storage and impoundment of water, 6. Technical, 7. Environmental, 8. Legal Aspects, 9. Land Ownership and 10. Economics.

It may be noted that many critical issues can be identified by simply visiting the site and carrying out a visual inspection. Such a site visit will also help to size and estimate the magnitude of the project and aid the production of a study plan. Having established the design criteria and identified the critical issues a number of the more favourable scheme arrangements may be found using various processes. An economic appraisal of this limited number of arrangements can then be undertaken in order to assess the initial feasibility of the development.

The objective of a feasibility study is to identify the most economically viable scheme from the investigations and results of the reconnaissance study. The main feasibility study is then used to carry out detailed analysis, design, assessment and costing for a number of the most favourable arrangements. The use of general design parameters and estimating is avoided and, wherever practicably possible, full design and analysis techniques are used. However, detailed design is often carried out in a subsequent stage, in order to postpone costs which the feasibility study might conclude are not justified. Before undertaking a feasibility study a detailed approach methodology must be formulated. The details of the proposed project methodology should be included within the study plan. Because every scheme is unique and the method of analysis, design and liaison is specific to the particular requirements of the site under investigation the formulation of a project methodology will be complex. Figure 5 shows how even the simplest of project studies can become complicated and involved. Good and adequate planning is therefore essential to ensure the success of a project.

SITE EVALUATION

The evaluation of a proposed small scale hydroelectric development site is one of the most important stages of a feasibility study. Such a study is used to assess the potential of a project and formulate the scope of work. The study is used to evaluate the current status of the site and to collate data and information to aid the feasibility study. The objective of a site evaluation is therefore to assess the overall picture to allow the feasibility study to be undertaken on a meaningful basis. Such a study provides the raw data from which both the technical and economic feasibility of a scheme will be determined. It should avoid wasteful effort by rejecting sites found clearly not to be economically feasible. (Technical problems can usually be solved - at a cost). The importance of an adequate and relevant site evaluation is therefore without question. A typical site evaluation will include the following processes; 1. Data extraction, 2. A walk over site survey, 3. Site layout assessment, 4. A map and mapping assessment or review, 5. A topological survey, 6. A geological site assessment, 7. An assessment of the catchment and 8. area and catchment characteristics.

Obtaining basic design data for a site is a fundamental step of both reconnaissance and feasibility studies. Such data is essential to the advancement and progression of a project and will include hydrological, topographical and geological information. A walk over visual inspection forms one of the main constituents of the site evaluation. This inspection is used to identify the existing situation, catchment characteristics, potential critical issues and possible development scenarios. The inspection will also include an assessment of existing structures at sites where such infrastructure already exists. Maps and mapping play an important role in many aspects of typical reconnaissance and feasibility studies. The information found on maps is used for hydrological, geological and planning purposes and in the identification of catchment areas. Where existing mapping data is inadequate a topographical survey may have to be undertaken to supplement existing information as appropriate - whether for the sizing of the reservoir, the limits of flooding, the layout of access roads and infrastructure or the design of structures. Having reviewed and assessed the information from a site evaluation, a detailed scope of works may be formulated. Such a scope of works can then be used in the production of the project study plan. If a study plan already exists, then the scope of works may be used to update the plan as appropriate.

DATA EXTRACTION

Data extraction forms the basis from which the technical and economic feasibility of a proposed small scale hydroelectric power scheme will be determined. Data can be found from a variety of sources in a number of different ways. Detailed planning of data extraction techniques should be included within the project study plan. Basic data is required for a number of different reasons including, technical analysis, design and costing. Broadly speaking data should be collected to assess the impact on a proposed scheme of the following; 1. Hydrological aspects, 2. Geological aspects, 3. Technical considerations, 4. Environmental effects, 5. Legal considerations, 6. Land Ownership and 7. Economics.

The sources of the basic data required for a typical hydroelectric feasibility study will vary widely from project to project. Every project is unique and careful consideration should be given to establish all data sources. The following list details organisations and authorities that may be capable of supplying relevant information; 1. Ordnance Survey, 2. British Geological Survey, 3. Institute of Hydrology, 4. The Institution of Civil Engineers, 5. Local and District Councils, 6. County Councils, 7. Natural Environment Research Council, 8. Local Libraries, 9. Local Environmental Groups, 10. Water plc, 11. National Environmental Agency, 12. Private and Public Land Owners, 13. The United Kingdom Land Registry and 14. Planning Authority.

This list is not exhaustive and a study should be initiated as part of a feasibility process to identify other potential sources of data. The amount of time and effort spent does, however, have to be scaled with regard to priorities dependent on the size and importance of the small hydro project and the expense justifiable.

SITE LAYOUT

The development of an effective site layout forms the basis of design (Figure 1 shows some typical small scale hydroelectric scheme layouts). The decisions regarding the location of plant and equipment are taken during both reconnaissance and feasibility studies. The walk over site survey also helps in the process of plant and equipment location. The design of a site layout is an iterative process that evolves during the process of a feasibility study and is governed by technical, environmental, legal and cost restraints.

HYDROLOGY

The hydrological study of a catchment is the prime activity of any feasibility study. The purpose of a hydrological study is to determine the optimum size of the main scheme components and to project the annual energy generating capacity and reliability of a scheme. The study ultimately determines whether a scheme is technically and economically feasible. Long term stream flow records form the basis for determining the feasibility of a hydropower scheme. The hydropower energy, power capacity and reliability calculations are derived from such records. The significance of accurate and reliable stream flow records is a major concern and it is important that such records be adjusted and validated wherever possible. Stream flow records are used to determine; 1. Flow-duration curves, 2. Power-duration curves, 3. Mass curves, 4. Flood frequency curves, 5. Low-flow frequency curves, 6. Pre-project and post-project water surface profiles, 7. Spillway design flood and 8. The amount of minimum discharges needed for local water users.

In most circumstances it is unlikely that the engineer will have good quality stream flow records available for the catchment area under consideration. It is therefore inevitable that some form of manipulation of flow records will be required. The degree of manipulation will depend upon the engineers confidence in the origins of such records. A stream flow record of at least twenty years is usually required to ensure adequate data for a representative analysis. If less than twenty years ot data is available, it will be necessary to extend the record. Stream flow records will normally consist of either daily, weekly or monthly records depending upon the information available. However, daily records should be used wherever possible as they give more representative results. Such records may be obtained, in the UK, from the Institute of Hydrology which maintains a National Water Archive at Wallingford, Oxfordshire.

Information with regard to flow extraction, discharge consents, land drainage, reservoir regulation etc. can be ascertained from a variety of sources. During the course of a feasibility study it is recommended that as many of the following organisation and authorities as possible be consulted; 1. Local Council, 2. County Council, 3. Local Libraries, 4. Environment Agency (National Rivers Authority), 5. Water plc and 6. Private and public owners.

This list is not exhaustive and all relevant parties should be contacted wherever possible.

Once the relevant stream flow data for the catchment has been collected, it should be reviewed to ensure its accuracy and reliability. Suspect data should be rejected unless an explanation for the inconsistency can be found. Stream flow records may be inconsistent for a variety of reasons. This is particularly true if records cover a long period of time. The main reasons for such inconsistencies include: 1. Poor recording or tampering with equipment, 2. Flow extraction, 3. Flow discharges, 4. Upstream land use changes and 5. Reservoir regulation.

Poor or inaccurate recording of stream flow information is a common cause of inconsistent data. Stream flow data should only be accepted as reliable if it originates from a reputable source or its accuracy can be proved.

Flow extraction and discharges are a common cause of inconsistent stream flow data. Their effects can be both long term, for example a discharge from a sewerage treatment works, or short term. Identification of all such extraction and discharges is difficult. Legally, any extraction and discharges must be licensed through the Environment Agency (formerly, National Rivers Authority - NRA) in England and Wales and through the water services departments of the regional councils in Scotland. These authorities should therefore be contacted to ascertain the location, size and period for any such extraction or discharges. The stream flow records should then be adjusted accordingly by a simple summation or deduction process as appropriate.

Changes in upstream land usage can also have a significant effect on stream flow records that extend over long periods of time. Changes in agricultural usage can significantly alter the flow and runoff characteristics of catchments. Past records of land usage should be thoroughly checked to ensure reasonable continuity of catchment usage. Such information can be obtained from a number of sources including local authorities, libraries and land owners. Impounding structures, such as dams and weirs, always have a major impact on the flow regimes in a river. Records are usually kept of regulating flows and stream flow records can easily be adjusted from this data. Such records can usually be obtained from the local water authority or managing organisation. The most reliable source of accurate stream flow data is obtained from stream gauging stations. However, in most cases it is unlikely that the area under study will contain such a gauging station. In this situation it is often recommended that the project should begin with the installation of a hydrological gauging network to collect the appropriate amount of data. This is impractical for most projects as a minimum record length of twenty years is normally required. Hence, the need for stream flow estimation arises. There are essentially three methods that can be adopted to provide stream flow extension. These are; 1. Regression and Correlation techniques, 2. Synthesis techniques and 3. Interpolation and extrapolation. Figure 6 indicates which technique should be used for a particular situation. Having validated, in filled and extended the appropriate flow data, the design process usually commences with the production of a flow duration curve at the proposed intake site. The flow duration curve is undoubtedly the most important tool used for assessing the feasibility of a proposed small scale hydroelectric scheme. A flow duration curve is essentially a plot of flow against the probability of such a flow being equalled or exceeded. The curve is therefore a fundamental representation of the hydrological characteristics of a catchment. Such a curve may be produced very simply by assessment of historical stream flows. Once established, a flow duration curve may be adjusted and analysed to assess the firm generating capacity of a proposed scheme and probabilities of failure. Ultimately the shape and profile of the flow duration curve, combined with the technical, legal and financial aspects of the scheme, will determine the feasibility of a project. One of the prime functions of the flow duration curve is its use in the calculation of the potential energy output of a scheme. This potential energy output is found from the area beneath the flow duration curve, with suitable adjustments for compensation flow and operating limits of turbine apparatus. Figure 7 shows the one day flow duration curve for the Plynlimon catchment with such adjustments. For the purposes of this example the compensation flow rate has been set at 0.1 m^3/s, the maximum turbine capacity is 3.0 m^3/s and the minimum turbine capacity is 0.1 m^3/s. The shaded area of the diagram therefore indicates the flow available for energy production on a run-of-river basis, in the absence of storage. Calculation of the energy potential is achieved by breaking the diagram down into a series of flow segments. The average flow in each segment is then converted into a power output and multiplied by the corresponding duration. Table 2 demonstrates the procedure involved using the flow duration curve shown in Figure 7. Given that the head difference, H, for the scheme being examined is 64m (this corresponds to a favourable scheme layout with an intake at location 9 and a power house at location 15, see Figure 9) the following calculation is applicable:

Table 2

Flow (m^3/s)	Average Flow (m^3/s)	Turbine efficiency η	Power Output $P = \eta\, g\, Q\, H$ (KW)	Percentage of time power can output (T) (%)	Energy (Kwh/year)
3.0	3.0	0.79	1488.0	1.6	20856
3.0 - 1.0	2.0	0.84	1054.8	10.4	960965
1.0 - 0.5	0.75	0.85	400.2	13.3	466265
0.5 - 0.2	0.35	0.83	182.4	22.7	362706
0.2 - 0.1	0.15	0.73	68.7	14.5	87263
----	----	----	----	Sum	1898055

The above figure (1898055 KWh/year) is normally multiplied by 0.9 to allow for the efficiency of the electrical generating equipment.

Flood flows can have a significant and highly detrimental effect on a proposed hydroelectric scheme. This is particularly true if a reservoir and impounding structure are to be included as part of the scheme. The design work undertaken as part of a feasibility study should address and cater for the effects of such flows. The costs of providing ancillary structures for the safe transition of high flows may then be incorporated into the feasibility study. Determination of flood flows is not a straight forward process and it is often problematic to determine accurate figures as little historical flow data may be available. At study sites where no historical records exist, flood flow magnitudes and probabilities must be based on catchment characteristics utilising statistical techniques.

One such method of analysis is given in the Flood Studies Report published by the NERC. This method is an empirical technique derived from the study of flows at numerous sites over many years. The analysis is based around a series of maps and equations and a procedure that results in the production of a hydrograph of either a specified return period or the estimated maximum. Such information can then be used to design many of the scheme components. One of the major advantages of this method is its adaptability. Where data or information regarding the site does exist, it can be incorporated within the basic procedure to give better site specific results. Applying the procedures to the case study yields the maximum flood flow hydrograph shown in Figure 8.

HYDROELECTRIC POWER POTENTIAL

A typical hydroelectric scheme will consist of an intake arrangement, a conduit system, a powerhouse arrangement, switching station and a transmission line. There may also be some form of storage reservoir or headpond depending upon the findings of a hydrological and a topographical investigation. In order to assess the ideal location for an intake arrangement and a powerhouse, a hydroelectric power study must be undertaken. The objective of this study is to identify the scheme arrangement which will yield the optimum power output. This process may not give the most economically feasible design solution; however, it will give the engineer a narrower field of investigation and yield a number of specific sites for further investigation.

The study is based around the fundamental equation for hydroelectric power:

$$P \quad = \quad \eta \times \rho \times g \times Q \times H \qquad \qquad \_\_\_\_\text{Eqn 1}$$

The efficiency (η) of typical generating equipment is normally around 80%, though this will vary depending upon the flow regime at the powerhouse and the type of turbine installed. It is noted that P will be the potential power output at the generator terminals when the flow rate through the turbine is Q working under a head H. It does not allow for transmission losses or outages.

Having selected the appropriate scale of map for the hydropower study, the following procedure can be adopted for the analysis:

- A point on the river, chosen for the feasibility study, is selected as a potential location for a powerhouse.

- A second point on the river is then selected upstream of the potential powerhouse location as a possible intake site.

- From the contoured map the catchment area of the river at the intake location is then determined.

- On the assumption that the runoff from the catchment is proportional to the area of the catchment an estimate of the flow at the intake can be determined.

- From the fundamental equation for hydroelectric power, the power potential for the combination of powerhouse and intake site can be found.

- By considering a series of points upstream of the powerhouse, the intake site which yields the maximum power can be found.

- By repeating this procedure at different powerhouse locations, the combination of intake site and powerhouse location that yields the highest potential power output may be determined.

Selection of both potential powerhouse and intake sites is somewhat arbitrary. As a guide the engineer should aim to select approximately 30 sites along the length of the main river. Each of these locations may then be considered as potential intake or powerhouse sites. The points should, unless circumstances facilitate a choice, be equally spaced along the length of the river. However, the engineer should be practical when choosing the spacing. If the engineer wishes to study particular areas in more detail then the addition of extra points at any location is allowed, though during the early stages of a feasibility study this is usually unnecessary. Figure 9 shows the location of the potential sites and their associated catchment areas on the Plynlimon case study.

So far this process has not allowed for any frictional head loss due to the flow of water. Where the combination of powerhouses and intake site requires a long pipeline (commonly referred to as penstock) the head loss due to flow could be significant. It is recommended that this head loss be allowed for in any calculations. In this case study, and in most feasibility studies, the use of approximately 30 powerhouse and intake sites yields a large number of combinations for analysis; with 30 potential locations there are some 450 combinations. This procedure is therefore ideally suited to analysis on a digital computer. The authors will be pleased to provide a program for undertaking such an analysis which will run on any IBM compatible machine. The computer program has been written to allow the many possible combinations of scheme arrangements to be examined as quickly and efficiently as possible. Two methods of analysis are possible using this program. Firstly, the method that ignores the frictional effects in the pipeline and penstocks is employed. Secondly, the method that includes an allowance of frictional head losses is employed, based upon the Darcy's equation for frictional losses in pipelines. The program assumes the pipeline length to be five times the head difference between the intake site and the power house location, for the initial calculation. This is because it has been assumed that the average fall on a small scale hydroelectric penstock is 1 in 5. Where necessary this figure can be adjusted to suit the actual site conditions.

The computer program generates the results in the form of a matrix, referred to as the power matrix. This power matrix then forms the basis of the reconnaissance or feasibility study. The power matrix is used to identify the most economically viable site and is therefore an extremely important design tool. The matrix is made up of possible combinations of intake sites, arranged down the side of the matrix, and possible power house locations, arranged along the top. Zero values within a power matrix indicate that the particular combination is not possible. An example is when the power house is higher than the intake site. Given a power matrix for a site under study, a certain degree of interpretation will be required before the project may be advanced to the next stage. For example, if it can be identified that a particular location included within the hydropower study cannot be used as an intake then the row associated with this intake should be deleted. Other project constraints, such as limiting the length of a leat for economic reasons, may be included as well. Table 3 shows how limiting the length of a leat to 3 kilometres and removing site 5 as a powerhouse site and site 20 as an intake site effects the power matrix. Shaded areas indicate combinations which are not possible by applying such restraints. It should be noted that, in this case, each of the potential intake and powerhouse locations are equally spaced at a distance of 500m. The combinations not affected by any of the restraints can now be scanned to ascertain the highest outputs.

ENVIRONMENTAL ASPECTS

A study of the potential environmental issues that are likely to affect a proposed hydroelectric development must be undertaken as part of a feasibility study. The type and nature of potential environmental issues can vary vastly depending upon the type of development, the location and the scale of the proposed project. Identification of potential problems at an early stage of the project will ensure that the appropriate steps can be taken to assess and mitigate against any adverse effects. The major impacts on small scale hydroelectric developments are usually associated with the construction of dams and reservoirs. The effects on land usage, flow regimes and fisheries should also be of prime concern within the feasibility report. The environmental effects of a hydroelectric development can be complex and can in some circumstances directly affect the feasibility of a scheme. Every care should therefore be taken to identify as many of the potential problems as possible so that their effects on the scheme as a whole can be assessed. Depending upon the type of scheme, each of the following environmental aspects should be addressed during a feasibility study; 1. Thermal stratification of reservoirs, 2. Dissolved oxygen levels, 3. Compensation flow and flow regimes, 4. Water level fluctuation in reservoirs, 5. Gas bubble disease, 6. Fisheries, 7. Dredging of reservoirs, 8. Blue-Green algae in reservoirs, 9. Pollution during construction, 10. Effects on land usage & property liable to be flooded, 11. Insect and wildlife and 12. Planning.

The environmental issues to be addressed as part of a small scale hydroelectric development are often considered a major barrier to the economic feasibility of a scheme. This is in fact not the case, as many of the issues can usually be addressed simply and efficiently. During the reconnaissance stage of the project the potential environmental effects of the project should be identified. Identification of all possible environmental effects can be achieved through a variety of techniques. These techniques include a desk study in which topographical maps and aerial photographs can be utilised for ascertaining specific problems. A site reconnaissance visit is also a recommended activity, as many of the environmental effects of the proposed project can be identified during a walk over survey. Local sources of information and regulatory authorities should also be contacted for specific information and problems unique to the site.

LEGAL RESTRAINTS

The legal restraints imposed on a potential hydroelectric scheme can be considerable. A feasibility study should address as many of the legal technicalities as possible and determine whether the scheme is practical to implement within the law. Broadly speaking the legal restraints governing hydroelectric schemes can be divided into the following categories; Generation of Energy, Abstraction, Impoundment, Prevention of Pollution, Land Drainage, Planning, Land Ownership, Fisheries and Environmental.

Although some of the legislation applies to the United Kingdom as a whole, there are significant differences in legislation between Scotland, Northern Ireland, England and Wales.

Historically, the development of small scale hydroelectric projects has always been difficult within the United Kingdom. In 1983 the Energy Act was passed by Parliament to address many of the problems encountered by developers of such schemes. One of the main problems associated with the development of a small scale scheme has been, in many cases, connection to the existing national grid or electricity distribution network. Under the requirements of the Energy Act 1983 the area electricity boards of England, Wales, Scotland and Northern Ireland were bound to purchase power from such developments.

In 1989 a new enabling act of parliament, the Electricity Act 1989, was introduced to the legislation of the United Kingdom. Under this Act the Central Electricity Generating Board (CEGB) and the area electricity boards in England and Wales were privatised. The Central Electricity Generating Board was split into four separate companies, the National Grid Company PLC, National Power PLC, Power Gen PLC and Nuclear Electric PLC.

One of the major outcomes resulting from this change in legislation, so far as small scale hydroelectric developments are concerned, is the new open market. Unlike previous United Kingdom legislation, the newly formed Regional Electricity Boards are free to purchase electricity from any source. In practice this will mean a number of factors may influence their decisions. These include cost, quantity, reliability and their non-fossil fuel obligation. The non-fossil fuel obligation is a section of the act that requires the Regional Electricity companies to obtain electricity from non-fossil fuel sources. The amount of electricity to be purchased is determined upon the limits set down by the Secretary of State for Energy under non-fossil fuel orders. To date there have been three such orders. The non-fossil fuels obligation clearly makes small scale hydroelectric developments attractive to the Regional Electricity Companies and it is noted that many developments are currently undergoing construction, feasibility and other investigations.

Hydropower schemes which include a physical abstraction from a river require an abstraction licence to be issued by the National Rivers Authority (NRA) under the Water Resources Act 1991. In cases where the installation is built into a weir or is directly in line with a river flow such that there is no diversion of flow, the scheme may be exempt from an abstraction licence.

Planning considerations are covered by the Town and Country Planning Acts. These acts of parliament dictate that the plans for a proposed small scale hydroelectric development should be submitted to the local planning authority as a planning application. Outline planning applications are made in the first instance to ascertain the planning authorities' general attitude to the proposed scheme. In most cases where the application is passed, certain conditions may also be attached. Anyone likely to be affected by the proposals contained within the planning application is invited to comment and to state whatever objections they may have. In the case of a hydropower scheme this would include local environmental groups as well as local residents and statutory bodies. In general, the local Council's Planning Officers will ensure that all environmental matters are covered and critically examined when they receive the Planning Application. The Local Planning Authority is also responsible for determining whether an Environmental Assessment will be required as a result of compliance with European Community Directive 85/337/EEC. This directive has been made law in England and Wales for projects requiring planning permission through the Town and Country Planning, England and Wales (Assessment of Environmental Effects) Regulations 1988. Similar conditions exist in Scotland and Northern Ireland.

Land ownership and the exchange of ownership is governed by the Land Registry Act. This act of parliament dictates that any change in land ownership is registered with the appropriate land registry office. As the exchange of land requires the undertaking of a number of legal searches and an exchange of contracts, the process is best handled by a specialist solicitor.

The NRA has a duty under the Water Resources Act 1991 to maintain and improve fisheries. If a hydropower installation is to be located at a site that already proves a barrier to migratory fish, the NRA may require a fish pass to be constructed. Where a new barrier is being created, the NRA will consider whether a facility for fish migration is required. Also, the Salmon and Freshwater Fisheries Act 1991 contains a number of conditions relating to fish passes.

ECONOMICS

Economics is the single most important factor governing the feasibility of a proposed small scale hydroelectric scheme. A scheme that cannot pay back the initial capital investment within a given period will not be feasible. When it can be proved that a proposed scheme will be financially feasible then consideration of the profit likely to be generated within the scheme's life span is estimated. Such economic and financial calculations not only help to assess the scheme's feasibility, but also encourage developers and other private investors to find the initial capital investment required for design and construction. Because of the importance of economics within a development, the financial status and projected feasibility of a scheme are constantly reviewed. The two most important initial assessments occur at the reconnaissance and feasibility stages of the project. At the reconnaissance stage of a project the data available to the engineer is often insufficient to be able to accurately assess the potential feasibility of a scheme. Despite this restriction, a decision regarding the potential of the scheme will be made on the basis of the reconnaissance study. During the feasibility stage of the project, sufficient data should be available to achieve accurate economic estimates and judge the overall feasibility of the scheme.

The largest cost experienced by a developer on a small scale hydroelectric scheme is the capital expenditure for construction. This cost is re-paid by the income earned through the generation of electricity. An estimate of the potential revenue that a scheme is likely to produce is therefore of prime importance. The economic feasibility of a scheme is a balance between the cost of the initial capital investment, the actual income from electrical generation and the operation and maintenance costs. For a scheme to be financially stable and successful, the net income must counter the initial capital investment within a stated pay back period.

Cost/benefit analysis is one of the fundamental methods for establishing the economic feasibility of a proposed development. The method essentially compares the initial capital cost against the benefits, or financial gains, of a scheme. The particular benefit/cost is then expressed as a ratio by dividing the financial benefit by the cost.

$$R \quad = \quad \frac{f_b}{f_c} \qquad \qquad \underline{\quad}Eqn\ 2$$

Where
R = Benefit/Cost Ratio
f_b = Financial benefit of the scheme
f_c = Financial cost of the scheme

The benefit- cost ratio may be calculated for any stage of a project. Often, calculations for the benefit-cost of a scheme are calculated yearly for the full period of the schemes design life, normally at least 25 years. Such calculations will indicate to the engineer the economic status of the scheme with respect to time. This is a useful tool for identifying pay back periods and profit margins. Schemes with a benefit-cost ratio of below one at the end of their design life will not be economically or financially feasible. In many cases, however, the benefit- cost ratio of many small scale hydroelectric developments is high, and final benefit- cost ratios in excess of two or three are very common.

The benefit gained from a hydroelectric development is mainly governed by the proposed end use of the power. There are two basic forms of scheme. In the first type of scheme the electricity is sold to the national grid at specified prices depending upon the time of day and season. The benefit therefore arises from the direct profit from the sale of the electricity. With the second type of scheme, all electricity is utilised by one consumer, usually the investor of the scheme and the benefit is gained by the investor who no longer needs to buy electricity from the national grid. Since the price of producing electricity through a locally developed scheme is small, the saving, or benefit, is very favourable. However, it must be remembered that this saving must be used, initially, to pay off the capital expense of the scheme.

As many of the benefits arising from a scheme occur at a relatively steady rate during the project's life, the effects of inflation must be taken into consideration. The effect of inflation is to gradually reduce the equivalent present value of any benefits. Therefore the present value of the benefit gained by selling electricity in the first year of operation will differ greatly from the benefit gained during the twenty fifth year. However, inflation is often ignored and current benefits and costs are used in the analysis, on the assumption that both benefits and costs will vary equally under the effects of future inflation. In order to calculate the present value at today's prices of a benefit arising from a proposed development, a technique referred to as discounted cash flow analysis should be employed. The principle is relatively simple. The value of the project, calculated from the present value of benefits earned during the expected design life of the project, are added together and the net present value at today's prices is found by subtracting the present value of costs. A comparison can then be made between this figure and the estimated capital cost. The advantage of this method of analysis is that it can be used to calculate a number of useful economic indicators. These include; 1. The benefit- cost ratio at any stage of the project, 2. The benefit - cost ratio at the end of the project's design life, 3. The pay back period of the project, and 4. The net present value at the end of the project's design life.

In order to calculate the net present value of a future benefit two economic parameters are required. The first parameter is the real cost of borrowing money. Over recent years this value has remained relatively static at around 8%. The second parameter is the real cost of electrical energy relative to the cost of living. This parameter over recent years has been very high, around 7%, however, a realistic value of 2% or more should be adopted in practice. In lieu of better information the engineer may adopt different figures as appropriate. The following is given as an example:

It is proposed to carry out a discounted cash flow analysis for a small scale hydroelectric scheme in mid-Wales as part of an economic appraisal. Cost estimating techniques have established that the capital cost of the scheme will be £160,000. The calculated annual income, at present values, is £27,400 and the annual operation and maintenance cost is £1,400. Assuming the real cost of borrowing to be 3% and the average escalation in electrical power costs to increase by 2% relative to the cost of living, the following equations apply:

$$PVC_n = \frac{1}{1.03^n} \hspace{4cm} \text{\_\_\_\_\_Eqn 3}$$

$$EC_n = (1.02)^n \hspace{4cm} \text{\_\_\_\_\_Eqn 4}$$

$$PV_n = PVC_n \times B_n \hspace{4cm} \text{\_\_\_\_\_Eqn 5}$$

$$E_n = EC_n \times 27400 \hspace{4cm} \text{\_\_\_\_\_Eqn 6}$$

The process of analysis is best demonstrated using Table 4.

A number of useful economic statistics may be obtained from the above table. These include the cost-benefit ratio and the pay back period. In this case the pay back period is approximately 6.5 years (Calculated by establishing the point at which the cost-benefit ratio is equal to one) and the final cost benefit ratio is 3.63, with a net present value of £420524.

Having established the economic feasibility of a proposed scheme the results and conclusions of the study are submitted to the developer. The financial feasibility of the scheme is then determined from the economic attractiveness of the proposed development and the level of available investment. The factors that influence the final feasibility of a scheme include the cost-benefit ratio and the pay back period. Often, the long pay back periods involved with such developments discourage investors. However, low running costs and the later high benefit-cost ratios help to counter other disadvantages, making many schemes financially feasible. To give definitive guidelines on what constitutes a financially feasible scheme is not possible within this text as each situation is unique. The feasibility of potential financial backing for a scheme is dependant upon many economic and social factors and each scheme should be considered on its own particular merits during financial reviews.

CONCLUSIONS

The procedures involved in undertaking a modern feasibility study of a hydroelectric project are amply illustrated by considering the planning aspects of the study of individual elements in detail. Also considered is the undeveloped hydroelectric potential of the United Kingdom. The principles of feasibility study are demonstrated using Plynlimon catchment area in Mid Wales as a case study. A computer program developed to select potential powerhouse sites and intakes gives a power matrix from which the most economically viable scheme with highest output power can be identified with ease. The potential environmental and economic issues together with legal restraints involved in the development of small scale hydroelectric power schemes in the United Kingdom together with social aspects are discussed with particular emphasis on planning and developing hydropower schemes in England, Wales, Northern Ireland and Scotland.

REFERENCES

1. Flood Studies Report. Volume II, Meteorological Studies., Natural Environment Research Council, 1975.

2. Design of Small Dams. United States Department of the Interior, 1977.

3. Sutcliffe, J.V., Methods of Flood Estimation: A Guide to the Flood Studies Report, Report No. 49, The Institute of Hydrology, Walligford, U.K., 1978.

4. Low flow studies. The Institute of Hydrology, Wallingford, U.K., 1980.

5. Report on the development of Small-Scale Hydropower Plants. Volume I, Technical Guide, Department of Energy, U.K., 1983.

6. Tolland, H.G., The Development of Small-Scale Hydroelectric Power Plant. Energy Technology Support Unit, U.K., 1984.

7. Small Scale Hydroelectric Generation Potential in the U.K. Volumes 1 - 3, Department of Energy, U.K., 1989.

8. Civil Engineering Guidelines for Planning and Designing, Hydroelectric Developments. Volume I-Planning, Design of Dams, and Related Topics, and Environment. The American Society of Civil Engineers, 1989.

9. Small Scale Hydropower Study of non-technical Barriers. Department of Energy, U.K., 1990.

10. Renewable Energy Sources - A Select Reading List. Office of Electricity Regulation, U.K., 1993.

11. Renewable Energy Information List No. 2 - Small Scale Hydropower. Department of Trade and Industry, U.K., 1994

12. Third Renewables Order for England and Wales. Office of Electricity Regulation, U.K., 1994.

13. First Scottish Renewables Order. Office of Electricity Regulation, U.K., 1994.

14. hydropower Developments and National Rivers Authority. The National Rivers Authority for England and Wales (now called Environment Agency), 1994.

15. Floods and Reservoir Safety Guide. Institution of Civil Engineers, U.K. 1996.

Table 3

Power House Location

Intake Location	1	2	3	4	5	6	7	8	9	10	11	12	13	14	15	16	17	18	19	20	21	22	23	24	25	26	27	28
1	0.0	2.8	7.9	10.3	14.4	15.7	18.0	19.4	20.3	21.1	21.5	23.0	23.9	25.2	25.6	27.0	27.7	28.0	28.7	29.3	29.6	30.1	30.5	30.9	31.2	31.5	31.7	32.0
2	0.0	0.0	12.7	36.3	38.7	42.6	49.6	54.0	56.8	59.0	60.2	64.7	67.4	71.6	72.6	77.1	79.1	80.1	82.0	83.8	84.8	86.2	87.5	88.7	89.7	90.5	91.2	92.2
3	0.0	0.0	0.0	45.5	78.4	70.6	85.3	94.7	100.4	105.1	107.8	117.2	122.9	131.8	133.9	143.3	147.5	149.6	153.8	157.5	159.5	162.7	165.3	167.9	170.0	171.6	173.1	175.2
4	0.0	0.0	0.0	0.0	46.6	61.5	78.3	93.3	102.5	110.0	114.2	129.2	138.3	152.5	155.8	170.8	177.5	180.8	187.5	193.3	196.6	201.6	205.8	210.0	213.2	215.8	218.3	221.6
5	0.0	0.0	0.0	0.0	0.0	17.3	45.4	63.9	75.3	84.6	89.7	93.6	103.4	118.5	122.1	138.1	145.2	148.5	155.9	162.1	165.7	171.1	175.5	180.0	183.5	186.2	188.9	192.5
6	0.0	0.0	0.0	0.0	0.0	0.0	39.0	64.1	79.4	91.9	90.9	113.9	128.0	149.8	154.9	177.9	188.2	193.3	203.5	212.5	217.6	225.3	231.7	238.1	243.2	247.1	250.9	256.0
7	0.0	0.0	0.0	0.0	0.0	0.0	0.0	31.1	49.0	64.2	72.7	98.4	116.1	143.6	150.0	157.9	169.3	175.0	186.4	196.4	202.1	210.6	217.7	224.8	230.5	234.8	239.1	244.8
8	0.0	0.0	0.0	0.0	0.0	0.0	0.0	0.0	27.1	48.7	59.6	123.4	143.4	162.3	171.4	212.6	230.8	217.1	233.7	248.1	256.4	268.8	279.1	289.5	297.8	304.8	310.2	318.4
9	0.0	0.0	0.0	0.0	0.0	0.0	0.0	0.0	0.0	37.0	57.2	171.0	231.0	246.8	295.7	316.3	326.7	341.2	376.3	395.7	410.2	434.7	453.6	468.4	482.4	493.7	504.6	519.1
10	0.0	0.0	0.0	0.0	0.0	0.0	0.0	0.0	0.0	0.0	21.9	57.2	142.6	213.9	222.2	264.0	295.7	347.9	388.9	393.1	405.7	432.0	450.2	455.5	468.2	480.7	491.4	506.5
11	0.0	0.0	0.0	0.0	0.0	0.0	0.0	0.0	0.0	0.0	0.0	97.8	155.9	242.8	264.0	347.9	388.9	409.3	450.2	478.1	507.1	531.2	555.4	574.7	589.1	603.7	—	623.0
12	0.0	0.0	0.0	0.0	0.0	0.0	0.0	0.0	0.0	0.0	0.0	0.0	68.0	168.0	192.2	264.0	347.9	410.7	433.9	456.4	490.2	518.4	536.0	569.5	586.0	602.9	—	625.4
13	0.0	0.0	0.0	0.0	0.0	0.0	0.0	0.0	0.0	0.0	0.0	0.0	0.0	0.0	107.5	131.5	240.6	290.0	355.0	397.3	421.3	457.4	466.9	495.8	518.4	536.1	553.4	576.5
14	0.0	0.0	0.0	0.0	0.0	0.0	0.0	0.0	0.0	0.0	0.0	0.0	0.0	0.0	0.0	124.9	178.3	205.8	260.6	303.9	329.9	370.3	404.0	424.5	455.5	474.0	492.8	517.7
15	0.0	0.0	0.0	0.0	0.0	0.0	0.0	0.0	0.0	0.0	0.0	0.0	0.0	0.0	0.0	26.8	143.3	195.4	246.0	295.7	339.7	368.3	399.3	424.0	450.5	470.2	489.8	516.0
16	0.0	0.0	0.0	0.0	0.0	0.0	0.0	0.0	0.0	0.0	0.0	0.0	0.0	0.0	0.0	0.0	59.0	88.2	144.9	195.7	221.8	264.7	300.5	329.9	357.6	378.6	399.7	427.7
17	0.0	0.0	0.0	0.0	0.0	0.0	0.0	0.0	0.0	0.0	0.0	0.0	0.0	0.0	0.0	0.0	0.0	32.7	96.8	152.3	184.4	230.1	269.8	304.7	335.9	359.9	382.8	414.0
18	0.0	0.0	0.0	0.0	0.0	0.0	0.0	0.0	0.0	0.0	0.0	0.0	0.0	0.0	0.0	0.0	0.0	0.0	66.4	123.5	155.4	204.4	242.7	283.2	310.5	334.3	358.3	390.1
19	0.0	0.0	0.0	0.0	0.0	0.0	0.0	0.0	0.0	0.0	0.0	0.0	0.0	0.0	0.0	0.0	0.0	0.0	0.0	59.4	92.9	142.9	183.6	222.8	255.8	280.6	305.3	332.6
20	0.0	0.0	0.0	0.0	0.0	0.0	0.0	0.0	0.0	0.0	0.0	0.0	0.0	0.0	0.0	0.0	0.0	0.0	0.0	0.0	43.8	108.5	162.1	214.9	255.7	287.6	319.6	362.2
21	0.0	0.0	0.0	0.0	0.0	0.0	0.0	0.0	0.0	0.0	0.0	0.0	0.0	0.0	0.0	0.0	0.0	0.0	0.0	0.0	0.0	67.2	122.5	177.3	220.1	253.3	283.3	327.5
22	0.0	0.0	0.0	0.0	0.0	0.0	0.0	0.0	0.0	0.0	0.0	0.0	0.0	0.0	0.0	0.0	0.0	0.0	0.0	0.0	0.0	0.0	57.5	114.5	159.7	193.7	225.6	270.7
23	0.0	0.0	0.0	0.0	0.0	0.0	0.0	0.0	0.0	0.0	0.0	0.0	0.0	0.0	0.0	0.0	0.0	0.0	0.0	0.0	0.0	0.0	0.0	58.3	104.4	138.7	172.6	218.6
24	0.0	0.0	0.0	0.0	0.0	0.0	0.0	0.0	0.0	0.0	0.0	0.0	0.0	0.0	0.0	0.0	0.0	0.0	0.0	0.0	0.0	0.0	0.0	0.0	57.6	100.5	143.3	199.2
25	0.0	0.0	0.0	0.0	0.0	0.0	0.0	0.0	0.0	0.0	0.0	0.0	0.0	0.0	0.0	0.0	0.0	0.0	0.0	0.0	0.0	0.0	0.0	0.0	0.0	45.7	91.0	150.9
26	0.0	0.0	0.0	0.0	0.0	0.0	0.0	0.0	0.0	0.0	0.0	0.0	0.0	0.0	0.0	0.0	0.0	0.0	0.0	0.0	0.0	0.0	0.0	0.0	0.0	0.0	46.8	108.6
27	0.0	0.0	0.0	0.0	0.0	0.0	0.0	0.0	0.0	0.0	0.0	0.0	0.0	0.0	0.0	0.0	0.0	0.0	0.0	0.0	0.0	0.0	0.0	0.0	0.0	0.0	0.0	64.1
28	0.0	0.0	0.0	0.0	0.0	0.0	0.0	0.0	0.0	0.0	0.0	0.0	0.0	0.0	0.0	0.0	0.0	0.0	0.0	0.0	0.0	0.0	0.0	0.0	0.0	0.0	0.0	0.0

Table 4

Yr	Income	EC	Benefit	O/M Costs	Net Benefit	PVC	PV	Σ	B/C
1	27400	1.02	27948	1400	26548	0.9709	25775	25775	0.1
2	27400	1.0404	28507	1400	27107	0.9426	25551	51326	0.3
3	27400	1.0612	29077	1400	27677	0.9151	25327	76653	0.4
4	27400	1.0824	29659	1400	28259	0.8885	25108	101761	0.6
5	27400	1.1041	30252	1400	28852	0.8626	24888	126649	0.7
6	27400	1.1262	30857	1400	29457	0.8375	24670	151319	0.9
7	27400	1.1487	31474	1400	30074	0.8131	24453	175772	1.1
8	27400	1.1717	32103	1400	30703	0.7894	24237	200009	1.2
9	27400	1.1951	32746	1400	31346	0.7664	24024	224033	1.4
10	27400	1.2190	33400	1400	32000	0.7441	23811	247844	1.5
11	27400	1.2434	34068	1400	32668	0.7224	23599	271443	1.7
12	27400	1.2682	34750	1400	33350	0.7014	23392	294835	1.8
13	27400	1.2936	35445	1400	34045	0.6810	23185	318020	1.9
14	27400	1.3195	36154	1400	34754	0.6611	22976	340996	2.1
15	27400	1.3459	36877	1400	35477	0.6419	22773	363769	2.2
16	27400	1.3728	37614	1400	36214	0.6232	22569	386338	2.4
17	27400	1.4002	38367	1400	36967	0.6050	22365	408703	2.5
18	27400	1.4282	39134	1400	37734	0.5874	22165	430868	2.6
19	27400	1.4568	39917	1400	38517	0.5703	21966	542834	2.8
20	27400	1.4859	40715	1400	39315	0.5537	21769	474603	2.9
21	27400	1.5157	41529	1400	40129	0.5375	21569	496172	3.1
22	27400	1.5460	42360	1400	40960	0.5219	21377	517549	3.2
23	27400	1.5769	43207	1400	41807	0.5067	21184	538733	3.3
24	27400	1.6084	44071	1400	42671	0.4919	20990	559723	3.5
25	27400	1.6406	44953	1400	43553	0.4776	20801	580524	3.6

Net present value　　=　　580524 - 160000　　=　　420524

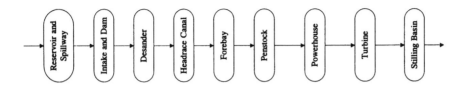

Reservoir and Spillway → Intake and Dam → Desander → Headrace Canal → Forebay → Penstock → Powerhouse → Turbine → Stilling Basin

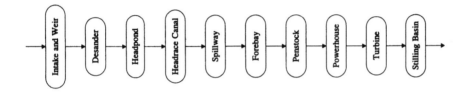

Intake and Weir → Desander → Headpond → Headrace Canal → Spillway → Forebay → Penstock → Powerhouse → Turbine → Stilling Basin

Figure 2

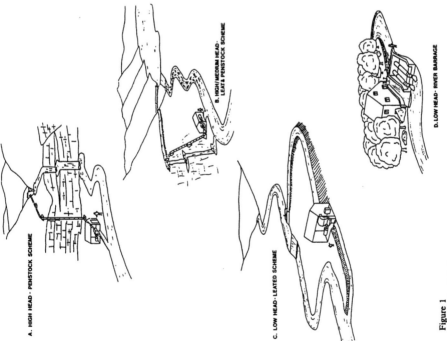

A. HIGH HEAD - PENSTOCK SCHEME

B. HIGH/MEDIUM HEAD - LEAT& PENSTOCK SCHEME

C. LOW HEAD - LEATED SCHEME

D. LOW HEAD - RIVER BARRAGE

Figure 1
Typical Small Scale
Hydroelectric Site Layouts

71

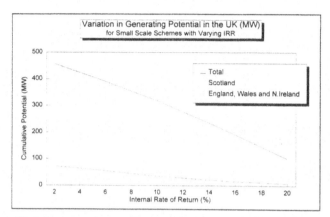

Figure 3

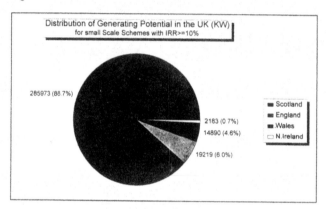

Figure 4

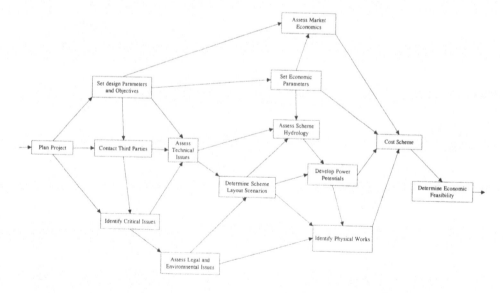

Figure 5

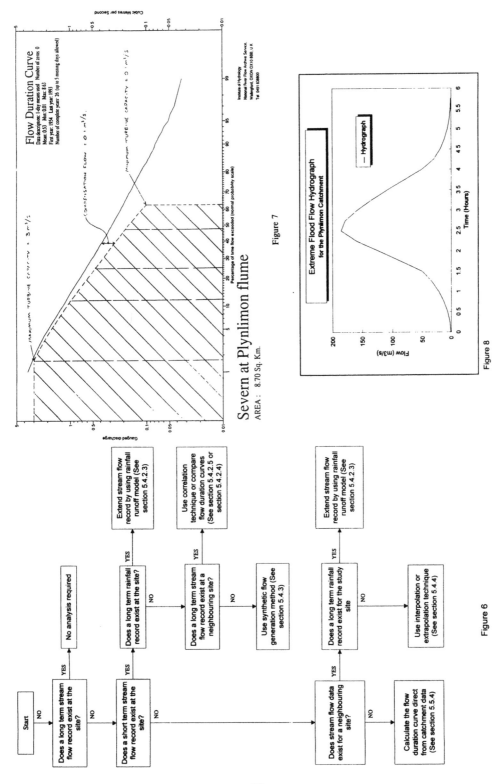

Flow Duration Curve

Data description: 1-day means used Number of zeros: 0
Mean: 0.33 Min: 0.01 Max: 8.63
First year: 1954 Last year: 1993
Number of complete years: 26 (up to 5 missing days allowed)

Severn at Plynlimon flume

AREA: 8.70 Sq. Km.

Cubic Metres per Second

Percentage of time flow exceeded (normal probability scale)

Gauged discharge

MAXIMUM TURBINE CAPACITY = 3 m³/s
MINIMUM TURBINE CAPACITY = 0.1 m³/s
COMPENSATION FLOW = 0.1 m³/s

Institute of Hydrology
National River Flow Archive Service.
Wallingford, OXON OX10 8BB, U.K.
Tel: 0491 838800

Figure 7

Extreme Flood Flow Hydrograph
for the Plynlimon Catchment

— Hydrograph

Flow (m3/s)

Time (Hours)

Figure 8

Start

Does a long term stream flow record exist at the site? — YES → No analysis required

NO

Does a short term stream flow record exist at the site? — YES → Does a long term rainfall record exist at the site? — YES → Extend stream flow record by using rainfall runoff model (See section 5.4.2.3)

NO (from Does a long term rainfall record exist at the site?)

Does a long term stream flow record exist at a neighbouring site? — YES → Use correlation technique or compare flow duration curves (See section 5.4.2.5 or section 5.4.2.4)

NO → Use synthetic flow generation method (See section 5.4.3)

Does stream flow data exist for a neighbouring site? — YES → Does a long term rainfall record exist for the study site — YES → Extend stream flow record by using rainfall runoff model (See section 5.4.2.3)

NO → Use interpolation or extrapolation technique (See section 5.4.4)

Calculate the flow duration curve direct from catchment data (See section 5.5.4)

Figure 6

73

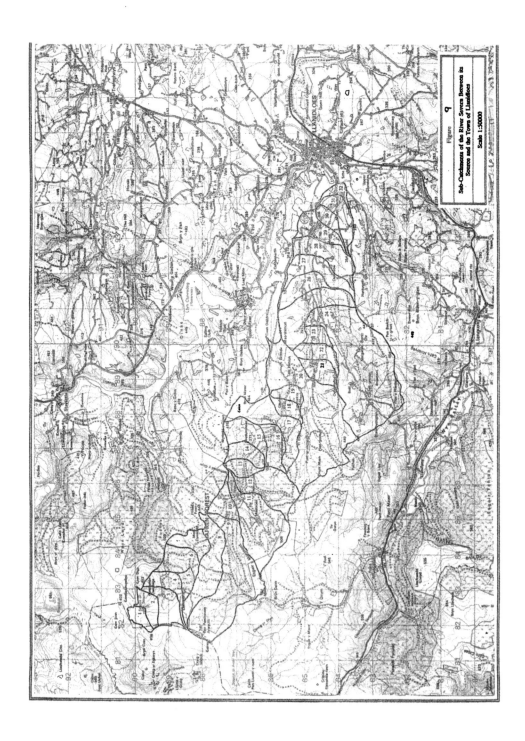

Figure 9

Sub-Catchments of the River Severn Between its
Source and the Town of Llanidloes

Scale 1:50000

74

REDUCING THE COSTS OF SMALL HYDRO GENERATION: THE UK PERSPECTIVE

M.A. Allington

ETSU, Harwell, Didcot
Oxfordshire, OX11 0RA, UK

ABSTRACT

Since the privatisation of the electricity industry in 1989, there has been a resurgence of interest in small scale hydropower (<5MW) in the UK. Near-market renewable technologies, including small hydro, have been encouraged by the Non-Fossil Fuel Obligation (NFFO) mechanism, which requires the privatised electricity companies to secure a specified amount of electricity supply from renewable sources. Successful proposers receive a 15 year duration, fixed-price contract for electricity purchase from their renewable scheme. The mechanism is competitive, and has resulted in a reduction in the maximum contracted selling price for small hydro generated electricity from 3.75Rs/kWh in 1990 to 2.42Rs/kWh in 1995 (2.08Rs/kWh in Scotland). This price convergence is set to continue with a further round of contracts offered in 1997. The maximum contracted price for small hydro in 1997 is expected to be around 2.0Rs/kWh. The main reasons for price convergence are better design, lower finance costs and improved procurement methods.

Between 1989 and 1995, 100 subsidised electricity purchase contracts were awarded under the NFFO and similar arrangements, to sites with a combined installed capacity of 73MW. Several of these sites were existing power stations, which required the subsidy to avoid closure. However, the development of new sites, and the refurbishment of abandoned sites has also been stimulated, and is now the main focus of the initiative.

The NFFO initiative has promoted growth in some sectors of the UK small hydro industry, particularly new project developers and consultants. Around 75 companies are directly involved in small hydro in the UK, including 13 turbine manufacturers, 12 professional developers and 50 consultants. In addition there are many manufacturers of ancillary equipment, such as generators, switchgear, control systems, sluice gates and pipes. Key development issues have emerged as finance, skills, abstraction licencing and environmental impact. Experienced developers, particularly those with access to equity finance, are recently beginning to dominate the competitive market. These developers, which include subsidiaries of the privatised water and electricity companies, and private 'entrepreneurial' companies, are now seeking to expand their activities overseas. Experienced partnerships, developed under the NFFO, can now offer a complete finance, skills and equipment supply packaged approach for rapidly developing markets, such as that in India.

INTRODUCTION

Small scale hydropower (defined in the UK as hydropower with installed capacity of less than 5MW per site) is an established technology, with little scope for significant price reductions through research and development. Several sites exist in the UK which have been operational since before the war. Many others fell into disuse when large fossil-fuelled power stations and widespread distribution via the electricity network made them uncompetitive. Heightened environmental concern over acid rain and global warming has caused interest in small-scale hydro generation to re-surface. Government policy, primarily via the Non-Fossil Fuel Obligation (NFFO) has encouraged the uptake of the technology.

This paper describes the progress made to date with small-scale hydropower ('small hydro') in the UK, analyses the current status of the industry and describes future prospects.

UK MARKET AND RESOURCE

The status of existing hydro capacity in the UK, excluding pumped storage sites, in June 1996 is shown in Table 1.

Table 1: Operating hydro capacity in the UK (June 1996)

	No. of sites	Total MW	Total GWh/year
Small scale			
<5MW Eng & Wales	29	14.48	41.3
<5MW NI	22	1.77	8.5
<5MW Scotland	41	61.35	206.6
Total <5MW	92	77.6	256.4
Large Scale			
>5MW Eng & Wales	5	156.95	687.4
>5MW NI	0	0	0
>5MW Scotland	42	1135.75	4974.6
Total >5MW	47	1292.7	5662
Total UK hydro	139	1370.3	5918.4

A total installed capacity of almost 78MW is currently operating in the UK at 92 small hydro sites each with installed capacity <5MW. In addition to these there are an unknown number (perhaps 200) of operational micro-hydro sites, with installed capacities <20kW. These are primarily used for self-supply. Approximately 50MW operates unsubsidised, managed by the main generators, particularly Scottish Hydro-Electric plc. A further 28MW has been supported by the various NFFO arrangements in all parts of the UK. Of this 20MW comprises operational sites whose medium-term future has been secured by the NFFO. The remaining 8MW (20 sites) are refurbishments of abandoned sites or green-field developments. The most recent rounds, NFFO-3 and SRO-1, concentrated on the latter, 'new' sites, contracting a further 33MW (30 sites) most of which are currently at the detailed design stage or are under construction.

The remaining practically feasible UK resource which might be commercially attractive (including the 33MW contracted under NFFO-3 and SRO-1) is estimated at between 40MW and 110MW (<5p/kWh (<2.5Rs/kWh @ 50Rs:£1) unit generation cost at 15% and 8% discount rate over 15 years respectively). Using the same economic parameters, the remaining resource at <10p/kWh (<5Rs/kWh) is between 300MW and 550MW. The resource in Scotland, at micro-hydro sites and within the water industry may however have been underestimated. While small-scale hydro can therefore make a useful contribution to the UK electricity supply, it is unlikely to be significant in terms of total demand.

PROGRESS UNDER THE NON-FOSSIL FUELS OBLIGATION

The majority of the operational sites (72 sites totalling 70MW) were running before the NFFO was introduced. Mostly these sites were operational before World War II. In some cases they are historic water power sites from the previous century, which were first used to generate mechanical shaft power and later for electricity generation, before cheap, grid-supplied electricity became widely available. The electricity generated from these sites was usually direct current (DC), and to no common standard or voltage. Some of the remaining sites were converted to generate mains frequency alternating current (AC) and received contracts under the 1983 Energy Act. The terms offered in 1983, essentially a must-take contract but without a premium price, were sufficient to prompt the refurbishment of two previously derelict sites and for construction to begin at one green field site in Scotland.

In 1989, the UK electricity supply industry was privatised. The Electricity Act 1989 introduced the Non-Fossil Fuel Obligation (NFFO) as a guaranteed premium market arrangement for the supply of electricity from renewable energy sources in England and Wales. Under NFFO, electricity distribution companies are required to secure specified amounts of electricity generating capacity from renewables. They meet their obligations by signing contracts with the renewable-based generators, who are paid a premium price for the electricity produced, funded ultimately by the final customers for that electricity. After three rounds of bidding for England and Wales, output from over 325MW of renewables-based capacity is now being delivered to the grid. Similar arrangements have been put into place for Scotland under the Scottish Renewables Obligation (SRO) and in Northern Ireland under the Northern Ireland NFFO.

UK Government policy is to stimulate the development of renewable energy sources, wherever they have prospects of being economically attractive and environmentally acceptable, in order to contribute to:
- diverse, secure and sustainable energy supplies
- reduction in the emission of pollutants
- encouragement of internationally competitive industries.

In March 1993, the Government announced its intention of working towards 1,500MW of new renewable electricity generating capacity in the UK by the year 2000, with the specific objectives of developing the market for electricity from renewables and providing a platform for the commercial demonstration of technologies. NFFO is the Government's main instrument for pursuing the development of this 1,500MW capacity. It will also contribute to the policy of aiming to return carbon dioide emissions to 1990 levels by the same year.

The purpose of the NFFO is to create an initial market, so that in the near future the most promising renewables can compete without financial support. That objective requires there to be steady convergence under successive Orders between the price paid under the relevant Order and the market price for electricity. This is achieved by allocating contracts on a competitive basis.

The first tranche of renewable energy covered by a NFFO order was set in 1990 for 102 MW of capacity and the second in 1991 for 457 MW. Twenty-six small-scale hydro schemes with a total capacity of 11.9 MW were offered contracts in the first tranche. Fourteen of these were at new or renovated sites with a capacity of 5.2 MW. In the second tranche 12 hydro projects were contracted at 6p/kWh (3Rs/kWh) with a total capacity of 10.86 MW. At the same time 23 hydropower projects in Scotland received contracts under a transitional arrangement. Further NFFO tranches for renewable energy were set in 1994 for 627MW of capacity in England & Wales, 15.6MW in Northern Ireland and 76MW in Scotland. the small hydro sites contracted are summarised in Table 2.

The figures for bids received reflect the bid status at the receipt of first prices. To illustrate the level of interest in small hydro, 194 requests for NFFO-3 tender packs were received. Ninety-one of these potential developers submitted initial prices, with only 72 proceeding to submit firm bids. Fifteen of the original 194 were contracted. The interest in NFFO support for small hydro is therefore substantial.

Overall, approximately 1 in 3 applicants submitting firm bids have been successful in securing a premium price contract. If the Scottish transitional arrangement, which was not competitive is excluded, this ratio falls to 1 in 4.

Table 2: Small hydro performance in NFFO (@ 31/3/95)

Tranche	Max. price p/kWh (Rs/kWh)	Bids received No.	Contracts		Commissioned	
			No.	MW	No.	MW
NFFO-1	7.5 (3.75)	62	26	11.85	21	10
NFFO-2	6 (3)	69	12	10.86	8	10.25
NFFO-3	4.85 (2.42)	91	15	14.48	2	0.26
STA	5.3 (2.65)	23	23	15.46	21	13.41
SRO-1	4.15 (2.08)	46	15	17.3	0	0
NI NFFO-1	6.25 (3.12)	21	9	2.79	2	0.28
Total		312	100	72.74	54	34.2

Data from Dept of Trade & Industry, NI Office, Scottish Office, Office of Electricity Regulation

The geographical distribution of contracted schemes, excepting the NI NFFO-1 and Scottish transitional arrangement sites, is shown in Figures 1 and 2.

It can be seen that NFFO sites are clustered on the western side of England and Wales, which receives greater rainfall, has lower potential evaporation and hence has greater prospects for hydro development. A few sites have been contracted away from the western side of the country. These tend to be low head sites on large lowland rivers such as the Trent, which include in their catchment area rainfall from the western upland regions.

Many contracts in the first two rounds of NFFO were awarded to sites in Devon and Cornwall. NFFO-3 has seen the emphasis shift to Wales and Derbyshire.

In Scotland also, the western highlands provide the most promising resource, and ten of the fifteen schemes contracted in SRO-1 are situated in west Scotland, particularly in the Highlands region.

Over 80% of schemes contracted in NFFO-1, 2 and the Scottish transitional arrangement are now operational. The 39 sites contracted in the NFFO-3, NI NFFO-1 and SRO-1 tranches are mostly at the detailed design stage, or are under construction. Four are operational.

The NFFO has successfully prevented the closure of existing sites and has stimulated the refurbishment of decommissioned sites and green field developments.

Industrial Impact

The arrangements have successfully stimulated the growth of some sectors of the UK small hydro industry, particularly new project developers and consultants, and have helped to develop their skills. The developers of the new and refurbished schemes fall into 4 broad groups:

- water companies, which have developed generating stations within their existing infrastructure, usually using reservoir compensation flow outlets or water supplies to treatment works. Some water companies have established subsidiaries responsible for industrial ventures outside their core business of water supply and treatment. One, Hyder Industrial (formerly Welsh Water Industrial Services) is expanding to consider sites outside of its own infrastructure, including overseas sites. Installed capacities vary from 0.2MW to 4.35MW, with a trend to larger capacity sites (>1MW)

- subsidiaries of electricity companies, such as Norweb plc, which has developed and operates hydro sites of average capacity 0.5MW, on land leased to the company. Like the water companies, these have the benefit of finance and resource backing from the holding company. The trend from this sector is also to invest in larger schemes (>1MW)

- small entrepreneurial groups whose main business is power generation from renewables, in some cases just from small hydro. Generally these companies involve staff experienced in the hydro industry, seeking to create a generation/site building business through the NFFO arrangements. Prominent entrepreneurial developers include Derwent Hydropower Ltd, Hydro Energy Developments Ltd, Garbhaig Hydro Power Ltd and Powerstream plc. Installed capacities tend to be small (<0.5MW, excepting GHP, which constructs larger sites), and sites are leased from their owners.

- private individuals or organisations, whose main business is not power generation or renewables, such as private estates or small businesses. In general these developers only pursue one project on their own land. The hydro project is pursued to generate revenue which is unlikely to be invested in further hydro projects. High returns are generally required on the capital invested.

The average scheme size reduces with each of these categories.

The water companies and Norweb Generation Ltd have been the most successful NFFO bidders, probably because larger sites tend to be cheaper to develop, and the companies have access to equity finance. There is a significant minority of entrepreneurial developers, which have achieved varying degrees of success. Although many private bids are received, few are successful.

The NFFO has provided a home market for UK hydro consultancies and equipment suppliers. A number of new consultancies have been established and have gained experience through the NFFO. Among the most successful (in winning NFFO business) have been Edinburgh Hydro Systems Ltd, Powerstream plc, Shawater Ltd, Wilson Energy Associates Ltd and Dougall Baillie Associates Ltd.

UK turbine manufacturers have gained 85% of the NFFO market by capacity. The main markets for small hydro turbine supply are however overseas and the manufacturers were established as exporters before the NFFO created a home market.

Monitoring of NFFO projects has revealed several key development issues:
- finance
- lack of expertise
- abstraction licensing
- environmental impact.

Finance

The water and electricity subsidiary companies have tended to use 100% equity funding to develop projects. Required rates of return have generally been between 8% and 12%, with the development considered as a long-term, secure company investment. In general, several sites have been initially considered prior to a NFFO tender round. These are usually within one or two regions, to minimise planning negotiations, and to allow grouped maintenance if the sites are built. This number has then been narrowed down to a few sites (perhaps 5), which are considered in greater detail, involving the preparation of feasibility studies, NFFO and planning applications. Those sites which receive contracts are then developed.

Entrepreneurial groups often require project finance from a lender, typically with a debt:equity ratio of around 3:1. In general the length of the loan is several years less than the NFFO contract - usually between 5 and 10 years, exceptionally 12. Some turbine suppliers can provide equipment leasing arrangements, with debt:equity not exceeding 4:1. Because of the requirement to repay loans, developers in this category usually need higher rates of return, in excess of 15%. The companies involved are all small in size and may have difficulty obtaining finance in some cases.

Private developers finance hydro projects either wholly from equity or, more usually, with a debt:equity ratio of around 5:2 or less, the loan being secured against other assets if available. Some are ethically motivated, others are hydro enthusiasts. Some are seeking a reliable income stream to offset other, unrelated business costs, such as reducing revenue from estate tenancies. In general, private developers require high returns (>15%). Equipment leasing organised by the turbine supplier is also used as a means of project finance by private developers.

Key issues which affect the unit price demanded for small hydro generation are contract length and the investment return rate required. The water companies typically require around 10%, with entrepreneur developers requiring 15% or more. The length of the electricity sales contract is critical. Without a contract of sufficient length comfortably to pay off the loan, entrepreneur developers would be unable to raise project finance. The water companies could proceed with equity funding, but would almost certainly require higher returns due to the increased risk. Private developers would be likely to require even higher returns, without long term contracts, because of the increased risk and since generation is not the developer's core business. Most schemes simply would not proceed without some security of income from electricity sales.

Capital costs are very site specific, but typically vary between £1,000/kW to £2,500/kW (0.5 to 1.25 lakh/kW) installed capacity. The cheapest sites to develop tend to be within the existing water supply infrastructure, where little civils work is required to accommodate the turbine and where controlled flows are available. Costs for refurbishment sites vary widely, depending on the amount of work required to re-instate generation. Often the original site was used for fairly low power mechanical operations only, such as milling. Refurbishment for profitable generation can involve extensive civil works, for example to enlarge the dimensions of the existing water channels to allow greater installed capacity. Green field sites also have widely varying costs, however it is noticeable that capital costs tend to decrease as the head increases (since cheaper turbines can be used and gearboxes avoided), and as the installed capacity increases (again mostly because larger turbines are cheaper). Sites in North Wales and Scotland, which are predominantly high head in nature, tend to be cheaper to build than sites of similar capacity in England and Northern Ireland, which are mostly low head.

Capital costs have generally risen with inflation in recent years, and there is little prospect for their reduction, particularly as most of the best known sites are already being exploited. It is likely however that many viable sites remain in Scotland, where less is known about the resource.

There is interest in and scope under current legislation for selling generated electricity to local factories, if necessary using the local electricity company network for transmission. The electricity selling price and contract length remain critical to the viability of such schemes.

Expertise

Occasional problems have been experienced which are related to a lack of expertise, particularly amongst private, 'one-off' developers, or from a reluctance to accept conventional practice. This has been exacerbated by some developers' reluctance to pay for quality advice at the outset of a project. Stimulation of best practice guidelines for developers, and increased market share for experienced, 'professional' developers should help to avoid recurrence of these problems.

Abstraction licencing

The abstraction of water from rivers in England and Wales is licensed by the Environment Agency (EA, formerly the National Rivers Authority). The preparation and assessment of small hydro abstraction licence applications in England and Wales has improved as the EA and developers gain experience. The production of national guidelines and policy statements by the EA, which clarify requirements for developers and abstraction officers, has helped this process. EA catchment plans may provide further guidance in future. Currently there is no requirement to licence abstractions in Scotland and Northern Ireland. In these countries only discharges are managed, but abstraction restrictions can be applied by the local authority in granting planning consent for schemes.

Environmental impact

Environmental objections have come to the fore during the planning process for NFFO schemes in North Wales and Northern Ireland. These relate principally to fisheries and the impact of developments on riverine ecology, particularly bryophytes in certain acidified mountain streams in North Wales. The EA is funding research in some of these areas, and a substantial amount of information already exists.

Hydropower development in the UK is at present closely allied to the NFFO process. Environmental research is generally long-term in nature and so meaningful results are unlikely to be available quickly enough to influence the UK Government policy of reaching 1,500 MW installed capacity by the year 2000.

The screening of intakes and allowance for passage of migratory fish are controlled by statute. The normal method involves mechanical screens. Alternative methods of intake and outfall screening are however allowed in the UK, using electric, acoustic or light barriers to divert fish, rather than mechanical screens. These forms of screening offer potential environmental and energy benefits, because fish damage and frictional head loss may both be avoided. At least one UK company is marketing an acoustic fish guidance system to address this barrier.

Price Convergence

Table 2 shows that significant convergence has been achieved so far, with NFFO maximum contract prices falling from 7.5p/kWh to 4.85p/kWh (3.75Rs/kWh to 2.42Rs/kWh) over three orders, and in Scotland, prices falling from 5.3p/kWh (2.65Rs/kWh) for the 1991 transitional arrangement to the maximum contract price in SRO-1 of 4.15p/kWh (2.08Rs/kWh). This convergence is however largely due to the extension of the contract period to 15 years, facilitating longer payback periods, and the inclusion of a 5-year planning period for NFFO-3 and SRO-1. The capital costs of hydro generation are well-understood and are unlikely to reduce significantly. Operating and maintenance costs are already very low for hydro equipment and further reductions are unlikely to be significant. Many of the best sites have already been contracted and the scope for further convergence is small.

Some further convergence is possible, but in the short term this is likely to stem from better design and lower finance costs, rather than from development of the technology.

It is likely therefore that experienced developers with access to equity finance will gain market share while private developers, who have to raise external finance, will face increasing difficulty. These key developers are likely to focus on Scotland where quality sites of appreciable capacity (>0.5MW) are more readily available.

UK SMALL HYDRO INDUSTRY

A minimum of about 75 UK companies are involved in small scale hydro, including 13 turbine manufacturers, 12 professional developers and 50 consultants. In addition, there are many manufacturers of other equipment, not specific to hydro, such as generators, switch gear, control and monitoring equipment, sluice gates and pipes. Civil works form a large proportion of any hydro project, and can be undertaken by several thousand civil contracting companies in the UK. In addition, the International Water Power and Dam Construction Handbook, a leading journal for construction of large scale dams and hydro plants, lists details of 133 UK-based organisations providing services and equipment to the large hydro market. Many of the disciplines required undertake other civil engineering projects, such as bridge-building, tunnelling, road-building and general construction as their main business.

The DTI support programme and the premium market established by NFFO have successfully stimulated development of the industry and enabled it to broaden its experience and skills. Water company and regional electricity company subsidiaries have emerged as prominent developers and are now actively researching opportunities to invest in small hydro overseas. Several entrepreneurial developers have also joined the market. DTI support for site-specific feasibility studies has encouraged developers to investigate sites more thoroughly. Consultancy and developer expertise has been improved and a recognised standard of study has emerged. Key developers now undertake feasibility studies as a routine part of their site investigations.

Turbine Manufacturers

UK turbine manufacturers generally include turbines alongside other product lines, such as pump manufacture, which provide their core business. Of the 13 companies, 3 are predominantly involved with large hydro turbines (>10MW), 5 manufacture turbines for the 0.1 to 10MW 'small hydro' market and a further 5 companies work mostly below 0.1MW ('micro-hydro'). The DTI support programme, including NFFO, has primarily been concerned with the second and, to a lesser extent with the third of these groups. The 5 small hydro manufacturers, and a few of the micro-hydro manufacturers are well-established and most have good export sales, representing the larger part of their hydro business. Several offer an in-house consultancy service, or have links with hydro consultancies.

Virtually all recent development in the UK has been undertaken under the NFFO arrangements. Many of the sites contracted were existing schemes trading under Energy Act terms. Some of these sites were refurbished to allow enhanced operation under the NFFO contract terms. Other sites installed new turbines, with the bulk of the new turbine contracts let to UK companies. The share of the home small hydro market (in terms of capacity) for known schemes is shown in Figure 3.

85% of the home market by capacity is currently supplied by UK manufacturers, with Gilkes, Biwater and Newmills dominating. Including civil engineering work, it is estimated that UK industry has won over 90% of the NFFO small hydro market by value. The three foreign companies represented all have agents in the UK, which carry out installation and maintenance work.

Trends for NFFO-3, NI NFFO-1 and SRO-1 indicate that Newmills, Gilkes and Biwater will continue to dominate the home market, with Powerstream, Flygt and Ossberger retaining a small share. The principal overseas competition is likely to be from ESAC, in the low head turbine market.

Consultants

For many of the 50 consultants (e.g. Balfour Beatty, Gibb, Halcrow), small hydro is only a small part of their business, which usually concerns large civil engineering projects, such as dam- or road-building. About 17 are small hydro specialists, offering advice in all disciplines required - hydrology, geology, civil engineering, electro-mechanical engineering, project management and economics. Environmental impact assessment tends to require separate specialist consultancy, especially at sensitive sites, and at least five companies have developed skills in this area. There are perhaps another 5 companies (e.g. Dulas, ITDG, IT Power) which specialise in small renewable energy projects, including small hydro. Some of these have expertise in technology transfer of innovative micro-hydro techniques to developing countries, including India.

Developers

As described earlier, the developers of new and refurbished schemes fall into 4 groups:

- water companies, which have tended to develop sites on their own land using equity finance;
- electricity company subsidiaries, which are actively developing equity-financed hydro schemes on other's land;
- entrepreneurial developers, generally small companies without access to equity finance whose core business is hydro development;
- private individuals or companies, which tend to pursue one hydro project only on their own land.

It is expected that developers supported by large company equity finance will have most success in both the home and export markets. This is primarily because larger capacity sites are cheaper to develop, provide greater income and the capital amounts involved require access to cheap finance backed by substantial security. It is desirable that opportunities are pursued in partnership with other sectors of the UK industry, such as the entrepreneurial developers, able consultants and equipment suppliers. This partnership approach is likely to reduce costs and establish proficient teams to develop projects in overseas markets.

Financial Institutions

No UK financial institutions yet are known to have particularly strong involvement in small hydro. The small hydro turbine suppliers generally can provide equipment lease terms in association with supporting banks. Typically these arrangements require a maximum debt:equity ratio of 4:1.

Professional Grouping

No specific representational body exists for companies involved in small hydro (or large hydro) as a business activity in the UK.

The National Association of Water Power Users (NAWPU) is the main UK interest grouping relevant to the hydro industry. NAWPU was established in 1975 to provide a common forum for information exchange between hydro developers and the industry. In 1994 there were over 250 members, predominantly small developers operating individual turbines or waterwheels. Companies engaged in hydro development as a business activity are in the minority. NFFO developers and applicants are well-represented, and several prominent consultancies and equipment manufacturers are registered as members.

Other organisations which are relevant to small hydro in the UK, but not as a main activity are the Association of Electricity Producers (AEP), the Association of Renewable Energy Producers in Northern Ireland (AREPS), the European Small Hydropower Association (ESHA), the Chartered Engineering Institutions and the Engineering Council.

THE FUTURE

The global market for small hydro development is estimated at 900MW of new capacity each year, 70% of this in the developing world. The remaining, economically feasible, world-wide small hydro resource is estimated at 180GW. The value of the market is estimated at between £900M and £1,350M (4,500 - 6,750 crores) each year. In addition there is a substantial market for plant and scheme refurbishments. Currently the UK has only a small share of this market, perhaps 1-2%. European competitors, principally from Norway, Germany, Switzerland and France account for around 75% of this business. The scope for market and market share expansion is therefore large, and UK companies and consortia, experienced through the NFFO, could play a significant part.

Some funding is provided from the international aid agencies for hydropower development in certain countries, and UK companies have experience of aid-funded projects. There is however an increasing market for privately-financed projects.

Problems have been experienced with construction of small hydro schemes in some developing countries. These primarily concern inadequate site investigation, inappropriate design, insufficient involvement of local effort, inadequate operation and maintenance training of local personnel and difficulties with local utilities and consumers. Positive benefits can include the retention of rural populations, rural wealth creation, improving social conditions in rural areas and reductions in emissions. However, small hydro generation mostly has only a marginal impact on the national economy and power supply situation, while it still demands adequate local institutional and operational organisation in order to function effectively. The essentially local nature of small schemes therefore can work to their detriment, except where environmental or social concerns are to the fore, or competing power sources are not feasible.

A 'packaged' approach to hydro site development can overcome these difficulties and ensure satisfactory continuity of supply. Mechanisms such as Build Own Operate (BOO) and Build Own Transfer (BOT) have become popular in recent years. The purpose of these mechanisms is to ensure that a new power plant is delivered into a receptive environment, and that it will continue to operate in this environment for a long time. The developer, contractor and equipment supplier all have direct vested interests in this arrangement, which contributes to its success. The package would include finance, design, equipment supply, installation and commissioning, along with facilitating measures, such as organisational guidelines, familiarisation and training. The knowledge, experience and partnerships developed between UK companies through the NFFO place the UK in a good position to assist other countries to deploy small hydro effectively via this 'packaged' approach. It may be mutually beneficial to combine the skills of smaller, specialist hydro companies with the resources of larger UK firms, to enter into joint venture agreements with local partners in the country of interest. For example, the development of low head hydro schemes on canal drops to power Indian industry, could be undertaken by a consortium involving UK investment, skills, equipment supply, procurement and project management techniques, along with Indian investment, skills and market knowledge.

Further work is needed to define key markets, UK opportunities and the most appropriate approach. This is best undertaken on a collaborative basis between Government and industry.

CONCLUSIONS

1. Small hydro unit generation costs can be reduced by better design, lower finance costs and improved procurement methods.
2. Electricity purchase price and electricity sale contract duration are key parameters in determining hydro scheme viability.
3. A competitive, market enablement scheme, such as the Non Fossil Fuel Obligation can prevent the closure of existing sites, stimulate the refurbishment of derelict sites and green field developments, reduce the electricity sale price required and promote industrial growth.
4. As a result of the NFFO arrangements, the UK can now offer a complete finance, skills and equipment supply 'packaged' approach to hydro development, based on proven project management and procurement methods. This is well suited to development of the Indian small hydro resource, jointly with Indian industrial partners.

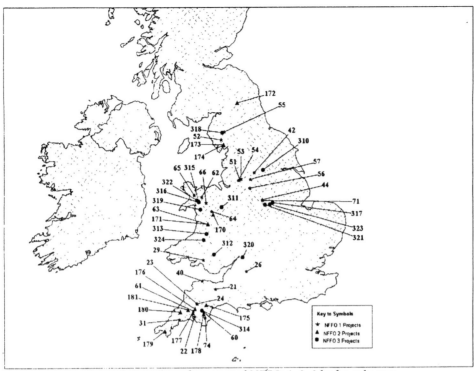

Figure 1: Location of contracted NFFO-1, 2, 3 hydro schemes

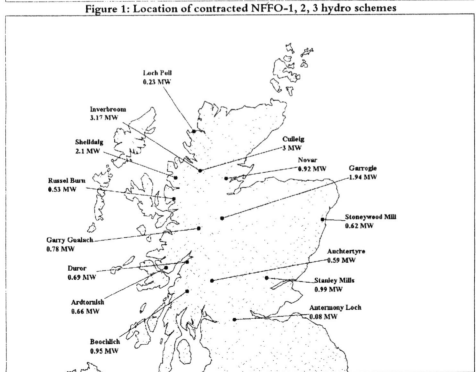

Figure 2: Location of contracted SRO-1 hydro schemes

Site no.	Capacity MW	Name	Site no.	Capacity MW	Name
NFFO-1			**NFFO-2**		
21	0.320	Maundon WTW, Somer.	170	0.200	Dolanog, Powys
22	0.425	Derriford WTW, Devon	171	0.500	Clywedog 1, Powys
23	0.490	Meldon Dam, Devon	172	5.95	Kielder res, Northumbria
24	0.950	Roadford Dam, Devon	173	0.075	Backbarrow, Cumbria
26	0.080	Reybridge Mill, Wilts.	174	0.268	Low Wood, Cumbria
29	0.150	Aberdulais Falls, S. Wales	175	0.016	Sowton Mill, Devon
31	0.432	Ponts Mill, Cornwall	176	2.605	Mary Tavy, Devon
40	0.300	Glen Lyn Gorge, Devon	177	0.638	Morwellham, Devon
42	0.090	Armitage Mill, Yorkshire	178	0.100	Kilbury Mill, Cornwall
44	0.400	Belper Mill, Derbyshire	179	0.175	Trelubbas, Cornwall
51	0.660	Barton Lock, Manchester	180	0.300	Delank, Cornwall
52	0.250	Church Beck, Cumbria	181	0.030	Trecarrel Mill, Cornwall
53	0.570	Mode Wheel, Manchester			
54	0.800	Irlam Locks, Manchester	**NFFO-3**		
55	0.410	Glenridding, Cumbria	310	0.386	Kirkthorpe, Yorkshire
56	0.105	Erwood res., Derbyshire	311	0.363	Oswestry WTW
57	0.180	Bottoms res., Manchester	312	0.367	Pontsticll, S. Wales
60	0.046	River Dart CP, Devon	313	2.950	Elan Valley, Powys
61	0.050	Kelly College, Devon	314	0.076	Old Walls, Devon
62	0.680	Garnedd, N. Wales	315	0.410	Cwmorthin, Gwynedd
63	0.170	Clywedog 2, Powys	316	1.750	Ffestiniog, Gwynedd
64	0.118	Vyrnwy Dam, Shropshire	317	0.890	Holme Pierrepont, Notts
65	0.120	Padarn Park, Gwynedd	318	0.176	Thirlmere, Cumbria
66	3.800	Llyn Celyn, Gwynedd	319	0.645	Ganllwyd, N. Wales
71	0.200	Milford Mill, Derbyshire	320	0.098	Ebley Mill, Glos.
74	0.055	Tuckenhay Mill, Devon	321	0.179	Borrowash Mill, Derbys.
			322	0.180	Gelli Iago, Gwynedd
			323	1.660	Beeston Weir, Notts.
			324	4.350	Llyn Brianne, Powys

Table 3: Index to Figure 1 NFFO contracted sites

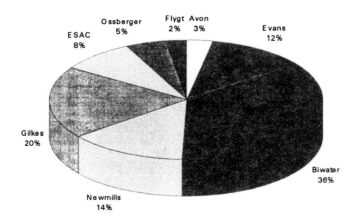

Figure 3: Recent UK small hydro turbine contracts

First International Conference on Renewable Energy–Small Hydro
3 – 7 February 1997, Hyderabad, India

SMALL HYDRO TECHNOLOGY AND DESIGN IN CZECH REPUBLIC

Radomir Zachoval[1] *and Vladmir Hebelka*[2]

[1] CKD Blansko Engineering a.s.
  Gellhornova 14, 67818 Blansko, Czech Republic
[2] CKD — Turbo Technics Rajec, Czech Republic

The traditional water turbine producer ČKD Blansko a.s., together with its daughter company ČKD-Turbotechnics Rájec, is presenting the comprehensive offer of small hydraulic water turbines. Their design, manufacturing and operation is based on the proper research activity and verified by model tests. The design exploits all positive and negative experience from operation and service of all foregoing units. The high level warranty of ecological quality and functional reliability has ensured.

ČKD Blansko a.s. has been staying the traditional producer of equipment for water energy machinery since the end of 19th century. This company has long time tradition in this branch and its historical roots have reached down to the year 1698. The present manufacturing program has been covering all types of water turbines including their accessories, turbine valves and equipment of water dam intake structures as well as hydrotechnical equipment for dams, weirs and lock chambers.

Small water turbine machinery is completely produced in new company ČKD - Turbo Technics which has been formed by the fusing of two independent subjects ČKD Blansko, small water turbine division, and Turbo Technics Brno.

Manufacturing program of this company includes all types of small water turbines in following range:

1. Turbines with Kaplan runner

1.1 Kaplan turbines placed either in concrete flume or in spiral case, horizontally arranged, for outputs from 20 up to 400 kW. In case of higher discharges and heads these turbines are offered in vertical arrangement from the output 50 kW and from the minimum head value about 1.5 m. The existing runner standard is covering diameters as follows: 300, 400, 560, 710, 800, 1000 and 1300 mm. Two latest diameter values are considered for vertically arranged units only.

This type of turbin is suitable especially for refurbishment of old low head power stations with open flume and with changing discharge and head values. These turbines are working with high efficiencies at wide range loading.

The turbines are normally delivered with speed increaser and high speed asynchronous generator for parallel operation into grid. Their operation is automatically controlled by water level governing system. The delivery can also include an electrical distributing system with all complete failure automatics.

1.2 Straight flow horizontal Kaplan block units with special cone wheel gear box fitted to high speed generator for outputs from 50 up to 600 kW and head range between 1.5 and 8 m. They are usually used for new built power stations. Their progressive compact design enables very fast and easy site erection into existing civil part. These units are designated for high discharge values. Their high efficiency is caused by optimum hydraulic profile shape with straight arranged draft tube.

1.3 Straight flow horizontal Kaplan "S" turbines with gearing fitted to high speed generator designed for outputs from 50 kW at heads from 1.5 up to 12 m. They are usually used for higher outputs and head range exceeding the possibilities of before mentioned block units. They enables very simple solution of power station civil part. The efficiency is rather decreased in comparison with former specified block units because of S-shaped draft tube. Anyway they can operate in very wide changeable discharge range.

The runner diameters row is 400, 560, 710, 850 and 1000 mm. The Kaplan "S" turbines are preferably equipped with the most progressive three or for blades runners developed in the hydraulic research department of ČKD Blansko a.s.. The highest parameters of these runners have been tested and verified on the model device in the laboratory. The efficiencies are fully comparable with other world water turbine top producers. The maximum discharge capacity of these runners allows minimising overall dimensions of complete turbine units with very good influence on general power station construction costs.

The units are equipped with diagonal wicket gate mechanism with adjustable or fixed vanes. Adjustable guide vanes are usually closed with help of counterweight. This solution enables omitting turbine intake valve installation. The using of fixed vanes wicket gate on the other hand saves the cost of the turbine mechanism but the operation valve instalment in front of turbine or in draft tube outlet is necessary.

1.4 As the extending of Kaplan turbine standard row the PIT turbines with three blade runner (diameters 1300 and 1500 mm) are offered. The energy transmission to generator is performed with epicyclical gear box designed in monoblock with horizontal generator in turbine pit. The second used solution includes turbine with right angle cone gear box and vertical generator situated above the turbine unit. The claw clutch, type "ESCO Gear", for connecting shafts is used.

1.5 The runner blades of all above mentioned individual types are usually made of stainless material Ni-Resist, Cr13 Ni1 or aluminium bronze. They are fitted in selflubricated bushes, type KU, with teflon or high-molecular polyethylene "Solidur 1000" lining. In case of higher specific pressure the selflubricated bushes "Iglidur", made by firm Hennlich, are used. The same type of selflubricated bushes is used in the bearings of wicket gate vanes. This is very important feature from the ecological point of view preventing grease leakage into flowing water.

The turbine shaft sealing box is designed in several variants. In case of Kaplan block and "S" turbines the sealing ring "Gufero" with hard stainless steel or ceramics bush is used. The flume turbine are equipped with labyrinth sealing ring and straight flow PIT turbines with soft sealing elements.

90

Under the condition of smaller range discharge variation, from 40 up to 100%, the simplified design of Kaplan turbine with fixed runner blades and adjustable distributor vanes can be used. Further the next simplifying can be taken into account in case of the long time period of changing discharge value. It consist in using runner with fixed blades adjustable by hand only. The wicket gate vanes can be adjustable or fixed.

The simplified design of Kaplan micro-turbine without wicket gate mechanism and with spiral case, type Reiffenstein, is possible to be used too. The efficiency of this turbine is a bit lower however it allows to operate with high range discharge between 40 and 100%.

2. Turbines with Francis runner

Francis turbines with spiral case, horizontally arranged, operated under the heads from 10 up to 120 m and for outputs up to 5000 kW. In case of larger units the vertical arrangement can be used. These turbines have high efficiencies however they can operate at lower discharge capacity range in comparison with Kaplan units.

The runners are made of stainless material Cr13Ni4 or Cr13Ni1 by method of precision casting "SHOW" with minimal allowance for grinding. The turbines are equipped with pressure spiral casing and adjustable guide vanes seated in selflubricated bushes. The wicket gate at smaller units is closed by weight. The generator can be connected directly to the turbine unit. Further the gear box or belt transmission can be used.

3. Turbines with Pelton runner

The simple horizontal arrangement of Pelton turbine units operated under the heads from 40 up to 200 m and for outputs from 30 up to 1800 kW are offered. The generator can be connected directly to the turbine unit as well as with gear box or belt transmission. These turbines have flat shaped efficiency curve and they can operate with relatively wide discharge capacity range.

The runners are made of stainless material Cr13Ni4 or Cr13Ni1. Horizontal shaft is seated in two antifriction bearings. All welded casing is equipped with one or two nozzles. The needle in nozzle is controlled with hydraulic or electric servomotors. The water stream can be declined with deviator.

4. Cross-flow turbines

This type of turbines due to their economical, simple and robust construction is to be very widely used. The most important advantage of this turbine is a wide range of heads overlapping those of Kaplan, Francis and Pelton. The total water flow capacity is usually controlled with by-pass valve installed on intake part.

Considering the experience gained so far from the market, operation and manufacturing the turbine has been permanently innovated. The optimum hydraulic shape of runner blade has been tested on model device in research laboratory. Three standard type series of the turbines with unified runner diameter and various rotor width has been produced nowadays as follows:

BK 45 - runner dia 450 mm, width range 150, 300, 450 and 600 mm

B 30 - runner dia 300mm, width range 110, 200, 350, 460, 600 and 1000 mm

B 15 - runner dia 150 mm, width range 100 and 300 mm

The turbine are manufactured and assembled in standard way in workshop. They are designed and delivered to the individual localities following all demands of the client. The using of typified components enables to reach very low price level. This is the reason why the cross-flow turbine is very suitable for private person activities.

Pic. 8 represents the operating ranges of all before mentioned types of turbines as a function of the head and the discharge. The head by itself constitutes the first criterion in the choice of the turbine to install.

When turbine and generator operate at the same speed they both can be directly coupled. No power losses are incurred and maintenance is minimal. Due to tolerating certain misalionme.it the application of flexible coupling is recommended. In many instances, particularly in the lowest power range, turbines run at less than 400 rpm, requiring a speed increaser to meet the 1000 up to 1500 rpm of standard alternators. Transmission of energy between turbine and generator in such case is made with multiplied V-belt drivers or gear box. Gear boxes increase the noise level in the power house and require additional maintenance.

Majority of customers demands the small turbine units for parallel operation into common public grid. Asynchronous generators are simple electric squirrel-cage induction motors, with no possibility of voltage regulation, which operate at a speed directly related to system frequency. They draw their excitation current from the grid, absorbing reactive energy. The absorbed reactive energy can be compensated for by adding a bank of capacitors. These units cannot generate when disconnected from the grid because they are incapable of providing their own excitation current.

When the operation into separated grid is required the synchronous generator involving voltage regulator and excitation system together with speed governor is delivered. Voltage, frequency and phase angle are to be controlled before the generator is connected to the grid. It supplies the reactive energy required by the power system when the generator is tied into the grid. On disconnection of the paralleled connection, the synchronous alternator can continue to generate at a voltage and frequency specified by its control equipment. Synchronous generators can run isolated from the grid and produce power since excitation power is not grid-dependent.

Control equipment involving speed governing system delivered to the turbine units is usually based on programmable module SAIA. All system consists of individual modules. The block system allows modifying governor according to the client s demands. This system can be designed either for complete power station control or for governing turbine speed only. In any time it can be extended in accordance with further additional requirements of client.

Radomír Zachoval, MSc., senior sales manager

ČKD Blansko Engineering, a.s.

Gellhornova 1

678 18 Blansko, CZECH REPUBLIC

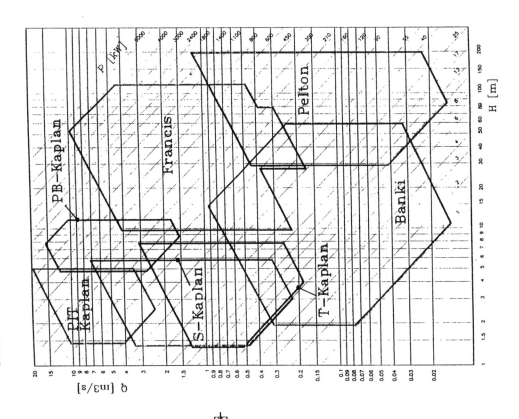

BANKI-MITCHELL TURBINES

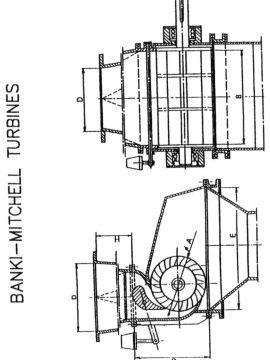

TYP	A	B.	C	D	E	H
B15	150	30–100	270	100–150	320	300–400
B30	300	110–1000	480	250–600	640	500–700
B45	450	150–600	740	300–600	960	700–800

93

PELTON HORIZONTAL TURBINES

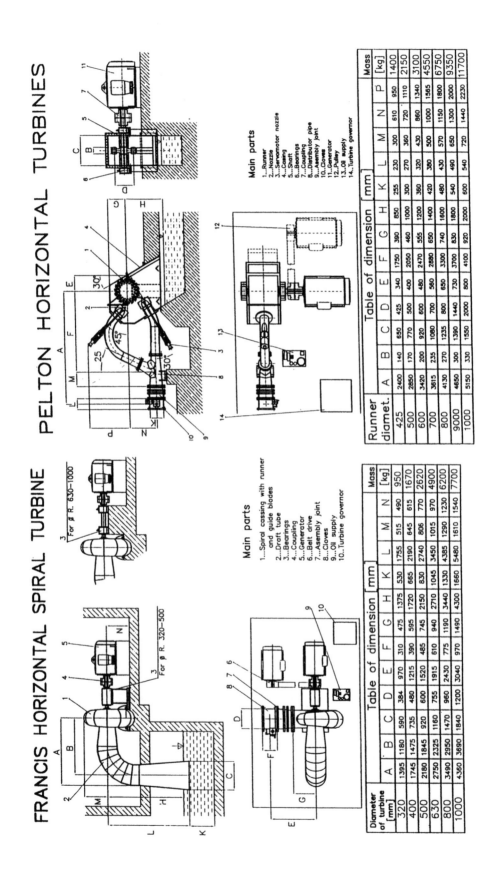

Main parts

1...Runner
2...Nozzle
3...Servomotor nozzle
4...Casing
5...Shaft
6...Bearings
7...Coupling
8...Distributor pipe
9...Assembly joint
10..Cloves
11..Generator
12..Pulley
13..Oil supply
14..Turbine governor

Runner diamet.	Table of dimension [mm]													Mass
	A	B	C	D	E	F	G	H	K	L	M	N	P	[kg]
425	2400	140	650	425	340	1750	390	850	255	230	300	610	950	1400
500	2850	170	770	500	400	2050	460	1000	300	270	360	720	1110	2150
600	3420	200	920	600	480	2470	555	1200	360	320	430	860	1340	3100
700	3615	235	1080	700	560	2880	650	1400	420	380	500	1000	1565	4550
800	4130	270	1235	800	650	3300	740	1600	480	430	570	1150	1800	6750
9000	4850	300	1390	1440	730	3700	830	1800	540	490	650	1300	2000	9350
1000	5150	330	1550	2000	800	4100	920	2000	600	540	720	1440	2230	11700

FRANCIS HORIZONTAL SPIRAL TURBINE

For ø R. 630—1000

For ø R. 320—500

Main parts

1...Spiral cassing with runner and guide blades
2...Draft tube
3...Bearings
4...Coupling
5...Generator
6...Belt drive
7...Assembly joint
8...Cloves
9...Oil supply
10..Turbine governor

Diameter of turbine [mm]	Table of dimension [mm]											Mass
	A	B	C	D	E	F	G	H	L	M	N	[kg]
320	1395	1180	590	384	970	310	475	1375	1755	515	490	950
400	1745	1475	735	480	1215	390	595	1720	2190	645	615	1670
500	2180	1845	920	600	1520	485	745	2150	2740	806	770	2620
630	2750	2325	1160	755	1915	610	940	2710	3450	1015	970	4900
800	3490	2950	1470	960	2430	775	1190	3440	4385	1290	1230	6200
1000	4360	3690	1840	1200	3040	970	1490	4300	5480	1610	1540	7700

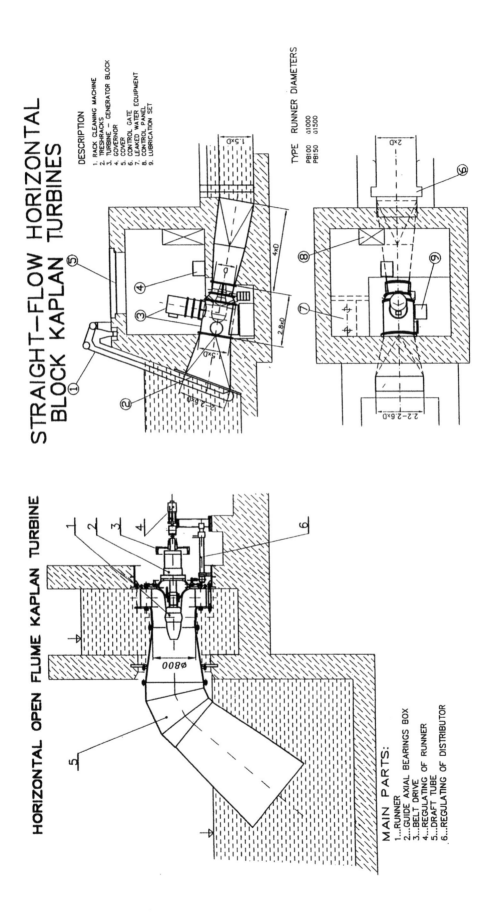

STRAIGHT–FLOW HORIZONTAL BLOCK KAPLAN TURBINES

DESCRIPTION

1. RACK CLEANING MACHINE
2. TRESHRACKS
3. TURBINE – GENERATOR BLOCK
4. GOVERNOR
5. COVER
6. CONTROL GATE
7. LEAKED WATER EQUIPMENT
8. CONTROL PANEL
9. LUBRICATION SET

RUNNER DIAMETERS

TYPE	RUNNER
PB100	Ø1000
PB150	Ø1500

HORIZONTAL OPEN FLUME KAPLAN TURBINE

Ø800

MAIN PARTS:

1...RUNNER
2...GUIDE AXIAL BEARINGS BOX
3...BELT DRIVE
4...REGULATING OF RUNNER
5...DRAFT TUBE
6...REGULATING OF DISTRIBUTOR

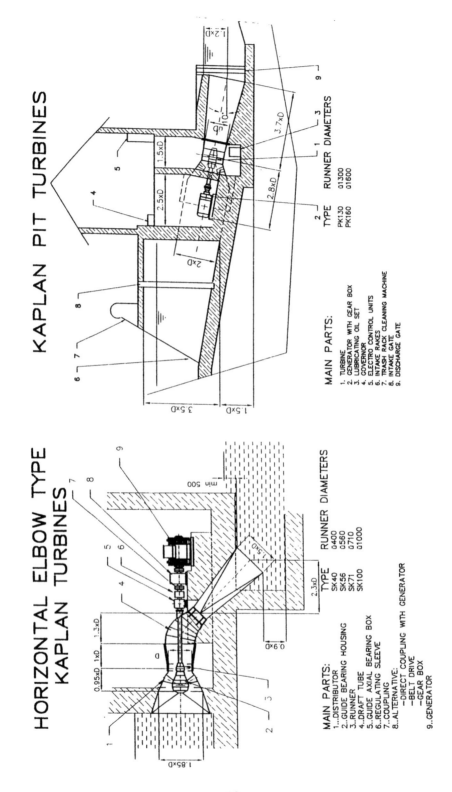

KAPLAN PIT TURBINES

MAIN PARTS:

1. TURBINE
2. GENERATOR WITH GEAR BOX
3. LUBRICATING OIL SET
4. GOVERNOR
5. ELECTRO CONTROL UNITS
6. INTAKE RAKES
7. TRASH RACK CLEANING MACHINE
8. INTAKE GATE
9. DISCHARGE GATE

RUNNER DIAMETERS

TYPE	RUNNER DIAMETERS
PK130	Ø1300
PK160	Ø1600

HORIZONTAL ELBOW TYPE KAPLAN TURBINES

MAIN PARTS:

1. ..DISTRIBUTOR
2. GUIDE BEARING HOUSING
3. RUNNER
4. DRAFT TUBE
5. GUIDE AXIAL BEARING BOX
6. REGULATING SLEEVE
7. COUPLING
8. ALTERNATIVE:
    −DIRECT COUPLING WITH GENERATOR
    −BELT DRIVE
    −GEAR BOX
9. .GENERATOR

RUNNER DIAMETERS

TYPE	RUNNER DIAMETERS
SK40	Ø400
SK56	Ø560
SK71	Ø710
SK100	Ø1000

SMALL HYDRO — A DECENTRALISED ENERGY OPTION FOR RURAL ELECTRIFICATION; THE EXPERIENCE OF VILLAGE HYDRO PROJECT IN SRI LANKA

Madhavi Malalgoda Ariyabandu

Intermediate Technology Development Group (ITDG)
15 B Alfred Place, Colombo 3, Sri Lanka

ABSTRACT[1]

In Sri Lanka the main supplier of electricity is the state owned national grid. At present the grid can supply only 45% of the households, mostly urban. The existing and planned Rural Electrification (RE) schemes are extensions of the national grid. RE schemes, although justified on the basis of social equity, are often a burden to the State. They are large scale and capital intensive schemes that are low on returns.

Village Micro Hydro projects, a pioneering effort of ITDG, generate electricity in remote villages, and are operated and managed by the village communities. At the end of a five year pilot phase, Village Hydro can be recommended as a promising option for decentralised supply of electricity for rural areas. A comparison with the grid extended RE schemes indicate that the primary objectives of rural electrification can be met with Village Hydro with no or minimum cost to the state.

The pilot phase has identified a few areas which require further improvements. The State can capitalise on the positive aspects of Village Hydro to address rural electrification problem, by recognising Village Hydro as a practical option for rural electrification, and by extending support for further research and development. The spread of Village Hydro will contribute towards achieving rural electrification goals at a faster pace with no burden to the State.

1. INTRODUCTION

In Sri Lanka State owned grid is the main supply source of electricity. About 45% of the country's households are served by the grid. The rural electrification rate is about 35%.

There are two options to supply electricity to the balance population who are over 12 million in number; extending the grid, and/or developing decentralised options of energy; such as mini and micro hydro, solar, wind and thermal power, and bio gas. Extension of the grid to the rural areas has its own limitations, the primary concerns are the cost, Sri Lanka Rupees 1.2 - 1.4 million per km, (1 US\$ = 55 SLRs) and returns on the investment, and the time span it will take to reach nearly 25,000 remote villages.

This paper intends to discuss the experience of Intermediate Technology Development Group (ITDG) in establishing micro-hydro power projects to electrify remote villages, in a decentralised manner. The paper draws parallels between this option and RE schemes, critically analyse the micro hydro option; its advantages and limitations. On this basis, micro-hydro development at village level is recommended as a suitable option for rural electrification.

---

[1]The author wishes to thank Sunith Fernando, Programme Manager - Energy, ITDG Sri Lanka for his valuable comments.

2. A PROFILE OF RURAL ELECTRIFICATION SCHEMES IN SRI LANKA

Some of the general notions about the RE Schemes are that they are expensive and bring low returns, which are the main constraints for expansion. Rural electrification is justified primarily on the basis of social equity (in relation to urban population) and improvements to the quality of life. It is widely recognised that electricity makes a significant contribution towards the overall improvement of the quality of life (QOL) of people. In particular, it is expected that supply of electricity will lead to better quality of lighting in houses facilitating education and literacy, will enrich social and cultural life by the improvements to public places such as school, community centres, temples, improve safety and security in the village and so on (1). The perception that electricity supply is automatically followed by growth in economic activities is however questionable, since electricity is only one contributory factor which induce economic growth.

In Sri Lanka, Ceylon Electricity Board (CEB) has the sole responsibility of generation and transmission of electricity. Up to the middle of this century, the grid mainly supplied the town and city centres. The first attempts of extension of grid to the rural areas came into effect in the 1960s, primarily to supply tea and rubber factories in the estates.

It is estimated that there are about 25,000 villages in Sri Lanka. At present only a little over 7,000 villages are supplied with electricity under RE schemes. Within the present structure of power sector extension of the grid is considered the best option for rural electrification. This view is based on the facts that the grid can meet all the demands of the households, and it has the potential for subsequent industrial and economic development in the villages. The main criteria for selection of villages for electrification include; distance to the grid from the village, potential for economic development, and political concerns. At present the CEB makes five year extension plans for the grid.

In the RE schemes, economic and financial benefits are determined by a number of factors. These factors essentially are: number and mix of consumers, their uses of energy prior and after rural electrification, growth rates of the number of consumers, and per consumer electricity usage (1, 2). Residential consumers comprise the majority of RE schemes, accounting for the major share of energy consumption. In an average scheme residential consumers make up to 88% of the total number consuming upto 81% of electricity. The mix of consumers, residential, commercial, and industrial depend on the economic activity in the area. It is observed, in the paddy and coconut areas electricity is put to relatively more productive use compared to rubber and tea economies, since there are more commercial consumers.

According to CEB surveys (1), on average 44, 26 and 41 percent of potential residential, commercial and industrial consumers respectively are electrified in RE schemes. On average, a residential consumer uses 43 kWh per month. The tariff paid is only about 1/2 the money spent on kerosene prior to obtaining electricity connection.

Further, in RE schemes on average only about 44% of "potential consumers" obtain connection. For commercial and industrial consumers this figure comes up to 26 and 41 per cent respectively (potential consumers are those permanent households, religious institutions etc. within 100 meters of the distribution line). Connections to the commercial and industrial consumers are vital for the economic and financial viability of RE schemes.

Economic and financial returns are low for remote and smaller schemes. Economic returns on average for remote small schemes are at 5%, while in large schemes with more productive uses can be as high as 16%. On average this figure is about 12%; which is considered as a reasonably sound figure (2).

Financial returns, like in many RE schemes in the Asia Pacific region remain low in Sri Lanka. Further, if the load factor is considered in RE schemes it is much lower compared to urban electrification schemes (which is about 55%) . The tariff is subsidised by the urban consumers, and industry who consume greater quantities of electricity. Considering all these factors, it is clear that RE schemes are justified mainly on a social equity basis.

3.   HISTORY OF MICRO HYDRO DEVELOPMENT IN SRI LANKA

History of harnessing water for power development in Sri Lanka goes back to early this century. The well distributed rainfall coupled with the hilly and rolling terrain offers ideal opportunities for harnessing micro hydro power in the upper reaches of the mountainous region of Sri Lanka. Beginning around the early twentieth century, the British planters exploited this potential widely to derive motive power for the rapidly expanding tea and rubber plantation industry. It is estimated that by the middle of the century about 500 tea and rubber estates had been operating micro hydro power plants.

With the extension of the national electricity grid to remote regions, beginning around late fifties, most plantations had abandoned their micro hydro plants in preference to the grid electricity supply. Since then, the use of micro hydro power in the plantation sector witnessed a rapid decline. Interest in micro hydro power, however, re-emerged in Sri Lanka in late seventies in response to the escalating grid electricity prices as well as international developments in the energy scene in the aftermath of 1974 "oil crisis" which favoured the development of renewable sources of energy. Revival of micro hydro power has, however, remained a slow process up to now.

ITDG played a major role in the revival of micro hydro power in Sri Lanka beginning from 1980's. With the establishment of the Sri Lanka office - ITSL, efforts were made to explore wider opportunities for application of micro hydro power. The organisation initiated its work in Sri Lanka with small hydro schemes in early 1980 by rehabilitating some of the abandoned schemes in the tea estates, with the objective of developing local technical capabilities for power generation in rural areas. These schemes were categorised under the mini hydro category, with capacities ranging from 60-150 kW. These efforts lead to the identification of some indigenous efforts of enterprising rural people who have been experimenting with self made hydro power devices to generate electricity for their own needs. This marked the birth of ITSL's Village Hydro Programme in 1990.

3.1   Village Hydro Project

Generating hydro power at micro scale in the villages to meet community requirements is a pioneering effort. At present there are 24 such schemes generating power, their capacities ranging from 0.9 to 35 kW. The power generated is primarily used for household lighting. There are a few schemes where power is utilised for economic activities; to charge batteries during the day, and one scheme which power a rice huller/polisher.

Technically, village hydro projects are built on the same principle of generating power with water wheels, at a micro scale. A typical scheme consist of a wier, a penstock, turbine, and a motor converted to a generator to generate electricity. The cost of generating 1 kW is estimated at 2000 US$. ITDG, through the village hydro schemes attempts to improve local technical capabilities related to manufacture of equipment, operation and maintenance skills of user community, and the local management structures.

One of the key requirements for hydro power generation at micro scale is streams which do not dry up during a most part of the year. Hence, all 24 completed schemes, and nine more in implementing stage are located in the wet zone of the country, where average annual rainfall is over 2,500mm.

Economies in these villages are agricultural, with subsistence paddy, smallholder tea, rubber and cinnamon. These are typical wet zone villages in Sri Lanka. Away from the nearest town, and the main grid (2-5 km), they lack good roads, clinics and schools with adequate facilities. Population varies, from about 10 households to 275 per village. Average annual income of the villagers fall within a range of Rs 18000 - Rs 75000 (5). Table 1 provides some background information on hydro villages.

Village hydro projects are a combined effort of a number of parties: village community, a funding source and ITDG. The community members are responsible for all the civil works and transmission. Civil works include constructing a wier, installing the penstock and building a power house. The community contributions in terms of cash, goods and labour which range from 25-40% of the total project cost. ITDG provide engineering consultancy services, which comprise of technical feasibility studies, installations and the study of socio- economic aspects. This component ranges between 15-25% of the total cost. The balance capital cost come from various funding sources, non-government organisations, varying government and private bodies. In addition to engineering consultancy services, during the pilot phase ITDG provided regular training on operation and maintenance aspects and institutional management of the Village Hydro projects.

Projects are initiated on the basis of requests form village communities. The approach to project implementation is participatory from the planning stage. The village community has the entire responsibility of the project implementation and management.

Background Information on Selected Village Hydro Projects.

Table 1

Project	Distance from nearest Town (km)	Distance to the grid (km)	Installed capacity	Economic Activity
Pathavita	10.0	2.0	3.8	Agriculture, labour, self employment
Dolapalledola	13.8	3.0	2.0	Agriculture, labour, self employment
Ihala Maliduwa	12.0	2.4	3.0	Agriculture, labour, small business
Umangedara	10.0	4.0	1.7	Agriculture, gem mining, labour
Katepola	9.0	4.0	24.0	Agriculture, labour, small business

Source : ITDG Socio-economic reports of individual Village Hydro projects, 1992-1995

In Village Hydro projects often not all the households in a village can be electrified from the power generated due to technical and socio-economic limitations. Table 2 shows the power consumption patterns in selected Village Hydro Projects. The technical and socio economic limitations are as follows:

Technical limitations

- Installed capacity depend on the 'head' (vertical height of the water source), and 'flow' (quantity of water flow per second)
- distribution from the point of generation is economically feasible only within a km radius[2].

Socio-economic limitations

- willingness of community members to be part of a community activity
- ability to contribute towards the project, in terms of capital, labour, and time
- ability and willingness to participate continuously in management and maintenance activities related to the project.

To minimise the impact of particularly technical limitations, a general recommendation made to ensure maximum possible equity in access to power is to limit the quantity consumed by a household to 100 Watts. This quantity allow five 20 W bulbs, or a combination of a few bulbs with a B/W TV and/or a radio cassette player.

---

[2]For transmission beyond 1 km radius, transformers are required. This escalates the cost of the project substantially.

Table 2

Power use pattern

Project	Population	No of Households	Connected Households (%)	Non-Connected Households	Power Consumption (Watt)	Times of Supply
Pathavita	1280	220	64 (29)	29 (30)	100 -140 - 10 houses 50 - 80 - 5 houses 40 - 36 houses 20 - 40 - 15 houses	5.30 p.m. - 7.30 a.m. - daily weekends and holidays - whole day
Dolapalledola	176	30	17 (57)	9 (35)	100 - 15 houses 50 - 2 houses	5.00 p.m. - 6.30 a.m. - daily weekends and holidays - whole day
Ihala Maliduwa	868	45	32 (71)	35 (50)	100 - 140 - 15 houses 80 - 3 houses 65 - 4 houses 40 - 10 houses	5.30 p.m. - 6.30 a.m. - daily during dry months 6.00 p.m. - 10.00 p.m. 4.00 a.m. to 7.00 a.m.
Umangedara	360 (est)	72	17 (24)	23 (57.5)	100 - 17 houses	week days - irregular weekends and holidays - whole day
Katepola	1050	210	89 (43)	121 (57)	100 - 89 houses	24 hrs daily

Source : ITDG Socio-economic reports of individual Village Hydro projects, 1992-1995

4.   IMPACT OF THE VILLAGE HYDRO PROJECTS; THE DIRECT, AND INDIRECT BENEFITS

Impact is the changes the project activity has brought about. Immediate and visible impact is observed during, and immediately after the project implementation. Some of the indirect and long term benefits are to be observed subsequently at different stages of project life.

Direct Benefits of the village hydro project are primarily consumption oriented. The main feature is illumination of the households. Lighting has given village communities a better quality of life, safety and security. Annual monitoring surveys of the project indicate that as a result of electrification communities are having greater access to information through TV and radio, household chores are made convenient and flexible particularly for women. Every kitchen has a light in electrified houses, women save time on cooking and other chores, and children are more enthusiastic on studies. A noteworthy comment made by women with young children is, that absence of kerosene lamps has made child minding much convenient and a less nervous task.

Fuel economy is another direct benefit of the project. Village Hydro Project Monitoring survey 1995 (3) indicate that there are substantial savings from cutting down the use of kerosene oil. Reduction in consumption ranges between 54-77%. The savings range from SLRs 44-96 a month. In a ten project sample, monthly savings on kerosene had been adequate to pay the monthly tariff in six projects. Savings are made on reduced torch battery consumption too. On average savings are made on 8 batteries a month, approximately amounting to Rs. 160/-.

As mentioned, not all the houses are electrified. The percentages of connected households in selected sample of projects is given in Table 2. However benefits of illumination (security, convenience), particularly public places such as school, community hall, clinic and access to TV are spread to the persons in unconnected households too. In addition, most projects have the facility of giving temporary connections to unconnected houses on special occasions such as funerals and weddings.

Productive use of electricity is not widespread for two reasons; lighting is the priority need, and in most projects installed capacity is relatively small to introduce productive uses. However, there are 4 projects where battery chargers are installed, and one scheme with a rice mill using the power during the day time. Battery chargers aid particularly non-connected households to have access to 'power', save them the trouble of taking batteries for re-charge (used car batteries re-charged for lighting, TV and radio) to city centres which are 2-5 km away. Similarly rice mill is a convenient service to the community, which benefit all households.

Positive changes in community in terms of capacity building are observed during and after project implementation. During the early stages community members get together, organise themselves to form a village institution to implement the project generally known as "Electricity Consumer Society" (ECS).

The ECS is responsible for planning, implementing and managing the project (4). The ECS is the key decision making body, and it liaise with ITDG and other institutions involved in project. This experience strengthen community members' capacity, and enhance their skills in a number of ways;

- organisational and decision making capacity (including information collection, assessment of situations, informal evaluation)
- technical skills in manufacturing, in operation and maintenance

- communication at village and institutional level
- management and leadership skills
- community sharing of work and benefits

Capacity building of this nature has resulted in strengthening and empowerment of communities. The communities have become stronger, they are able to deal better with outside institutions, and work towards positive changes. This process, initiated at project implementation continues afterwards, beyond the project.

Further, the project has set in motion a skill development process; both for local manufacturers and beneficiaries. Most of the equipment required for these schemes are manufactured locally, in local workshops. This include turbines, Induction Generator Controllers and other electronic items. With a little enhancement of knowledge the local manufacturers are capable of producing the equipment to the required standards. The Village Hydro project, which was initiated as an experiment has resulted in strengthening about 10 local turbine manufacturers. Further increase in this figure will depend on how the project would spread.

Community members by participation in installation and related activities develop skills which were foreign to them before. ITDG has trained selected community members to attend to basic operation and maintenance needs of the schemes. Further, project management brings out the leadership qualities in community members, which is refined through this experience. Potential community members who demonstrate leadership qualities are provided with training to further their skills.

In totality, Village Hydro projects are able to illuminate rural households, to strengthen and to empower village communities to a considerable extent, make a marginal contribution towards economic activity, and to develop local manufacturing capabilities. From an environmental point of view, the technology does not make any adverse effect on the surroundings; water resources, fauna and flora. The communities in project areas ensure that trees in the catchment area is protected. Efforts are made to plant additional trees to ensure a sufficient water flow throughout the year.

5.   VILLAGE HYDRO -AN OPTION FOR RURAL ELECTRIFICATION?

A comparative analysis of benefits and limitations of village hydro with RE indicate that benefits from both schemes are similar to a large degree. While it is acknowledged that the scale of benefits are less in Village Hydro, the improvements to the quality of life as a result of lighting, and fuel economy can consider to be on par with RE schemes.

In both schemes consumption is at a low level, 48 kWh and 18 kWh /month respectively for RE and Village Hydro and the electrification rate is below 50% (Table 3). The mix of consumers is similar with a large number of residential consumers, and few of commercial and industrial consumers. As a result the total consumption is low and the load factor is considerably low for both schemes. In RE schemes this is about 7% on average (1), and in Village Hydro it is lower than this, since the consumption is limited to about 6 hrs/day in most projects.

Table 3

Comparison of Basic Characteristics of RE and Village Hydro Schemes[3]

Characteristic	RE Schemes	Village Hydro Schemes[4]
Consumption (residential consumers)	48 Kwh/month	18 Kwh/month
Cost of Service connection	Rs. 7,100*	Rs. 7000
Cost of Wiring	Rs. 9,900*	s. 2,000 - 5,000
Electrification rate	about 43%[5]	about 40%[6](limited to 1 km radius)
Restrictions on consumption	No restrictions	Only up to about 100 Wt
Mix of consumers	High residential, low commercial and industrial	High residential, low commercial, very limited industrial
Load factor	low	low
Management	Uniform (by the State)	Individual for schemes, space open for conflicts, sabotage etc.
Demand for more schemes	high	high
Capital and maintenance cost	entirely by the state	shared by community and other funders, no cost to the state

* 1987 prices in the original source (1) deflated to 1996.

---

[3]Sources:1, 3, 5
[4]Calculated on the basis of 100 W used on average 6 hrs per day

[5]The balance are not connected mainly for financial reasons
[6]Not connected due to technical , financial and social reasons

Most marked disadvantage of village hydro power in comparison to the RE schemes is the restrictions imposed on consumption, which allows up to a maximum of 100W, severely limiting the use of any appliances·other than a few light bulbs, a B/W TV, radio/cassette player[7]. This clashes with the expectations of most rural consumers who wish to have access to the grid. However, in the existing village hydro schemes the consumers who can afford appliances such as fridges, colour TV, fans are extremely few.[8] For many, initial priority need is household lighting. Once this need is met, there is a tendency to demand further improvements and more quantity.

In terms of capital, maintenance costs and returns, the cost of extending the grid by 1 km is SLRs. 1.2 - 1.4 million. The entire cost of RE schemes, and subsequent maintenance costs are met by the State. The tariff is subsidised[9], overall consumption is low (since a majority of the consumers are residential), therefore the cost recovery rates of RE schemes are at a very low level. Returns for smaller, more remote schemes are as low as 5% (1). The financial performance of RE schemes is reported to be always weaker than the economic performance and, according to the Ministry of Power and Energy Evaluation (1) they impose a financial burden on CEB. However, the demand for more RE schemes are on the rise. The State is pressurised to allocate finance for RE despite low returns due to political and social equity reasons.

The capital costs of the village hydro projects vary according to the magnitude of the civil works required, the costs of power generation and the costs of distribution systems. It is estimated that generating 1 kW by village hydro costs approximately 2,000 US$. The consumers of electricity pay a monthly tariff for the use of power. The amount is an independent decision of the respective ECS, therefore the tarriff levels vary. Money collected is credited to a fund, which is utilised for operation and maintenance purposes. In general practice the recommended tariff is Rs. 1/- per Watt, but, this can be as low as cents 50 in some schemes. These tariff levels are not realistic considering the capital costs, and monthly revenues vary drastically. As a result, similar to RE schemes financial performance of Village Hydro is poor, projects have unusually long pay back periods, on average about 30 years (3).

Village Hydro however, imposes no cost to the State. Capital and maintenance costs in most projects are borne by the community, non-governmental organisations, and private bodies. (there are a number of projects part financed by different government bodies; Energy Conservation Fund and Government de-centralised budget, on an experimental basis). The maintenance is entirely carried out by the user community. With the present limitations on consumption and access, there is a considerable demand for more village hydro schemes. At present there are over 200 requests received by ITDG. They come from remote locations where extension of the grid is unlikely for a few more years. In totality, there are positive and negative facts about village hydro when considered as a rural electricity option (Table 4).

---

[7]This apply only under the present system of Village Hydro Project implementation, carried out with partial community financing. Technically there is no such limitation. Removal of this limitation will raise the cost of project.

[8]There is a relatively greater demand for the use of irons. In some village hydro projects this demand is met by having an arrangement of common irons which are shared.

[9]According to Siyambalapitiya 96) a vast majority of residential consumers in Sri Lanka pay at rates well below the average price. Policies on electricity tariff subsidises the households and penalise industries.

Table 4

A Comparison of Positive and Negative Aspects of Village Hydro

Positive	Negative
Can improve quality of life of rural people	consumption is limited
Can generate electricity in most remote locations	connections limited to a 1 km radius (refer Footnote1)
No capital cost to the State	Industrial use difficult in small capacity projects
No maintenance cost and losses to the State	possibility of management related problems
savings on kerosene oil and batteries	poor standards in transmission (due to high costs), which compromises safety
managed by the users	
contribute to rural community development	

Given the situation that the eventual expectation of the rural consumer is a grid connection (or an equivalent), and weighing the pluses and minuses of the village hydro in comparison to RE schemes, village hydro can be recommended as a suitable option for "pre-electrification" of rural areas till such time the grid reaches them. An alternative option is to invest to improve the quality of Village Hydro to an equivalent level of the service provided by the grid (which is technically possible), and include in the rural electrification plans as a decentralised option.

This will contribute to village and community development, meet the demand for lighting in rural areas, while saving much needed State funds. However, given that Village Hydro is considered as a viable option for rural electrification, the State has an important role to play.

Firstly, it is necessary to recognise the potential of the Village Hydro concept as a decentralised energy option for rural electrification. The State recognition will contribute to the wider acceptance of the concept, and boost the local manufacturing and implementing capacity. More organisations will take on projects, the spread will be faster. Secondly, Village Hydro just completed a phase of pilot testing with ITDG initiative. During this period a number of areas which require further work have been identified. This include technical as well as institutional aspects. It is felt that the State can support further research and development activities to improve the quality of the future projects.

Thirdly, as mentioned in Table 3, transmission is one component of village hydro which require relatively higher investment. Therefore, often transmission lines are of poor quality, and the safety aspects are questionable. For future expansion, or as a decentralised option, the State can invest on village hydro transmission to a level of accepted standards.

6. CONCLUSION

Village Hydro concept is a promising option to meet rural electricity demands in Sri Lanka which can be implemented in a decentralised manner. A pilot phase initiated by ITDG has demonstrated that it's pluses outweigh the minuses when analysed in the light of the priority electricity needs of the communities. The pilot phase has identified areas which need further attention to minimise the negative aspects. Village Hydro in its present style of implementation cannot meet the energy requirements of the consumers fully[10], whose yardstick is the service of the main grid owned and managed by the State. Grid connected RE schemes are capital intensive, and levels of revenue are low, but the State has the responsibility of providing electricity to the rural areas. Village Hydro is recommended for "pre-electrification", or as a decentralised option, which can raise the quality of rural life in a relatively shorter time span compared to long term grid extension. Since this can be achieved at marginal capital and maintenance cost to the State, this is an opportunity to be capitalised.

The pilot phase has shown that further work is required in refining the implementation of the concept, particularly safety and quality. The State can step in by giving due recognition to the potential of this concept, and investing in research and development, and on safety aspects.

REFERENCES

1. Ministry of Power and Energy, 1988. Evaluation Report, Rural Electrification Schemes in Sri Lanka.

2. Tissera, A.M. 1990, Rural Electrification in Sri Lanka, In *Power Systems in Asia and the Pacific, with Emphasis on Rural Electrification*. United Nations, New York ,pp507-515.

3. ITDG, 1995, Micro-Hydro Monitoring - Synthesis Report.

4. Ariyabandu, M.M. 1996, Formal and Informal Evaluation; its use in Institutional Strengthening: the Case of Village Hydro Project in Sri Lanka. Pending publication.

5.ITDG Socio-economic Reports of Individual Village Hydro Projects, 1992-1995.

6. Siyambalapitiya, T. 1996, Policy Issues in the Electricity Sector. Seminar Presentation.

---

[10]Limitation on consumption of energy in Village Hydro projects can be entirely removed technically. But, this raises the project cost, and therefore to ensure community participation the additional financing need to be on a subsidy basis.

First International Conference on Renewable Energy–Small Hydro
3 – 7 February 1997, Hyderabad, India

WATER RESOURCES IN SUDAN

Abdeen Mustafa Omer

National Water Equipment Manufacturing Company Ltd.
P.O. Box 15007, Khartoum, Sudan

Abstract:

Since water is the critical component for sustainable development in Sudan, this paper deals with the analysis of the present water resources, and use for different purposes.

Comprehensive water resource management is necessity for Sudan.

Human resource development should be the cornerstone of national development policies in Sudan, and emphasize education rooted in religion.

Introduction:

Sudan is the largest country in Africa with geographical area around 250 million hectares and a population of 26.5 million. It is endowed with vast natural resources and a special geopolitical location bridging the Arab world to Africa. It has plenty of sunshine, vast productive land estimated at 84 million hectares while the actual area under crops has not exceeded 12%. It has also plenty of water resources from the Nile system, rainfall, and ground water.

Extending from the arid north to the wet tropics in the south. Sudan has several agro-ecological zones with a variety of climatic conditions, rainfall, soil and vegetation. These agro-ecological zones and the large mass of arable land support a variety of food and industrial crops, vast natural pastures and forests support a large herd of livestock including cattle, sheep, goats and camels.

Water Resources:

Water resources in the Sudan include surface water and ground water spread over large parts of the country. Both resources are utilized for agricultural production by the use of canal system from dams, pumps, embankments (flood), and wells.

Rainfall:

Rain is the major water resource for agriculture followed by the Nile system. Annual rainfall varies from almost nill in the arid hot north to more than 1000 mm in the tropical zone of the south. The rains fall during the summer months of May-September. The duration and amount of rainfall increases from north to south. Rainfall varies from year to year. This variation is very crucial for rainfall farming.

Annual rainfall is estimated at:

< 100 mm	41.7 Mm3	
100 - 300	76.5 Mm3	
300 - 600	199.5 Mm3	
600 - 1000	515.5 Mm3	
> 1000	261.0 Mm3	
Total	1094.2	Milliard cubic meters

A great potential for increased production of food crops and livestock lies in the rainfed sector. Its realization will require attractive policies, generation and use of proper technology and improved physical infra-structure.

The Nile System:

The Nile system travers Sudan from south to north. The Nile basin system in Sudan comprises the Blue Nile, Rahad, Dender system. The White Nile, Bahr El Gabal, Zaraf, Bahr El Ghazal and Sobat. In addition, the Atbra System.

Bhar El Gabal	26000	million m3
Bahr El Ghazal	15000	
Sobat at Malakal	13000	
Total	54000	
Losses in Swamps	- 27000	
White Nile at Malakal	26000	
Blue Nile at Junction	53400	
Atbra	11600	
Losses along river	- 77000	
Net available at Aswan	84000	

In 1959 the Nile water agreement was concluded between Egypt and Sudan. The agreement is based on average Nile yield of 84 Mm3 (measured at Aswan). After construction of High Aswan Dam; (saving 22 Mm3); 14.5 Mm3 for Sudan and 7.5 Mm3 for Egypt. As a result Sudan share has become 18.5 Mm3, and 55.5 Mm3 for Egypt, leaving 10 Mm3 for evaporation from Aswan High Dam reservoir.

Ground Water:

The ground water quality is suitable for animal and human consumption as well as for agriculture and other uses.

The potential renewable ground water suitable is estimated at 6 Mm3. The main aquifer is Nubian sandstone covering 28% of surface of the country. Ground water is mainly used to satisfy animal and human needs in rural areas while small areas are irrigated from ground water.

About 5 Mm3 may be used for agriculture during the coming ten years. Current use for both agriculture and drinking water is about 1.2 Mm3. The ground water potential is believed to be much more than presently estimated.

Uses of Water in Sudan:

The utilization of water in Sudan is widely estimated in agriculture, human use, domestic, animal uses, industrial, hydropower generation and navigation uses.

The agricultural sector is the major source of water consumption in Sudan. Sudan is presently utilizing 15.6 Mm3 annually from its share in irrigated agriculture sub-sector, currently covering an area of 1.7 million hectares. The potential is three-fold irrigated crops include cotton, wheat, sorghum. groundnuts, sugar-cane, vegetables and fruits.

Mechanized and traditional rainfed farming sectors cover an area estimated annually at about 8-10 million hectares. The total area cropped in the rainfed areas vary from year to year depending on rainfall. Crops grown in the rainfall sub-sector include sorghum, millet, sesame, sunflower and groundnuts. The potential in the rainfall sector is more than five fold.

The southern tropical zone of Sudan is distinct with its red lateritic soil and heavy rainfall ranging between 1000-1500 mm over 7-8 months. Agriculture is largely subsistance with a wide range of food crops including maize, sorghum, millet, root crops, banana, pulses, tea , coffee, and tobacco.

Because of the marked fluctuation between the flood discharge and the low season period in the Nile system, storage reservoirs in Sennar, Roseiris and Girba were constructed to ensure the availability of water during the recession period. These dams are used for irrigation and for hydro-power generation.

Drinking water supplies have been provided for people and animals in most of the urban and rural areas, but still more than 40% of the rural areas, and more than 25% in urban areas are in need of safe drinking water supplies.

Water use for industry and sanitation is still very limited. The demand for water in industries at the present estimated as about 0.24 Mm3 per year; most of which from the surface water.

Management of Water:-

Water is a substance of paramount ecological, economic and social importance. Interrelationships inherent in water use should encourage integrated water management. Water resource is to be better managed to:

- ensure more reliable water availability and efficient water use in the agriculture sector.

- mitigate flood damage.

- control water pollution.

- Prevent development of soil salinity and water logging.

- reduce the spread of water-borne diseases.

Environmentally - sound water management is an absolute necessity and should address and try to resolve all the above issues simultaneously.

The following are the main constraints of environmentally - sound water management in developing countries (including Sudan) :

- Debts and financial deterioration; lack of funds or substantial delays in allocating funds for essential requirements such as operation and maintenance of irrigation and drainage projects, deterioration in data collection activities .. etc.

- Lack of appropriate and consistent policies for water development for both large - and small - scale projects.

- Serious delays in completing water projects after major investments like dams and other hydraulic structures and main secondary canals are completed. Thus potential benefits are not fully realized.

- Absence or inadequate of monitoring, evaluation and feedbacks at both national and international levels.

- Lack of proper policies on cost recovery and water pricing or, if policies exist, absence of their implementation.

- Shortage of professional and technical manpower and training facilities.

- Inadequacy of knowledge, and absence of appropriate research to develop new technologies and approaches, and absence of incentives to adopt them.

- Lack of beneficiary participation in planning, implementation and operation of projects.

- General institutional weaknesses and lack of coordination between various ministries such as water, agriculture, environment, planning ... etc.

- Lack of donor coordination resulting in differing approaches and methodologies, and thus conflicting advice.

- Inappropriate project development by donor agencies, e.g. irrigation development without drainage, supporting projects which should not have been supported.

The emerging water crisis, in terms of both water quantity and quality, requires new approaches and action. Priority areas needing concerted action in various sectors are :

- water use efficiency.
- flood control.
- salinity and drainage.
- scarce water resource management and provision of safe drinking water.
- coordination and integration of various aspects of water management, and water management with other related resources and societal concerns.

Human Resource Development in Water Sector:

Water is one of the most fundamental of natural resources that a country must harness in its efforts for rapid economic development. The role of water in the development process cannot be over - emphasized. The demand for water in Sudan has increased tremendously over the years and will continue to increase in view of the accelerating pace of population growth, urbanization and industrialization. Comprehensive water resource management is a necessity. Human resource development should be the cornerstone of national development policies in Sudan. There are many approaches and programs through which human resource can be properly developed for the water sector. These programs are not mutually exclusive and include:

- Non-formal education through media and non-governmental groups
- Formal education through universities and technical colleges
- On - the - job training and regular refresher courses

Sudan has to build a bigger and better base of science and technology. Besides strengthening the national science and technology systems, Sudan should collaborate, and achieve collective strength within the framework of the perennial Islamic values and ideals.

Education and training play key role in human resource development in water and other sectors. Education has to inculcate the moral, spiritual and ethical values of Islam. These values promote peace, contentment and inner happiness which are not dependent on material wealth and prosperity.

Religion is the fountain which irrigates the tree of virtue, goodness and truth. Human resource development in Sudan should emphasize education rooted in religion to inculcate and promote fellow-feelings, love and respect for all creatures, provide succour and support to those who are poor and destitute, and discourage hoarding and exploitation. Education based on Islamic teachings goads to simple living making minimal demands on natural resources. In the case of water, attention must pay due to how it is used and how it can be reused.

Future Water Resource Development Priorities:

The country has to work along the following strategic programs priorities:

1- Developing non-renewable ground water resources on a sustainable basis.

2- Raise the overall water use efficiency to the maximum limit.

3- Reuse all the possible amount of agricultural drainage water using proper technological means.

4- The conjunctive use and management of the dams and the underground water reservoir in the Nile valley with consideration to drought conditions.

Recommendations:

1- Use new economical technology of sea water desalination.

2- Water harvesting of rainfall on desert areas and make full use of torrential streams and flash floods.

3- Laws should be reconsidered to match with the required development and the existing scarcity.

4- Raise public awareness about water resource scarcity and government management plans.

Conclusion:

1. Sudan has plenty of Sunshine vast productive land yet to be utilized, and plenty of water resources from the Nile system, rainfall, and ground water not yet fully tapped.

2. Water resources available to Sudan from the Nile system, and ground water resources provide a potential for three fold increase in the irrigated subsector. There are also

opportunities for increased hydropower generation.

3. The physical and human resources base in Sudan can provide for sustainable agricultural growth and food security for itself and for others. Its failure to do so in the past derives from several causes and constraints that are manageable. These included misguided policies, poor infrastructure, low level of technology and political instability.

4. The strategy of Sudan at the national level aims at the multipurpose use of water resources to ensure water security for attaining food security, industrial security, navigation, waste disposal and the security at the regional levels within an environmentally sustainable development context and in harmony with the promotion of basin-wide integrated development of the shared water resources.

References:

(1) A. K. Biswas, Objectives and Concepts of Environmentally - Sound Water Management, Delhi, 1990.

(2) FAO, Yearbook 1989 : Production, Rome, 1990.

(3) FAO Report, Towards Agricultural Development in Sudan, Rome 1990.

(4) B. Mitchell, Integrated Water Management : International Experiences and Perspectives, London, 1990.

PROPOSITION FOR MASTER PLAN CADASTRE STUDY OF SHPP

Radovan M. Miljanovic

Energoproject-Hydroinzenjering Consulting Engineers Co. Ltd.
11 070, Novi Beogard, Bul. Lenjina 12, Yugoslavia

1.   INTRODUCTION

The cadastral register of the small-scale Hydro power plant includes the hydropower plant which installed power amounts to $100 \, KW \leq Ni \leq 10000 \, KW$.

The hydro power plants which power which is less than 100 KW are classified as a micro power plants category and as such they shall not be considered, except in special cases.

Considering the activities planned in this study, the cadastre of Small Scale Hydropower Plants shall treat and contain the following items:

- a) Desk work
- b) Field reconnaissance
- c) Cadaster processing
- d) Selection of typical solution

All enumerated activities shall be synchronized in order to prepare the study on high professional level.

2.   STUDY INVESTIGATIONS

2.1.   Data bases

The available data which have been published or which are kept in the files (professional library of the Employer, or the like, as for instance:

- topographical surveying data
- geological data

- hydrological data
- torrential erosion and a sediments data
- other data (transmission - distribution network, power consumption, infrastructure, needs in water, water pollutants and the like) are used for preparation of small-scale HPP cadastral register.

2.2. Analysis of the collected documentation and data

The men who process the cadastral register of small-scale hydro power plant throuht the analysis of the existing documentation, shall make the following:
- identification of small-scale hydro power plant for which there are the designs and the other technical solutions,
- identification of the possible consumers of electric energy and their demands, within time
- analysis of the conditions under which some regions are provided with electric energy
- identification of the possible water consumers out of electrical industry and establishment of their needs,
- identification of the issues and problems arisen during the analysis of the existing documentation and data.

2.3. Identification of the new locations for small-scale hydro power plant and of the reservoir

Within the study investigation, the preceding identification of the new locations is made for the small-scale hydro power plant in the following way:

a)   on the basis of topographical map (in scale 1:50000 or 1:25000 or ....) division the principal stream's (attachment area is made to the catchment areas of larger tributaries. The size of those sub-catchments is different and the approximate criterion is the lower installed power limit of the plant that would be constructed at the catchment.

b)   Markind of chainage of the main watercourse and the tribuatries nets shall be provided on maps. Longitudinal profile of the watercourse and the tributaries shall be drawn in a suitable scale.

c)   Use of the maps of specific run-off for the considered catchments. If there are no maps of such a kind they should be made. The estimate of run-off mens the guideline of the expected average discharge from the catchment area to the one who processes and it is used only in the phase of preliminary identification of the new locations.

120

d) Through usage of the effected processing under a), b) and c) as auxiliary documentation engineering analysis shall be taken up as regards possible solution of power utilization of the given watercourse. Within this activity, alternative recognition to the following shall be performed:
- position of the principal structures
- water diversion during construction
- hydraulic scheme of the plant
- preliminary determination assessment of the plant's installed capacity
- assessment of the conditions for completion of the S.H.P.P. (geology, alluvium, sediments, infrastructure, consumers and the like)

e) Notice natural advantages for formation of the reservoir should be connected with:
- presence of several water consumers
- demand for formation of the reservoir (tourism, fishery flood control, recreation), which is besides water management.

2.4. Preparation for field investigations

This preparation involves the following:

- Teams composing from the experts of different specialities (civil engineers, geologists, forest engineers specialized in sedimentation torrents, hydrologists etc.),

- work program for each team, provided with the schedule of visits of particular locations, collection of additional data, etc.),

- preparation of the material which teams should take along to the future site and which consists of topographical, geological and other data bases that have to be proved on site.

3. FIELD RECONNAISSANCE
3.1. Collection of geological and geotechnical data for the location of small scale hydropower plant

The geologist from the investigations team shall carry out the following field works:
Study of available documentation

-   Site visits and visual prospection to establish the geological features of the terrain (type of rocks, stability of slopes, expected strength of alluvial deposits, etc.)

3.2.   Collection of torrent-erosion data for the catchment area of the location

The erosion specialist shall collect the same data at the field as under Item 3.1. but from the erosion and torrent aspects.

3.3.   Other field investigations

During the field works a team of specialists will be assigned on the following tasks:
-   Contacts with competent institutions to establish all interests and requirements for water supply and electrical energy generation in the small scale hydropower plant.
-   Collect data on infrastructure (roads, settlements, electricity supply, etc.) being of importance for the plant location.
-   Collect data on water quality in the river and possibly, if any, pollution source in the catchment area.
-   Collection data of the construction material (quarry, borrow areas).
-   General earthquake data.

4.   MASTER PLAN (CADASTRE PREPARATION)

After completed site visits the available data shall be supplemented with information obtained from such visits.

-   These relates to modifications in geological, topographic and other data, the mean discharge will be determined (Qi), the 100 year return period flood computed, etc.
-   The final location of the small scale hydropower plant will be selected together with the processing of parameters to be included into the Cadastre sheets. The hydrological scheme will be prepared.
-   The installed capacity shall be computed according to the gross head of the plant Hg (m) and the adopted installed discharge Qi (m3/sec) by the following pattern:

$$N. = n \times Q \times Hg \ (kW)$$

-   The estimate of average annual output is obtained from the average duration of discharge curve and the evaluation of the lifetime of the hydropower plant.

\- The investments estimate will be carried on the basis of the costs for small scale hydropower plant with or without reservoir considering empirical data and analysis being available to the Consultant.

5. CREATION OF CADASTRE

The cadaster of small scale hydropower plants is consisting of
\- Lists of hydropower plants per catchment areas
\- List of hydropower plants per regions
\- Cadaster sheets

5.1. Lists of hydropower plants per catchment areas and regions

It will be established on the basis of cadaster sheets of small scale hydropower plants (SHPP) for main catchment areas divided by sub-catchments. Indications and numbers for catchment areas shall be made in agreement with the Client.

The hydropower plants are numbered from the downstream to the upstream part of the catchment area.

5.2. Cadastre sheets

For each denoted small scale hydropower plant a cadaster sheet will be issued containing all necessary data such as :

\- general
\- hydrological (average, floods and min. discharges)
\- reservoir, type
\- energy (power and output connection)
\- economic indexes (cost of 1 KW and 1 KWh)
\- infrastructure
\- geological and sediments, general earthquake
\- brief technical descriptions.

5.3. Annexes to Cadaster Sheets

These annexes shell provide more detailed outline or supplement the data included in the cadaster sheets (longitudinal sections of water courses, layouts, maps of catchment areas, etc.).

6. APPLICATION OF TYPICAL SOLUTIONS

Within this activity all aspects of the technical solution will be studied for the small scale hydropower plant, such as

- civil structures: dam, intake, spillway, outlet, pipelines, machine houses, etc.
- hydraulic equipment:gates, trashracks, valves, surge tanks, etc.
- electro-mechanical equipment: turbines, generators, automatic control, etc.
- construction of access road, river diversion, availability of construction materials, earthquake degree, construction management, etc.
- maintenance of hydropower plant : maintenance service, operation control.

7. CONCLUSION

The preparation of such cadaster would enable the review of all small scale hydropower plants thus the prospective Client could easily choose the plant he intends to construct.

The construction of such plants is advantageous since their construction is not requiring large investments as for the large hydropower plants. The estimated costs of SHPP are from 0.8 to 1.2 mln $ per 1 MW in dependence from tail water head, civil works, equipment, etc.

The local population can be employed in the construction of the plant.

Attention shall be paid when preparing the urban plans and care shall be taken when creating the reservoirs to serve for touristic and other purposes.

The consulting and contracting activities could be paid by barter arrangements, local and hard currency, depending on the costs.

From the preparation of such Water Resources Master Plan (Cadaster) of Small Scale Hydropower Plants many advantages will be obtained:

1. Attract the foreign financing institutions for investments in your country
2. Increase the trade exchange by barter arrangements
3. Provide better living conditions to the population
4. Possible development of light industries

5. Planned direction of financial assets
6. Permanent employment of labour during construction and later on maintenance during operation
7. The creation of small lakes would improve the touristic resources of your country
8. Construction of small scale hydropower plants would not deteriorate the environment
9. Provision of steady supply of electric energy
10. Independent operation of the plant
11. High level of professional documentation preparation and skillful construction
12. Training of your specialists for designing and construction of small plants
13. Short term of SHPP completion (few months to 2 years at least).

Finally, the small scale hydropower plants are not jeopardizing the environment and such project would certainly be supported by the banks.

8.   CADASTRE SHEETS

1.   USER NAME
2.   SHPP
3.   SORT OF DOCUMENTATION: FEASIBILITY STUDY, FINAL DESIGN
4.   GENERAL DATA: (LOCATION, TYPE OF DAMS, TYPE OF STRUCTURES, CATCHMENT AREA, STREAM FLOW)
5.   HYDROLOGICAL DATA: CATCHMENT AREA, MEAN ANNUAL, PRECIPITATIONS, MEAN ANNUAL, DISCHARGES ANNUAL, INFLOW, SPECIFIC DISCHARGE, FLOODS DISCHARGES)
6.   RESERVOIRS DATA: (TITLE, LIVE AND TOTAL STORAGE CAPACITIES, TOTAL VOLUME, STORAGE WATER LEVEL, TYPICAL REGULATION)
7.   CONDUITS DATA
8.   ENERGY DATA (TAIL, WATER LEVEL, HEAD, RATED DISCHARGE, TYPE OF TURBINES, NUMBER OF UNITS, INSTALLED CAPACITY, POWER GENERATION)
9.   ECONOMICAL DATA: (ESTIMATED COST)
10.   INFRASTRUCTURE DATA
11.   GEOLOGICAL DATA
12.   NECESSARY DRAWINGS: (MAP, SECTIONS - S 1:500, 1:1000, 1:5000/5000)

9. CONSTRUCTION OF SINGLE SMALL SCALE HYDROPOWER PLANTS

Such scheme has been proposed in the cadastre study and it could be applied for construction of individual, known locations of small scale hydropower plants.

This particularly relates to remoted, populated places, where it is possible to erect small sawmills and smaller manufacturing units, depending on the financial resources of the potential clients.

To all prospective clients investors all information could be provided at any time at the shortest terms possible, with most particulars on selected small scale hydropower plants and approximate cost estimates of the structure.

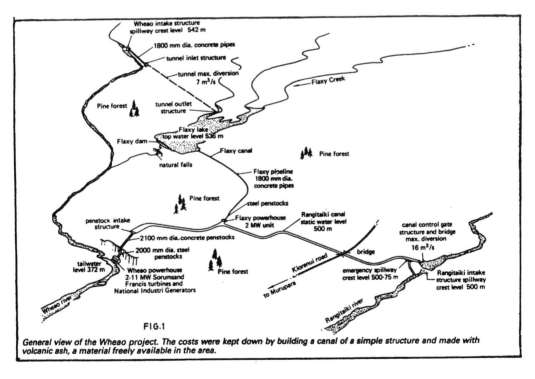

FIG.1

General view of the Wheao project. The costs were kept down by building a canal of a simple structure and made with volcanic ash, a material freely available in the area.

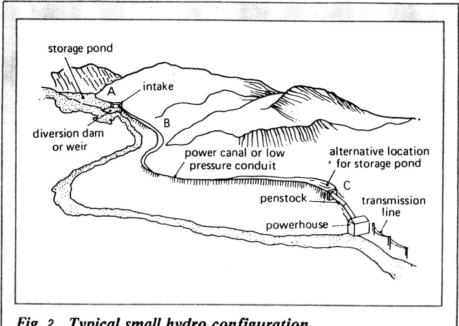

Fig. 2 . *Typical small hydro configuration.*

storage pond

intake

A

B

diversion dam
or weir

power canal or low
pressure conduit

alternative location
for storage pond

C

penstock

transmission
line

powerhouse

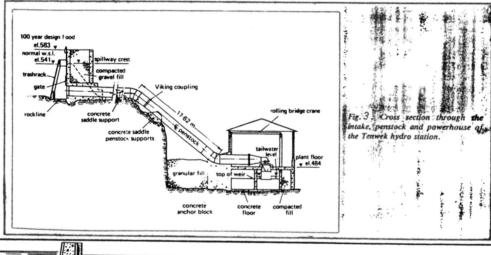

100 year design flood
el.583

normal w.s.l.
el.541

spillway crest

trashrack

compacted
gravel fill

gate

Viking coupling

rockline

concrete
saddle support

17.62 m

rolling bridge crane

concrete saddle
penstock supports

tailwater
level

plant floor
el.484

granular fill

top of weir

concrete
anchor block

concrete
floor

compacted
fill

*Fig. 3 . Cross section through the
intake, penstock and powerhouse of
the Tenwek hydro station.*

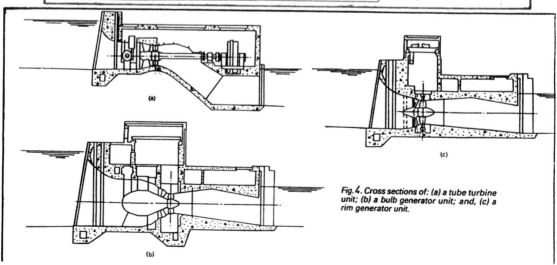

(a)

(c)

(b)

**Fig. 4. Cross sections of: (a) a tube turbine
unit; (b) a bulb generator unit; and, (c) a
rim generator unit.**

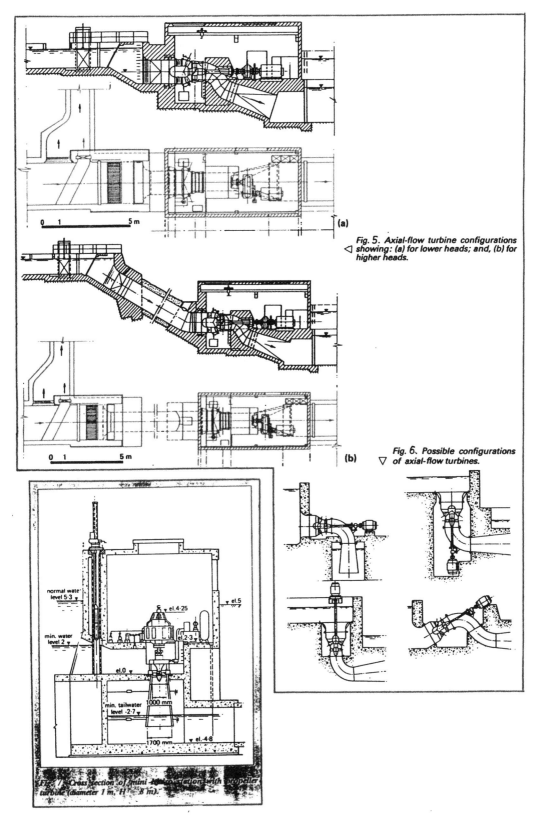

Fig. 5. Axial-flow turbine configurations ◁ showing: (a) for lower heads; and, (b) for higher heads.

(a)

(b)

Fig. 6. Possible configurations ▽ of axial-flow turbines.

normal water level 5·3

min. water level 2

min. tailwater level -2·7

el.5

el.4·25

el.2·3

el.0

1000 mm

1700 mm

el.-4·8

Fig. 7. Cross-section of mini installation with propeller turbine (diameter 1 m, H = 8 m).

129

Water level control for small hydro plants

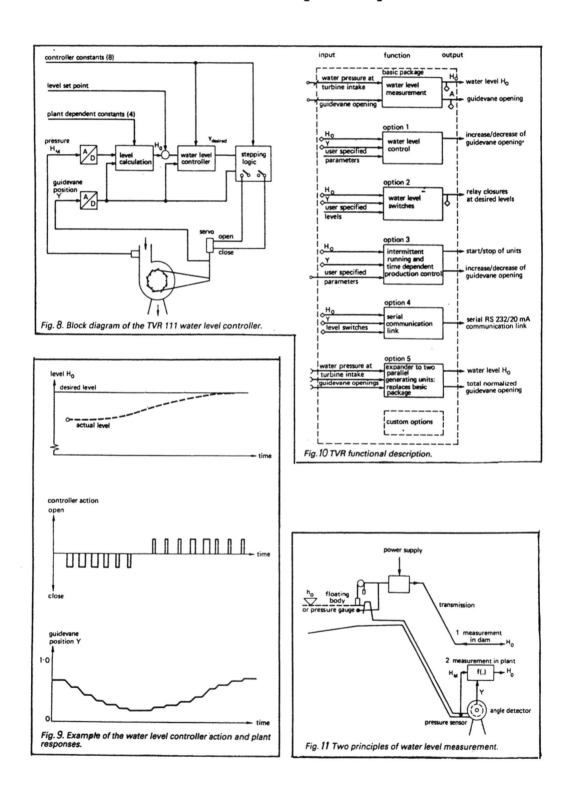

Fig. 8. Block diagram of the TVR 111 water level controller.

Fig. 9. Example of the water level controller action and plant responses.

Fig. 10 TVR functional description.

Fig. 11 Two principles of water level measurement.

ENERGOPROJEKT, BELGRADE
By: R.Miljanovic,B.S.C.E.

MASTER PLAN OF SMALL HYDRO AND THEIR
INFLUENCE ON DISTANT REGIONS OF SERBIA

As a result of the three-year work, the cadastre of small hydro on the territory of Serbia of about
36,500 km^2 is made three years ago. The inventory presented 850 hydropower plants with installed
capacity exceed 100 kW and is not higher than 10,000 kW. A cadastral sheet has been made for each
small hydro including the basic data (coordinates, small dams, catchment areas, discharge, flood, min.
flow, elevations, installed capacity power generation, general data, geological report, layout and
longitudinal profile).

Small hydro, which have been investigated at the territory of Serbia, are mainly designed on small
rivers and are located in underdeveloped and regressive regions.

Access to the power houses with appurtenant structures in majority of cases is generally bad, which
makes the costly construction of small hydropower plants still more expensive. According to the
rough estimates, the cost of 1 MW of a small run-of river power plant would amount to about 1
million US dollars.

In case to decide the construction of a SHPP with a dam and reservoir, the cost of 1 MW will be still
increased in dependence of the dam size, but the output and power shall be larger.

The installed capacity of small hydro in Serbia amounts to 443 MW, which represents 13% of the
total installed capacity of hydrpower plants in Yugoslavia. Annual power generation would amount
to about 1.55 billion kWh, which makes 14% of the total annual production obtained from Yugoslav
rivers.

It means that the calculated installed capacity and power generation of small hydro on the territory of Serbia is equal to the power production of a small nuclear power plant.

Whereas the cost of a nuclear power plant would amount to about 6 billion US dollars and represent a permanent danger to the environment, the construction of all small hydro and dams, in the territory mentioned would be similar the price above.

Therefore, it would be more purposeful to construct all small hydro first, while the nuclear power plants construction will never be late.

On the occasion of construction of small hydro a lot of labour would be employed, as well as afterwards - in the plant operation period. Besides, the new road will be constructed and better living and work conditioned in the village will be created. People would be interested to a greater extent to remain in their villages, and in this way dying out of rural population would be stopped. Agriculture and cattle raising would start to develop again, which would be the best method to enable Serbia and Yugoslavia to get out of underdevelopment and to increase its living standards.

First International Conference on Renewable Energy–Small Hydro
3–7 February 1997, Hyderabad, India

SMALL HYDROPOWER DEVELOPMENT IN NEPAL —
A CASE STUDY

S.N. Mishra

Nepal Electricity Authority
Darbar Marg, Kathmandu, Nepal

1. Back-ground

First Hydropower generating station in Nepal, PHARPING HYDROPOWER HOUSE, was established in 1911, (when Kathmandu had no road link even with India). It is the third oldest hydro power generating station in Asia. In spite of this historical back ground hydro power development growth remained so slow , that even after 85 years, today Nepal has a total 270 MW installed capacity, which is not sufficient to meet the internal power demand of the country.

Development of Hydropower is so capital intensive, that Nepal, having a Hydropower potential of 83000 MW could not explore her potential more than 270 MW today due to financial constraints. At the same time Nepal is a country with a very wide geographical and or topographical nature. Population density is very high in cities or municipalities and southern part of the country, where as middle and Northern Nepal, has a very scattered population with low density. This is why, rural electrification of hilly and mountain regions are not so easy. Financially most of the projects are not feasible. Grid supply is also not possible to be given to all parts of this country.

Seeing all these difficulties, HMG/N decided to encourage the micro-and small Hydropower generation and distribution in isolated (from grid) way for the benefit of the local people, and the result is very much satisfactory.

2. Hydropower development policy :

As Nepal has constraint of its own financial resources for the development of Hydro power potential of Nepal, HMG/N announced the Hydropower development policy, 1992, to attract the private investors to invest in this power sector. The policy is very liberal and a lot of facilities and benefits have been announced for the private investors- both national or international or jointly organised

In this regard HMG/N has come up with broad views and openess to develop the micro and small Hydropower in Nepal. Government has announced different schemes for the National investor in this small Hydro power sector , under which investor can get upto 75% of the total investment as grant by HMG/N through the local Nepalese Bank . Higher percentage of grant covers for the plants to be developed through joint (collective) efforts in the very remote areas. Any individual can also get a grant upto 50% of the investment. As a result a lot of micro (upto 100 KW) and small scale Hydropower (upto 5 MW) projects have come up and others are to come. A few big projects (more than 5MW) are also in the construction phase due to this new Hydropower development policy.

3. Financial Assistance from International and National Investors:

HMG/N has increased the sectoral allocation of Development Expenditure on Electricity from US$ 241 mill (7th plan 1986 - 91) to US$ 418 mill (8th plan 1992 - 97) by 19% of total Development Expenditure, but the need of financial resource of a single "Kali Gandaki "A" Hydropower project (capacity 144 MW) has been estimated US$ 458 mill. Thus HMG/N can not make available the required financial resources to meet the power demand of Nepal. This is why, majority of finance for investment in development project continued to be obtained from donors as foreign aid channeled through HMG in the form of Subsidiary Loan Agreement (SLA). Donors from the international scenario continued to show interest in developing Nepal's power sector.

As mentioned above Grid supply can not be made available at all places (within Nepal) due to financial non-feasibility and also constraints of financial resources, some international donors/Agencies have come to help Nepal in the areas of micro and small Hydropower development. UNDP and some others donor countries like Sweden, Germany,Denmark,Switzerland etc. are helping in the development of small Hydropower projects and UNESCO with the involvement of ICIMOD (International Center for Integrated Mountain Development), INGO (International non governmental organisation s) , ADB/N (Agriculture Development Bank of Nepal) , and other Local financial agencies are making investments in micro Hydropower (MHP) sector.

UNDP has signed recently an agreement to invest in SHP sector US$ 2.26 mill. This amount will be spent in five districts of Nepal over a period of 32 months. Till date about 40   SHP plants are generating power in Nepal. Investment in micro Hydro power plants is also very encouraging. Till date more than 1000 MHP plants are generating about 12 MW Hydropower in Nepal.

4. Potential growth of local enterpreneurship.

If 20 (twenty) years back when MHP or SHP plants were under development, only a few expertese were available in Nepal, today mostly all (vital) parts and equipment for the plants are being manufactured here. Demand for the machinery/equipments forced the local industries to manufacture cheap and good quality products. Balaju Yantrashala and Butwal Techinical Institute have played a leading role in this regard. Today about nine potential manufacturers are supplying good quality of micro and small Hydropower plants in Nepal at a very competetive rate. Thus, development of this small Hydropower sector has played a very positive role not only in partially solving the power shortage, but in the development of technical know how of the local people and creation of substantial job opportunities for them.

5. CHALLENGES REGARDING HARSENING OF SMALL HYDRO POWER IN NEPAL:

For the countries, like Nepal, where land is not flat and major portion is covered by hills and mountains (which are blessings for the Hydropower generation from one side, creat a lot of problems for the development of the power generating projects from other side) harsening of small hydropower (in Nepal) has taught (by experience) us a lot due to the following factors:

Geography and topography (means high construction cost)
Low density of population (means high distribution cost)
Low and scattered load (means high distribution cost)
Poor paying capacity of people (means less revenue generation)
Difficult and costly grid connection (means power development scope is limited)
Shortage of skilled man power in the remote areas (means Less reliability of power supply)
Transportation problem (means construction as well as operation and maintenance cost high)
Maintenance very difficult (as manufacturers find difficult to respond to calls for repair from far places) costly and uneconomical also. Due to this reason some users of power from water have replaced the power from diesel engine. Besides the above mentioned problems the following are also very serious:

Problem of sedimentation (in Himalayan rivers) decreases the life of machines.
Problem of utilisation of water from river or sources and reliable flow discharge ,
Poor financial investment,
Projects are selected on political pressures, not on financial feasibility.
Poor coordination between organisation and line agencies.
Tartff not realistic.

6. PROSPECT OF SMALL HYDROPOWER IN NEPAL

Efforts made by HMG/N and independent/private power producers, only about 12% population of Nepal has got opportunity to be benefited from electrical power. Main reason behind this scenario is the topography of the country. Lack of motorable road and scattered population on hills with difficult inter-community communication affects the speedy development of electrification in these areas. Main grid connection is not financially viable. Thus there are questions to be answered:

Whose responsibility is it to make electricity available in these villages or areas?
When will the villages or areas to be electrified ?
Should the villages or areas wait for the grid or build their own small or micro power plants ?

And everybody knows that rural development is impossible without electricity. Today concept about electricity is also changing in the society. It (electricity) is not regarded as a luxuary but as an essential infrastructure to fulfill the basic needs of the people.

As commercial Organisations of power sector cannot fulfill the power need of all the rural and hilly people of Nepal and grid supply is also not economical for the electrification in such areas, only option remains open to explore the local hydro power resources and electrify the areas to the extent of possibility. The plan and policies of HMG/N will play a very vital role in the Electrification of this part of the country. People's participation in the community development through rural electrification is very significant in this regard.

Supporters of the solar energy try to convince that it would be the best solution for the rural electrification. Today it is very costly and from the reliability point of view seeing the maintenance problems, it would not be so suitable at least in the present status of development of Nepal, to go for solar energy in the remote hilly areas. But this option will remain valid for those, where chance of local Hydro power development is rare or not available. This is why , for the rural development of Nepal the only option remains to explore the small Hydro power potential of Nepal to the maximum extent, no matter, how costly is the power generation for the particular place. In long run, it will be cheaper type of energy and certainly will aid positive inputs for the development of the areas and country.

RESOURCE ASSESSMENT AND PLANNING

First International Conference on Renewable Energy–Small Hydro
3–7 February 1997, Hyderabad, India

SOME GUIDELINES FOR INSTALLING POWER STATIONS IN IRRIGATION CANALS

V.V. Badareenarayana

Central Water Commission
Sewa Bhawan, R.K. Puram, New Delhi 110 066, India

SYNOPSIS

The importance of developing canal falls as source of hydro power generation in India has been well acknowledged. A number of power stations of the type have been commissioned and more are in the pipe line. The canal fall HE schemes are blessed with the advantages of small gestation periods, and are devoid of submergence, resettlement and other problems of environment and ecology.

Optimum scheme of hydro development with the best possible techno economic solutions, can be arrived in a canal system if the system is in its planning stage. However, certain limitations are faced while introducing the power component in the already existing canals. Site specific studies, with all possible alternatives are to be carried out to come out with the most economic layouts.

The most important feature of a canal fall hydro scheme is the automatic bypass arrangement to divert the water in case of load rejection(partial or full) at the power station. A fool proof arrangement is required to ensure uninterrupted flow in the canal system and to avoid the danger of overtopping due to upsurge. There are many ways to achieve this objective, but an appropriate method is to be adopted depending on the existing features, to arrive at an effective and economical solution. Self actuating gates which operate automatically according to a pre-determined design are very commonly adopted at the regulators for the canal fall power houses. Sometimes the gates at the canal regulators already existing before introducing the power house, may have to be utilised for serving the new purpose, after bringing out necessary modifications in them.

An analysis of the experience gained in the Central Water Commission, while rendering consultancy to more than 20 power houses helped in bringing out some useful suggestions in layout, planning, design and construction, for the benefit of future schemes of this type.

1.   INTRODUCTION

1.1   Development of the technology and equipment to exploit very low heads at reasonable cost made the canal fall type hydro power plants not only to come into existence, but also made them popular. A number of such schemes are already in operation in different parts of the country,   and many more are under construction/formulation. The canal fall hydro electric projects have the advantages of small gestation periods and simplicity in project layouts and are practically devoid of submergence, resettlement and other problems of environment and ecology. It is expected that more and more small hydro projects of the type will be taken up in future to cover almost every   possible site in the existing canal systems and to come out as a part of those canal systems under planning.

1.2   It is possible to optimise the scheme of hydro development in a new canal system by adopting the best techno-economic solutions. However certain practical limitations are required to be taken care of, while introducing power schemes in existing canal systems. The design features of a canal power house will depend upon the size of the canal, its discharge variations over the year, the head available at the fall and ground configuration in the canal reach under consideration. It may not be possible to standardise the layout or design of these power stations, but certain guidelines can be set, based on fundamental principles, and the experience gained from the performance of the numerous existing small hydro projects of the type. The most appropriate solutions will, however, have to be evolved after conducting site specific alternative studies, including their relative costs.

2.   LAYOUT

2.1   The essential requirements to be met while planning hydro power development in an irrigation canal are to ensure that (i) the water flow in the canal is not disrupted even for short durations and (ii) the upsurge created in the canal due to tripping of power house is well within the safe limits. A regulator is generally provided to achieve these requirements. When the power house is incorporated in a new canal, the best proposition would be to combine the power house and regulator structures, and provide them in the irrigation canal itself at a suitable location. Such an arrangement is shown in Fig.1.

2.2   When a power house is contemplated in a running canal, it would not be possible to construct it inside the canal. It therefore becomes obligatory to construct a separate power canal, bypassing the main canal in certain reach, for locating the power house. The arrangement shown in Fig:2 is one such example. The existing irrigation canal could still serve to carry the flow when the power house is not in operation. Normally a cross regulator exists in the irrigation canal at a fall, which may serve to divert/control the flow even subsequent to the introduction of the power scheme. Otherwise, one has to be constructed at a suitable point. As the regulators shown in the Figures 1 and 2 serve to bypass the flow during the power house closures, they are known as bypass regulators. They play a very important role in efficiently operating the power house and can serve different functions at varying situations.

2.3   Falls are provided in irrigation canals to suit the topography along the canal alignment. The size and location of falls are determined to achieve maximum economy

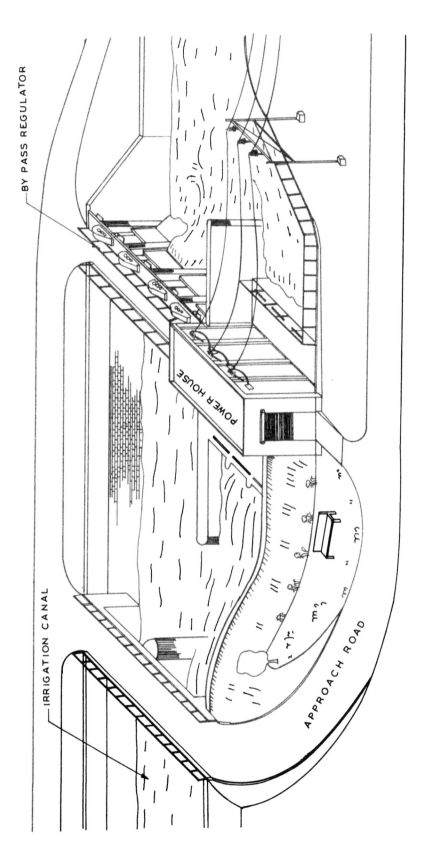

BY PASS REGULATOR

IRRIGATION CANAL

POWER HOUSE

APPROACH ROAD

FIG. I. COMBINED POWER HOUSE AND BYPASS REGULATOR IN A NEW CANAL

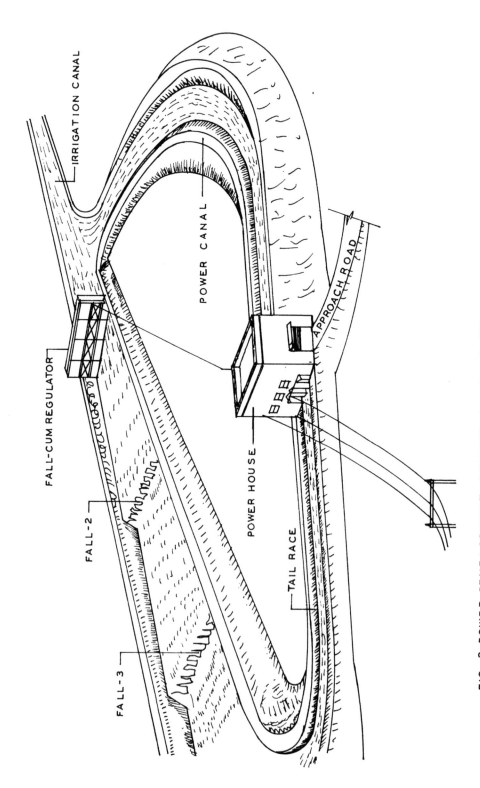

FIG. 2 POWER DEVELOPMENT SCHEME IN AN EXISTING CANAL

on the overall cost of the canal in the reach under consideration. Usually individual falls are not kept more than 3m. It may therefore be necessary sometimes to jointly utilise two or more falls to make the hydro development scheme viable. This may make the power canal very long depending on the relative locations of the falls which are proposed to be combined. As the land on both sides of the original irrigation canal would have been well developed, it becomes usually difficult to obtain land to construct the new canal. Under such circumstances it would be prudent to regrade the existing irrigation canal to combine some or all the falls under consideration. A typical example of such an arrangement is shown in Fig.3. Here, three falls are combined by raising the canal banks between the first and second falls, and lowering the bed of the canal between second and third falls. This has resulted in a big reduction in the length of power canal from alternative-1 to alternative-2. There can be a number of alternatives to choose the profile of the regraded channel. The process involves cutting in a part length and filling in some length of the canal. Different combinations have to be studied and the quantities of excavation and filling worked out in each case to arrive at the most economical proposition. Regrading works on the irrigation canal will have to be properly planned and meticulously executed during the annual closures of the canal, so as not to affect its function in the running period.

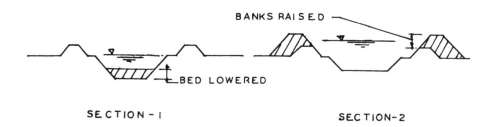

SECTION-1   SECTION-2

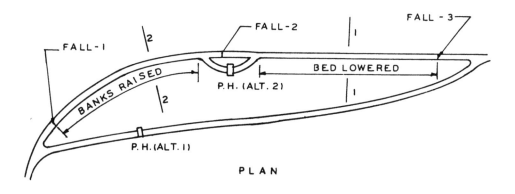

PLAN

FIG. 3 COMBINING MORE FALLS

3. BYPASS REGULATOR

3.1   The main functions of the bypass regulator are (i) to divert the required quantity of water towards the power house and pass the balance flow down the canal and (ii) to instantaneously come into action to pass the water down the canal, when the power house trips.   The second function is of great importance as it means not only the prevention of upsurge and consequent danger of overtopping of the canal banks on the upstream, but also maintaining flow continuity in the canal downstream. The regulator and its control gates should be designed according to the operational requirements of the scheme.

3.2   The bypass regulators are best equipped with self-activated automatically operated gates.   These gates open as soon as the upstream water raises over the predetermined level, and slowly get back to the closing position as the water level starts receding.   Thus the water level in the forebay is maintained nearly constant.   Many designs of the self activated gates are in vogue, but these can be broadly classified as counter weight operated and float operated gates. A detailed version on the various types of bypass gates and their selection are contained in the article `Bypass Gates and Their Selection for HE Projects' by the author included in the proceedings of the Workshop on Hydraulic Gates and Hoists in Water Resources Projects, CBI&P, Bangalore, June, 1995.

3.3   Sometimes it may be required to utilise the existing gates in the regulator in preference to constructing a new regulator with self activating gates.   This may be possible with suitable improvements in the hoisting system, to make them operate quickly and automatically.   The operation can be electrical or hydraulic, activated by a system of sensors, which act as soon as the water raises above a predetermined level. Usually, the level is fixed a few centimetres above the FSL in the forebay. Though this is the simplest way of improving the regulator causing the least changes to the existing structure, there are certain disadvantages in this arrangement.   A hundred percent availability of emergency standby power has to be ensured for opening the gates. The hoists can be designed to lift the gates only by predetermined fixed amounts, in each operation.   Also special efforts are to be made for automatically lowering the gates when water level recedes. Thus the system may not be successful in maintaining a constant water level in the forebay during the load fluctuations.

3.4   It may be possible to replace the existing gates in the regulator with the counterweight operated gates with minimum modifications to the civil structure. It is worth the effort and should be tried as an alternative to build an entirely new regulator. It may also be possible to manage by having the automatic gates only in some of the existing regulator bays. This facility of immediately releasing part discharge will significantly increase the time before the surge reaches the danger level, and the balance gates which are not automatic, could also be opened in the meantime. Detailed surge analysis will have to be carried out while attempting this alternative solution, for which computer programmes like MIKE 11 can be conveniently used.

4. OTHER MEANS OF BYPASSING

4.1   It may also be possible in some cases to get away with the costly arrangement of providing regulators with automatic gates. One way is to design the hydro electric machines for allowing flow of water through the units under no load condition. This would enable surplussing of the water through the units themselves. Such a facility has been incorporated in both the Sone Canal Power Houses. Here the flow is controlled by hydraulically operated draft tube gates and sluicing upto 80 percent of the design discharge is permitted.

4.2   Where the canal discharges are not very high, it is possible to divert the surplus water over one of the forebay walls. The top of wall for the required length is provided as the crest of a side channel spillway, with the crest level corresponding to FSL in the forebay. As and when the water level in the forebay raises due to load rejection at the power station, the surplus water flows over the spillway and falls into the tail race channel through the side channel. This simple type of arrangement has been adopted in the Birsalpur Canal Power House(Fig.4). The maximum depth of flow over the spillway has to be restricted in order not to unduly increase the free board and the height of canal embankments and lining. A flow depth of 50 to 60 cm could be a reasonable limit. If the discharge is too high to be passed over the forebay wall, one may even consider converting some length of the embankment of the power canal into a broad crested weir with suitable protection on the outer face of the embankment to act as a spill channel, which can be suitably connected to the main irrigation canal. Such an arrangement is shown in Fig.5, and has been adopted in a number of canal power houses in Rajasthan.

4.3   Providing syphon spillways in place of gated regulators is worth considering. They do not have any movable parts but at the same time are capable of operating themselves automatically when the water level rises beyond a certain limit. The syphon action utilises the entire head available at the fall and therefore the crest length is considerably reduced. Though syphon spillways are popularly adopted in the developed countries these have not been tried so far in our country. Serious thought is required to be given on this aspect and adequate designs of this type of arrangement should be evolved to suit the Indian conditions. Hydraulic model studies will guide in arriving at a proper design and also give us the confidence on the performance.

5. SELECTION OF TYPE OF BYPASS

It is not necessary that one single arrangement of bypassing the surplus water, should be adopted in any power house. Due to site specific considerations it may be advantageous and also economical to go in for a combination of the available methods. Thus, float operated automatic gates can be provided only for a part of the design discharge allowing certain raise in water levels, or the balance flow allowed to pass over the forebay walls, and so on. A judicious single arrangement or a combination will have to be adopted after a thorough study of all the possible alternatives, keeping in view of both performance and cost. Such studies would involve numerous calculations to determine the transient flow conditions and maximum surges in the canal for different combinations, which can be conveniently done on computers.

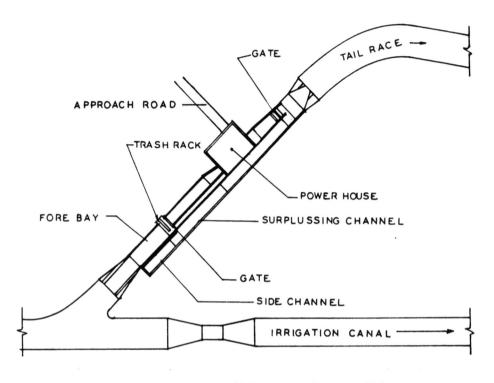

FIG. 4 TYPICAL SIDE CHANNEL BYPASS

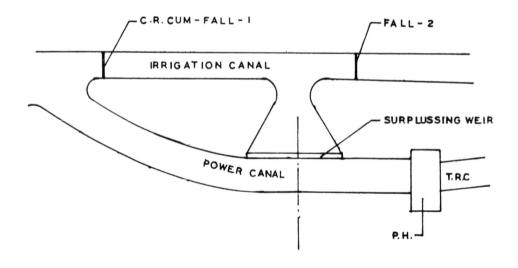

FIG. 5 TYPICAL SURPLUSSING OVER CANAL BANK

6.   ESCAPE FACILITY

6.1   Usually the canal power houses are planned to generate power according to irrigation flows, available round the year in the canal,. It means that the power house will have forced closures, when the irrigation requirements diminish temporarily, though there is enough water in the river. It could be possible to continue power generation even during such 'dry spells' if the water from the power house is disposed without passing it down the canal. An escape in the canal at a suitable point downstream of the power house, as close to it as possible, is a solution to permit this facility. The benefits are best achieved when the power scheme is located in the head reaches of the irrigation canal, and the escape can be joined into the river itself. Carrying capacity of the escape may be equal to the total discharge from the power house, but lesser capacities can also be decided, depending upon the relative economics. Cost analysis carried out for a few specific cases, indicated that the additional revenues accrued by way of the secondary power generated, compensate more than the cost of providing the escape. It is therefore suggested that an escape should be provided, wherever possible, to make the canal power houses more profitable.

6.2   The escape provision involves construction of an escape regulator on one of the canal banks, and a cross regulator in the canal, immediately downstream of the escape regulator. The location of these structures will have to be decided to get the best results. If a cross regulator is already existing, or is planned on other considerations, its advantage should be taken to combine the use. If the provision is planned afresh, taking the escape from the tail race channel is worth considering. A typical layout of an escape facility is shown in Fig.6.

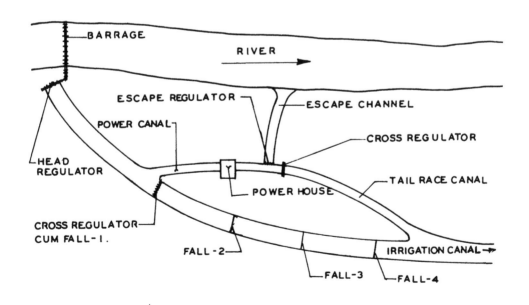

FIG. 6 TYPICAL ESCAPE FACILITY

149

7. SILT CONTROL

7.1   The entry of silt into the canal system is controlled by providing silt excluders in the diversion weir or barrage, and silt ejectors/desilting chambers in the canal. Normally the extent of silt control provided for the irrigation flows in the canal would be adequate and no additional provisions are required for incorporating power generation. The silt laden water does not cause any problems normally, due to the low heads. However, in extra-ordinary cases where the concentration of silt in the canal water is excessive the turbine blades may have to be given additional protective coating. The type and extent of protection depends on the concentration of silt, and size and hardness of the silt particles.

7.2   Sometimes silt gets deposited in front of the power house intake due to low velocity flow, particularly when the power house is not running or partly running. However, such silt usually gets washed out through the turbines, when the power house runs back to its full capacity. It is possible that the silt deposition becomes heavy, if the duration of the power house closure is more than few weeks. In such a situation the operation of the intake gate may also get affected. Keeping a provision for blowing the silt by compressed air or water under pressure and providing upstream seals on the gates can take care of such situations. Occassional removal of the silt deposits may however be required, which can be achieved either manually or by excavating machines.

8. MISCELLANEOUS

8.1   Construction of the power house involves deep excavations, depending upon the extent of fall contemplated and the ground levels. The scheme of dewatering for making and maintaining the power house pit will have to be properly designed to suit the ground water condition and permeability of the soils encountered. Extra caution is called for while dealing with sandy soils, and fine sands in particular. The problem becomes more acute when the construction is to be done adjacent to the existing canal. Sheet piles or diaphragm walls may be required in specific cases, to control the seepage into the excavated pit or to maintain safe gradient. Making the sump well may pose additional problems, as the depth of excavation would be maximum here. It may be prudent to construct the well by sinking process instead of adopting open excavation.

8.2   The power house upstream wall is subjected to the forebay water pressure. It is important to ensure that water doesnot leak across this wall, as such leakages will not only keep the inner face of this wall damp and ugly, but may also damage the cables etc. located on the wall. Special care should be taken while concreting the areas around the gate grooves and other embedments in the wall, as these are the potential leakage paths. It is desirable to add water proofing admixtures in the concrete, and/or treat the outer face of the wall with water repelling agents. Any leakages observed after construction should be sealed by polygrouts.

8.3   Some canals carry unusually large quantities of floating debris, which cannot be passed through the machines. The debris is intercepted by the trashracks and gets accumulated there, choking the water passage with consequent head loss. The trashracks therefore need frequent and sometimes a continuous cleaning. As per the current practice, the trash racks are provided near the intake gates. It is very difficult to clean these deep seated trash racks, unless some mechanised cleaning system is adopted. The problem can be alleviated to some extent by shifting the trash racks to a point at the beginning of the forebay where the water depth is comparatively much less, and removal of the trash would be easier. Similarly the stoplogs of the intake gates could also be shifted upstream to reduce their cost. If a road bridge is contemplated across the power canal it may be convenient to combine the trash racks and stoplogs with it.

First International Conference on Renewable Energy–Small Hydro
3–7 February 1997, Hyderabad, India

EVALUATION OF SMALL HYDRO POWER

B. Pelikan

Institute of Water Management, Hydrology and Hydraulic Engineering
University of Natural Resources, A-1190 Vienna, Austria

Recent restrictions force small hydropower to aim at high quality levels concerning economy, technic and ecology in order to proof further acceptance and continous exploitation. To evaluate the individual quality of projects or already existing plants the described tool, containing a complex system of small scale features, may serve. Within the evaluation process the achievment of goals in comparism with the best possible situation is to be decided but also interpreted in an overall context. The results will support administrative procedures, environmental impact recognition and the design process. Nvertheless a high degree of engineering experience and environmental feeling will be the assumption of a succesful performance.

1.INTRODUCTION

There is absolutely no doubt, that the resource of hydro power represents a significant, in mountainous regions even predominant part of the so-called renewables and consequently an energy potential of highest quality. Not only in Europe and some other countries, where the exploitation of hydro power has been carried out intensively whilst the last decades and the ecological sensitivity has increased as well, but all over the world, where hydro power is part of energy policy, the uncritisized position and blind acceptance of water power has been degraded to a historical fact and each new project has to recover singular consent. The only suitable procedure for future developement seems to be the proof of overall quality. The operative tool is the definition of a series of goals and the evaluation regarding to them.

The facility to realise ecologic as well as esthetic targets within a project depends predominantly on the scope of design respectively engineering on the one hand as well as the size of the station on the other hand. The connection with the rules of human range is obvios and favours the small scale twice. Therefore this article is dedicated to SHP.[1,2,3]

2. QUALITY

Quality is in case of hydro power a special kind of efficiency. The numerator shows the contents already achieved and the denominator presents the achievable maximum. Considering the complexity of a hydro power project the targets have to been devided up into the following special branches.[2,7]

  * economy (national view, individual view)
  * technical state of the art (system, material, design, machinery, equipment)
  * environmental sociability (ecology, recreation, balance of claims)

2.1 Economy

The discuss on is to be held on two levels. From the national superior point of view within the frame of energy policy the recent state is rather discourageing and unsatisfactory due to the fact of low prices particularily of gas. This temporary unbalanced and distorted system requires long termed decisions apart from quick profits respecting existential interests of the succeeding generations in a national as well as a world wide context. Some so-called generation - contracts have to be concluded even under close economic conditions.

In Austria for example a lot of medium size projects have already passed technical and even environmental protecting governmental exams and procedures but have been set back based upon politcal considerations. Under such circumstances this kind of quality is hard to be defined and therefore more or less unevaluable.[3]

From the individual economic point of view quality concentrates on the best possible benefit cost ration as a result of optimisation procedures particularly concerning rated head and rated discharge in a first step. Additional·a consequent low cost strategy with minimal losses in overall efficiency will support these efforts. The connections to the second area become obvious.

There might occur some conflicts in the field of economy between individual and national interests to be shown in the following two examples.
The highest stage of individual economic quality can be combined with low national economic quality if foreign products are used predominantly.
National energy resources should not be wasted by low efficiency equipment or low cost determined suboptimal exploitation. In both cases the government is called as a corrective.[11]

2.2 Technical state of the art

Concerning the technical quality there should be divided up into design and construction. A lot of significant contents of design like the fundamental idea, the degree of exploitation, hydrology and hydraulic calculations are proofable in a state of design as well as after realisation. Their extraordinary importance during the phase of operation does not change. Quality can be measured by the scales of scientific literature as well as the comparism with documentations of stations already established.[7,8]

Constructing details may be based upon a professional design but carried out amateurish, impairing the functions intended. Typical for this kind of conflict is the segment of micro design, not fit for representation within a project like a lot of ecological based contents for example concerning fish ladders or morphologic structures of channels. On the other hand there is an important potential of „last minute design" which should be activated in case of suboptimal macro design. Quality can be criticized exclusively by means of the reality.

Additional any judgement has to consider the variability of the input data like hydrology, efficiency, material but also environment and environmental conditions, leading to the following remarks.

2.3 Environmental sociability

Nature means living space not only for animals but also for mankind. Human well being depends on ecological conditions like climate or acoustic impressions but also on esthetic perceptions determined by the balance between order and variety. Consequently the term of environmental sociability respecting humen being includes ecologic as well as esthetic parameters. Nevertheless this must not express, that sectoral qualities are exchangeable and „compensation deals" are permitted. On the contrary the knowledge of a dangerous mixture forces to a very consequent separation in order to avoid foggy decisions.[4, 9, 10]

2.4. Restrictions in general

To define quality at all it is necessary to describe targets as level to be compared with. Besides some attributes which are numerable to be expressed or can be limited precisely by their technical functionalty there exist some topics beyond any classical units. Individual and specific new scales render a qualitative graduation.

In a simple way of interpretation it could be assumed that 100% of total quality has to be divided up into the three levels of quality and benefits on the one hand will cause necessarily losses on the other hand. Those almost classical connections well known from the ecologic - economic conflicts may occur in part but not in general.

155

Fig.1: Evaluation of Small hydro power stations - general level

river:			river - km:	river-typological attachment:				
station:			type:	date of review:				
top water level:	head:	discharge:	turbines:	name of reviewer:				
power:	annual energy:	degree of exploitation	designed by:					

	features	achievment of goals				
		perfect	extensive	partly	little	non
1.ECONOMY	best possible use of natural water yield					
	covering of one's own requirement of energy					
	little specific cost of power					
	little specific cost of energy					
	best possible financing					
	big portion of national products and performance					
	high value of energy produced, summer-winter diversion					
	including of secondary usage claims					
	no restriction of neighboring use					
	perfect operation conditions					
2.TECHNIC	optimised head					
	optimised discharge					
	production forecasting for normal and extreme years					
	design according to masterplans					
	use of reliable hydrological data					
	minimal construction cubature					
	perfect engineering					
	optimal adjustment to terrain conditions					
	river engineering according to the state of the art					
	architectural attractivity of the buildings					
	funktioning and adequate to the claims					
	optimal and fail save flood management					
	hydraulic calculation acc. to the state of the art					
	optimal safety in operation					
3. ECOLOGY	preservation of the river continuum					
	little ecological quality of the river section exploited					
	leastpossible change of limnological character					
	leastpossible technical impacts					
	attachment to already existing buildings					
	passability beyond limnological bypass systems					
	channel usable as aquatic living area					
	limited enlargement of depht in the back water area					
	limited reduction of flow velocity					
	measures to increase the morphological variety					
	use of soil bioengineering materials					
	ecological bank protection andn shaping measures					
	planting measures appropriate to zhe site					
	creation of new ecological attractive areas					
4. ADDITIONAL REMARKS						

Fig.2: Evaluation of the passability of fish bypass-systems

river:	mean flow:	reviewer:
station:	rated discharge:	date of review:
head to get over:	construction type:	river - km:
back water level:	operation discharge	river-typological attachment:

	features	achievment of goals				
		perfect	extensive	partly	little	non
1. OUTLET	no danger of silting up					
	near the scour					
	reachable even at low water					
	easy to find due to flow					
2. FISHWAY	variable flow velocities					
	continous line of maximum velocity					
	acceptable maximum velocities					
	acceptable minimum depths					
	no turbulences					
	no circle flow					
	acceptable height of small ramps					
	no fan out of water					
	flow velocity smaller then critical stage					
	submerged weir conditions					
	areas to hide					
	adequate shading					
	natural bedload					
	natural light					
	difficult accessible to man					
3. BASINS	adequate depth					
	adequate width					
	adequate lenght					
	resting basins					
	no aeration at overflow sections					
	acceptable height of drops					
	no danger of hurt					
4. INTAKE	adjustable discharge					
	unselective passability					
	protection against flood					
	protection against floating matter					
	protection against to much bedload					
5.MATERIAL	morphological adequacy					
	physical resistance					
6. ADDITIONAL REMARKS						

Fig.3: Evaluation of turbine water channels as aquatic living area

river:		river - km:		reviewer:
station:		date of review:		
lenght of top water channel:	flow velocity.	rated discharge:		
lenght of tail water channel:	flow velocity:	river-typological attachment:		

	features	achievment of goals				
		perfect	extensive	partly	little	non
1.CHANNEL	variable slopes					
	rounded transitions of slopes					
	variable heights of embankment					
	natural stabilized ways for maintenance					
	variable structure of bed load					
	little variable cross sections					
	little variable longitudinal gradient					
	environmental harmonised shaping					
	no surface sealing (head water)					
	dovetailed structures of the banks					
	vegetation cover on the inside slope					
	vegetation cover on the outside slope					
	vegetation to improve structures					
	border zones with little flow					
	adequate shading					
	passability of water intake					
	passability of water outlet					
	connection to groundwater (tail water)					
	withdrawal areas for fish					
	areas of shallow water					
	no waste water inflow					
2. PIPE	line of construction adapted to landform					
	digged					
	complete recultivated					
	additional impacts avoided					
	pipe not avoidable					
3. ENVIRONS	connections to existing vegetation					
	accessible to animals					
	recreation for people					
	careful connection to natural surroundings					

4. ADDITIONAL REMARKS

3. EVALUATION

3.1 System

In the first step, following the division into the fields mentioned above, a lot of attributes or features has been selected. They should almost have similar weight and are defined in its best case. A perfect balance among al the attributes is fairly unattainable. The consequences will be discussed on an upper stage.

The main problem to be overcome was the balance between completeness and independency of the characters to avoid some multiple evaluation or neglection of significant topics. A second aim was to prevent the compensation of uncompareable and nonequal attributes leading to an unbalanced result. It is to concede, that the solution found and presented now is not the only one or even the best. Some testing with quite different projects will show disadvantages or qualities. The evaluation defect possible has two roots - the unperfect features in the evaluation forms and the human individuality of the evaluators in experience and competence. (Fig. 1)

In all three sections it has become obvious, that the general catalogue of criterions does not render satisfying results yet and a special catalogue refining the first step concentrating on singular aspects has to be created in a similar way. Special examples for the necessity of refining are:

rentability, energy consumption and financing [*economy*]
optimisation of head and discharge, production calculation [*technical state of the art*]
fish ladder, residual flow, turbine water channels [*environmental aspects*]

The necessity of a second refining level is inhomogenous and depends on the subject. Some characters dont need any further refining. In the course of application it was rather surprising, that the alternative character of the attributes within the first general level did not necessarily lead to total opposing so-called either/or situations. In many cases a significant approach f.i.between ecologic and economic aims could be achieved provided an aggreement concerning some modifications. Nevertheless naturally some attributes remain incompatible. As examples for detailling you may find in Fig.2: Evaluation of the passability of fish bypass-systems and Fig.3: Evaluation of turbine water channels as aquatic living area.[4, 6, 9]

3.2 Applicability

As mentioned above the catalogue of features claims to be expressive and significant but not complete. In spite of the individuality of each project superposing various typologic differences a limitation exams is expedient. Even within the actual selection some characters stay unevaluable. The border-line of evaluability will be demonstrated by means of two examples:

159

Under limited conditions of discharge or head available in the case of an existing channel any optimisation work is lost time. The design parameters are more or less fixed within the existing situation. In a low head concept the ecologic functionality of the channel is to be considered. In case of a pipe the ecologic point of view is irrelevant. The evaluation may concentrate on esthetic parameters. The concluding statement is the request for an analogous individual adaption of the tool avoiding nonsensical supposed accuracy.

In general the recent tool is applicable for all types of small hydro power projects. In a further step of development it may be useful to devide into high head and low head plants.

3.3 Achievement of goals and interpretation

For the evaluation whether a project has fullfilled the goals demanded, the presented tool offers five steps with the following descriptions: *perfect - extensive - partly - little - non*. This verbal description lets come into being realistic pictures and makes the attchment easy. We should interpret this strategy as a human support to a scientific procedure. Practical experiences already confirm this reflection.

The scaling is very individual and hard to be compared. This leads to the following step of interpretation. A total evaluation must not be seen in calculating a mean value of the variety of marks neither on the general nor on the special level. This procedure would uncorrectly provide a full compatibility between the thematical groups as well as inside of them. It would not make any sense to try a compensation of to little residual flow with a remarkable architectural design of the power house. Wetlands in the backwater area will not compensate the wrong choice of the rated discharge.

Instead of calculating mean values the weak point of the chain is to be discovered and to be explored if there is any key function connected with. If not the next of the weak points should be examined in the same way. The result of this research has predominant influence on the total result. This „weak-point-system" has the advantage that efforts concentrating on them will cause sudden improvement of total quality.

Usually the variety of attributes does not demonstrate only singular points of defects but agglomerations showing centres of gravity respective centres of bad marks. The evaluator has to locate those areas and consequently identify the total character of the project considering the connections between the singular attributes. Douptless this evaluation and interpretation work needs a lot of experience and sensibility and should be carried out with responsibility.

3.4 Long range usability

Aiming at best possible projects it is proposed to devide up into repairable and not repairable defects. Of course in the project state any content is variable. But especially the application of this tool on already existing plants needs this distinction offering the facility of continuing improvement. The recognitions in the course of an accompanying control of stations in operation enable the owner to qualify investments and to find out thematical efficiencies of them.

Best efficiency can be defined in all the three groups. The enumeration of the single topics within the groups is not complete but exemplaric and symbolic.

> *economical efficiency = investment / improved income*
> based upon better buy back rates or credit conditions, better internal organisation and
> management, lowering the expenses for maintainance, partly automation

> *technical efficiency = investment / improved output*
> based upon better hydraulics, better governing, higher turbine efficiencies, avoidance of
> hydraulic and electrical losses

> *ecological efficiency = investment / reduced environmental impact*
> based upon river continuum, micro climatic conditions, additional recreation use.

As said at the beginning the change of input data, operation conditions, public opinion, scientific knowledge or the economic frame requests periodical uprating. Within the European Community the idea of an „ecological auditing" is part of European law. This procedure and instrument has been invented especially for industrial structures but will be applicable also for hydro power stations. the tool presented may be significant part af this adaption.

Further on the result can be used for several services:
Within the *governmental granting procedure* the projects have to described and commented precisely. The completed evaluation forms will help the government to check the single features of the project on a concentrated level. This application will focus on the technical section.
Concerning projects within sensible areas usually there is a need for environmental impact studies. Their contents can be tied up to the list of characters dedicated to environmental aspects.

Both the technical as well as the ecological subjects can be used as controlling system of the engineering process. This may support primarily the individual orderer of the project having less experience and specific knowledge.

161

Sometimes the placing of financial support or even crediting requires a precise presentation of economic figures, production expected and other interesting facts. The evaluation form will make the analysis easy and shorten up the necessary procedure.

4. ACCOMPANYING COMMENT TO THE FEATURES

Even the „detail level" of the evaluation form is not sufficient to execute the evaluation work. Each character has to be described in detail and limit values considering different situations and surroundings have to be provided. By means of examples the different degrees of quality are explained. The evaluation of the first projects will need a lot of time gathering the experience based upon the recent state of the art. Later on these additional descriptions will serve only in cases of doubt. Some short examples, one of each field, may serve as demonstration:

Regarding to economy the feature „inclusion of secondary usage claims" needs some explanation. First of all you have to take a spatial as well as a substantial dividing up. In a second step we have to find out and define different types of usage claims like smooth recreation, boating, swimming, fisheries, bridges, natural reservates, forestry, agriculture, hunting, scientific interests or governmental interests like increased flood control or groundwater management. The third step has to classify the aspects and claims regarding to their economic relevance which means direct profit, cost sharing, eventual losses, operation restrictions, third party financing or governmental support. Summing up the total result of evaluation shows the economical impact compared with the claims achieved.

Regarding to the technical range some explanations concerning „forecasting of production" which does not mean hydrological forecasting but calculation by means of hydrological data available. The most simple but unprecise way is the use of monthly mean values and fixing head and efficiencies. The usual way is by using the duration curve with some flexibility of head and efficiency approximated in classes. The most precise way is the calculation by means of daily values of discharge and adequate calculated values of top water and tail water levels, efficiences of the turbines considering the type and the number, the minimal discharge and some more details. (Fig. 4: sheet of results)

The ecological has been demonstrated already with the evaluation sheets in Fig.2 and 3.

5. CONCLUSION

The complexity of the evaluation tool demonstrates the complexity and diversity of small hydro. This feature is to be seen as a chance concerning further exploitation. Small hydro is able to be adapted best to changing marginal conditions excluding economical pressure from the non-renewables. Real long termed quality will be the only alternative to overcome even this critical period.

Fig.4: Calculation sheet - energy output - losses

river:	**Ybbs**		calculated year:	**normal3**	**1982**
power station:	**Kemmelbach**		rated discharge:	**30,0** m³/s	
turbine:	**1 Kaplan double**		rated head:	**4,57** m	

Var.:	**5**	Specification:			

Q_{res} (absolut/relativ):		**absolut**	number of days Q_{act}.< Q_{rated}		**275**

	absolut	m³/s		m³/s		m³/s	relativ	%
January	**1,00**	May	**4,00**	Septembe	**2,50**	% of Q_{act}	**15,0**
February	**2,50**	June	**4,00**	October	**2,00**		m³/s
March	**3,00**	July	**3,00**	November	**1,50**	Q_{min}	**1,50**
April	**3,00**	August	**3,00**	December	**1,00**	Q_{max}	**5,00**

consumption of the weir		consumption tailwater		efficiencies	
topwater level:	**266,80** m a.Sl.	bottom:	**261,20** m a.Sl.	generator:	**0,93**
fixed weir crest:	**265,00** m a.Sl.	bed width:	**30,00** m	gear:	**0,98**
width:	**45,00** m	n:	**1,00**	transformer	**0,98**
coefficient μ:	**0,55**	long.slope:	**0,001**	other:	**1,00**
		$k_{Strickler}$:	**30,00**	total:	**0,89**
topwater$_{max}$:	**267,82** m a.Sl.	tailwater$_{min}$	**261,77** m a.Sl.		

exploitation of total annual discharge load:	**69,2** %		
exploitation of useable discharge load:	**88,4** %		
capacity:		**1074** kW	
annual output:	year:	**6.144.226** kWh	%
	summer:	**3.231.795** kWh	52,6
	winter:	**2.912.432** kWh	47,4
loss of production due to Q_{res}:	year:	**580.607** kWh	8,6
	summer:	**333.338** kWh	
	winter:	**247.269** kWh	
theoretical amount of hours of maximum capacity:		**5720** h	
rate of exploitation:		**1,53**	

returns [in national currency]:		ATS	**3.939.360**
loss of returns due to residual flow [in national currency]:		ATS	**376.023**

summer	high	**1.538.950** kWh	**0,68** ATS/kWh	ATS	**1.046.486**
summer	low	**1.692.845** kWh	**0,48** ATS/kWh	ATS	**812.565**
winter	high	**1.386.872** kWh	**0,85** ATS/kWh	ATS	**1.178.841**
winter	low	**1.525.559** kWh	**0,65** ATS/kWh	ATS	**901.467**

calculated mean value of buy back rates:	**0,641** ATS/kWh	

days of weir overflow:	**90**		
minimum discharge for the turbine (% von Qa):	**15** % =	**4,5** m³/s
days out of operation:	**0**		

Bibliography

1. DIRECTORATE GENERAL FOR ENERGY, COMMISSION FOR THE EUROPEAN COMMUNITIES AND EUROPEAN SMALL HYDROPOWER ASSOCIATION:
   „Layman´s guidebook on how to develop a small hydro site", Part 1 and 2

2. NORWEGIAN INSTITUTE OF TECHNOLOGY, DIVISION OF HYDRAULIC ENGINEERING:
   „Hydropower development", Vol.4 and 5(1992), Vol.6 (1993) and Vol 7 (1995), Trondheim

3. ÖVE, SEV und VDE:
   „Wasserkraft, regenerative Energie für heute und morgen", Wien 1992

4. DVWK Merkblätter 204/1984:
   „Ökologische Aspekte bei Ausbau und Unterhaltung von Fließgewässern", Verlag Paul Parey

5. WASSERWIRTSCHAFTSAMT IN BAYERN, BAYRISCHES STAATSMINISTERIUM DES INNEREN:
   „Flüsse und Bäche, erhalten entwickeln gestalten"

6. JUNGWIRTH, M und PELIKAN B.:
   „Zur Problematik von Fischaufstiegshilfen", Österr. Wasserwirtschaft Heft 5/6, Wien 1993

7. RADLER, S. (Editor):
   Symposium on project design and installation of small hydro power plants", Wien, 1981

8. WATER POWER AND DAM CONSTRUCTION (PUBLISHER):
   „Uprating & Refurbishment Hydro powerplants V - Conference Proceedings"

9. LANDESAMT FÜR WASSER UND ABFALL NORDRHEIN - WESTFALEN:
   „Fließgewässer, Richtlinie für naturnahen Ausbau und Unterhaltung"

10. EUROPEAN SMALL HYDRO POWER ASSOCIATION and FEDERATION OF SCIENTIFIC AND TECHNICAL ASSOCIATIONS:
    „Conference Proceedings Hydroenergia 95", Milan

11. EIDGENÖSSISCHES VERKEHRS UND ENERGIEWIRTSCHAFTSDEPARTEMENT, BUNDESAMT FÜR WASSERWIRTSCHAFT:
    „Kleinwasserkraftwerke in der Schweiz", 1987

First International Conference on Renewable Energy–Small Hydro
3–7 February 1997, Hyderabad, India

CONDITIONS FOR THE CONSTRUCTION OF SMALL HYDRO IN CROATIA

Boris Berakovic

Faculty of Civil Engineering, University of Zagreb, Hydraulic Engineering Department
10000 Zagreb, Kacicieva 26, Croatia

1. INTRODUCTION

The construction and exploitation of small hydro besides the disposable energy in the nature and energy requirements depends considerably on the state policy on energy exploitation. The state policy is implemented through legislation, environmental protection, financial support and price policy. In this paper shawn are legislation and other measures related to small hydro in Croatia. It is discussed how succesful are the legislation and other measures and how they could be improved.

2. AVAILABLE ENERGY IN NATURE

In this paper is considerd the energy from the nature, which is used in electrical power generation. For electrical power generation in Croatia today are used fosil fuels (43%), water power (40%) and nuclear power (17%). Croatia imports the major parts of fuels. Nuclear fuel is not produced in Croatia and the nuclear plant used by Craotia is in Slovenia. Therefore Croatia must relye primarily on water power, which has been used for over 100 years. The total water power is estimated to 12,1 TWh/yr, from which 6,6 TWh/yr is being used (52%). Incoming period it is planned to increase the exploitation for about 2,1 TWh/yr by the construction of power plants larger than 5 MW. Based on the existing studies which do not comprise all the available resources it is estimated that with small power plants (around 50, with installed capacity of less than 5MW) about 0,09 TWh/yr could be exploited.

3. POWER REQUIREMENTS AND THE POSSIBILITIES OF SALE

Since the power requirements exceed the available capaties it is necessary to provide more and more power for the users. In Croatia the consumption before the war (1991-1995) was 15,8 TWh/yr and the production was 10,7 TWh/yr. The difference is beeing imported. From this it follows that there is no problem in selling the produced power in Croatia as long as the quality and acceptable prices are assured. In this respect the Croatian public power enterprise (HEP) is ready to accept all power which could be produced in small power plants in private ownership.

Beside that possibillity of selling electrical power to the public power network there is a possibility of using the produced power of small power plants for own requirements which means developing some other activity which will consume the produced power i.e. the power production would be only a secondary activity. Such solution in Croatia today are rare (negligable). Today almost all electrical power produced in the country is being overtaken by HEP . HEP itself is oriented on large power plants and is not interested to build the small ones. This leaves the space of activity for small un - dertakers in the power industry.

4. LEGISLATIVE REGULATION

The Croatian laws allow for private construction of small power plants. As with all other con - structions it is necesaryto fulfill some requrements and collect necessary permissions.

4.1 Ownership Question

All waters in Croatia are public good and they are controlled by some institutions in the name of the state. The use of waters and objects on them is subject to concessions given through public so - liciting for funders. For now it is still unknown for which time period the concessions are to be giv - en, but it is expected it will be 30 years, with the upper limit of 99 years. The use of water power re - quires indemnity payments. The potential undertaker has to solve himself the question of the land for the construction site and get permissions for construction on it.

The land along the rivers is still or was formerly mostly private property. For those which are still private there is the only problem of negotiation with the known owner, but those which were nation- alized and are to be returned to their former owners or their successors it make take time to clear the papers.

4.2 Location Permission

This is the basic document showing the conditions under which an object could be built and used. Especially unphesized are conditions for nature protection and water use.

4.3 Concession Contract

With this contact the contactor gets the right to use the water for electrical power generation. The contract is given by the State Board for Waters after a public soliciting for tenders.

4.4 Other Permissions

There are some other permissions which must be provided before the construstion starts like con- struction permission, water management permission etc. They do not increase the expenses but to get them some time is required.

5. NATURE CONSERVATION

Todays high nature protection conciousness influences strongly also the construction of small pow- er plnts. For locations which are not specially protected construction conditions regarding nature pro- tection are also prescribed.

5.1 Protected Areas

In National parks no construction is allowed. In other protected areas like nature reserves, nature monuments, nature park etc with some limitations in some places it is allowed to build small power plant provided that limitations concerning opertion resulting environmental changes are observed.

5.2 Biological Minimum

The biological minimum is the quantity of water which must be preserved in the river on which a power plant is to be built. It is the flow which still preserves life in the water. In protected areas this minimum is increased up to medium flow with special operation regimes (day-night etc). The biological minimum is defined by the responsible authority . This limitation exists with derivative power plants.

5.3 Fish Migrations

Water catchment for a small power plant is usually of a small height and it does not prevent fish migrations. If this could happen in some location, a fish path must be provided.

6. FINANCING

The financing is left in full to the private iniciative and the state has no incentive measures in this respect.

7. PRICES

The purchasing price of the electrical power is bound with the selling price and today it is between 75 % of the selling price for plants with capacity under 500 kW and 65 % of the selling price for plants above 500 kW, which is 0,09 DEM/kWh.

8. CONCLUSION

There is a considerable unused potential for small power plants in Croatia. In last five years only one was built (30kW) and that one for private use only The market for electrical power exist and there is no problem to find a purchaser , but the price of 0,08 - 0,09 DEM/kWh does not seem to be too atractive.

Though there are several locations suitable for power plants with capaties above 500 kW for which the above price could be acceptable.

The additional dificulty in some places could be buying of land because of the disputed owner - ships (nationalized land). Though rules for environmental protection are aproximately some as in some other countries.

First International Conference on Renewable Energy–Small Hydro
3–7 February 1997, Hyderabad, India

PRESENT SCENARIO FOR EXPLOITATION OF HYDROPOTENTIAL OF THE RIVER CHANGED ACCORDING TO ENVIRONMENTAL AND LEGAL ASPECTS

Svetlana Stevovic

Energoprojekt-Hydroinzenjering Consulting Engineers, Co. Ltd.
11 070 Novi Beograd, Bul Lenjina 12, Yugoslavia

ABSTRACT

Large population migrations are arising as well as national turmoils and territorial divisions. River valleys are populated and once adopted technical solutions of the river hydropotential exploitation are underestimated. Therefore, it is feasible and only advisable to build several small hydroelectric power plants, observing the environmental and legal aspects. The problem is to prove this mathematically. Their reservoirs shall mostly remain in the river bed, while the hydroelectric power plants shall be most often (if head are approximately the same) cascades in series provided with uniform equipment.

Key words: Environmental, legal aspects, hydroelectric power plants.

1. INTRODUCTION

Each country or region should have an elaborated water economics plan. Water streams and their valleys should be provided with the adopted land use plans of project areas which should be observed when hydraulic structures are erected. In practice,it may appear different, especially in economically underdeveloped countries. Large population migrations are arising as well as national turmoils and territorial divisions. River valleys are populated and once adopted technical solutions of the river hydropotential exploitation are underestimated . Therefore, it is feasible and only advisable to build several small hydroelectric power plants whereat they were once planned and whereat it was justified from techno-economic point of view to build one dam and hydroelectric power plant (having submerged a considerable part of valley)

169

due to newly formed settlements and observing the environmental and legal aspects. The problem is to prove this mathematically. Their reservoirs shall mostly remain in the river bed, while the hydroelectric power plants shall be most often (if head are approximately the same) cascades in series provided with uniform equipment.

The adopted technical solution of the Drina river hydropotential exploitation at the stretch extending from the town of Foca to the town of Gorazde has undergone similar modifications. Within this part of the river according to the available Preliminary Design adopted even in 1978, a dam provided with Gorazde HPP barrage power station and the reservoir situated at 375.00 m a.s.l. was foreseen.

Development of the area initiated elaboration of the report on "The study of Gorazde HPP effect" in 1985. The conclusion of the study was drawn taking into account the results obtained through the analysis of Gorazde HPP effect on land use plan of the riparian zone development. It still means the proposal for construction of one step at the profile of Gorazde, i.e. backwater is still at 375.00 m a.s.l. but on condition that the settlements situated on the banks of future reservoir are to be connected by means of communication lines and that water level fluctuations are to be reduced to minimum and that the Drina riparian zone improvement has to be worked out. Further development of Foca and Gorazde municipalities requires finding out of new technical solutions of hydropotential exploitation of the current Drina river stretch. In addition to techno-economic criterion it was necessary to consider social, environmental, urban and human factors and appraise their particular importance. This was made applying the method of multicriteria comprise ranking.

Selection of the optimal solution of Foca-Gorazde hydroelectric power system has been made provided that a group of six possible alternative technical solutions has been formed. Complete design was worked out for each of them. The first alternative was a dam with the reservoir which had submerged almost the whole valley (the existing solution), while the last alternative meant the series of four small dams and hydroelectric power plants with reservoirs in the river bed of the same stretch. Criteria functions essential for making a final decision were: the present value of profit expressed in (Din.), the present value of costs (Din.), the total power output (GWh), the total peak energy (GWh), the number of households which had to be submerged or endangered (pcs.), the amount of hectares of agricultural areas that had to be submerged (ha), the number of settlements or areas that had to be submerged or endangered (pcs.), and the effect on the environment (quantified by the mark from 1 to 5).

Evaluation of all six alternatives had been made applying quantitative indexes which were the results of economic analyses, engineering calculations and various measurements. Evaluation had been also made using the qualitative indexes which were the result of expert appraisal. The purpose determined through optimization was to select the solution which required rather small investments due to present economic situation in the country (criterion "cost" should be minimized), and at the same time it should not disturb the vicinity from the environmental, perspective-urban and social aspects (the other criteria functions should be maximized).

Multi-criteria optimization as a part of modern approach to system engineering mathematics decision-making. It is possible to quantify certain criteria as environmental and legal aspects, essential for making a final decision and to put them into the ranking calculation using their numerical values. When computation techniques of operation researches had been put into effect, a part of criteria existed only descriptively in the mind of a planner, Employer and decision maker in the municipality, republic or region. Also, one can say that the priority is given, in this way, to a series of small hydroelectric power plants as regards one large for the hydropotential exploitation of the same river stretch.

2. PRACTICAL EXAMPLE

It is really difficult today to find a suitable profile for construction of large dams. Most of the river valleys are populated. If not it is quite complicated to find capital investments, while the electric energy requirements are constantly growing. In many countries different movements for environmental protection, for safeguarding cultural and historical monuments, for preservation of various plant and wildlife species are more influential. The optimal technical and economical solution is not the only one sought when designing and constructing hydropower structures. The most recent schemes should respect both ecological and legal aspects, otherwise the adopted technically and economically proved alternatives will not be implemented.

This was the case with the scheme for utilization the water potential of the Drina river, at the tract between the towns Foca and Gorazde. The optimal solution techno-economically was a dam at the Gorazde II profile with normal water level elevation of 375.00 m asl. The adopted solution dealt with flooding the highway along the river, many fertile ploughlands and rural settlements. The fertile lands in this region are existing only there, in the river valley. Inundation of villages would create problems in the already delicate national aspects. All the above mentioned confirmed the idea that the adopted solution would remain "on paper" many years ahead. Through design task for preparation Preliminary Design implementation, as well as through the demand itself to analyze once more the reach of the Drina considered before 12 years, the need was underlined to evaluation profitability of the structures with respect to the new relation to the surroundings and the environment. For this purpose, multicriteria optimization was applied within the methodology for selection of the optimal scheme.

Multicriteria optimization as a part of modern approach to system engineering solves decision-making mathematically. Through the method of multicriteria compromise ranking, it is possible to quantify certain criteria indispensable for making a final decision and enter them with their numerical wights in the ranking computation. Not until the appearance of computation techniques necessary for carrying operation researches, a part of criteria which have effected the decision, existed only descriptively in the mind of the planner, employer and decision-maker in the municipality or the republic.

The practice has proved that construction of many structures could not have commenced although the standard feasibility study has shown that coefficient of their public payability is justified. The river valleys had been colonized and the construction postponed sometimes by the moment of complete loss of the area intended for the river hydropotential utilization. As to obstruct such an incident even with the Drina hydropotential utilization at Foca-Gorazde reach, it is necessary to make a decision on the optimal scheme selection as soon as possible, taking into consideration techno-economic, social, urban, ecological and human factors and to assign them the belonging importance. Because these are just the elements due to which once proposed solution could not be further considered in the course of this realization. (1)

3. MATHEMATICAL MODEL

The mathematical model is using the program package VICOR for multicriteria compromise ranking. The decision is mathematically presented detecting and elaborating first the possible solutions. This part of the work could be very extensive. The reason for this is that every possible solution should be developed at the relevant and uniform technical level. Practically it would be necessary to prepare as much designs for the hydropotential utilization of the river course as required and as existing. Then upon consultations with the Client (Clients) the essential criteria for decision (objective) are emphasized.

So, multicriteria optimization develops in several stages of phases as for instance:
- Designing of alternative solutions of the system,
- Defining of the criteria and criteria functions for evaluation of the alternatives (economic, technical, social, ecological, etc.)
- Evaluation of all the alternatives per each criterion individually.
  Evaluation can be made applying quantitative indicators (which are the results of economic analyses, engineering calculations or different measurements) or using the quantitative indicators (which are the result of expert opinion).
- Multicriteria ranking of the alternatives
- Adopting of the final (multicriteria optimal) solution.

Multicriteria optimization of Foca-Gorazde hydroelectric power system was made applying the method of multicriteria compromise ranking (2). Ranking of alternative solutions was made using the program package VIKOR which is described in the paper (3).

4. POSSIBLE SOLUTIONS

Alternative solutions have been determined in the form of the variant solutions of the system provided with a different number of the reservoirs or different levels of normal retention water. Possible reservoirs of this hydroelectric power system are presented on the diagrams within the figure provided with the group of alternative solutions.

On the basis of the elements of hydroelectric power system considered within the study (4), a group of six alternative solutions has been formed for multicriteria optimization, such as follows:

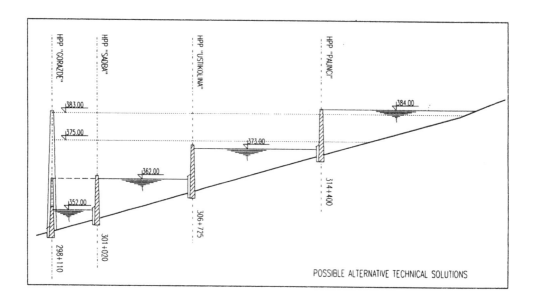

POSSIBLE ALTERNATIVE TECHNICAL SOLUTIONS

A1 - <u>Variant A</u> (GORAZDE II HPP, NHWL = 375) represents hydropotential utilization of the Drina at the Foca-Gorazde reach, characterized with one dam and retention water level at 375,00 proposed by Energoinvest Co. in their Preliminary Design from 1987.

A2 - <u>Variant B</u> (GORAZDE II HPP, NHWL = 383) means also a solution with only one dam, but with retention water level at 383,00 for which an order was made by the inspection team to be further analyzed and effected, but rejected in the study on the impact of Gorazde II HPP (Urban planning Institute - Bosnia and Herzegovina - Sarajevo, 1985).

A3 - <u>Variant C</u> means a series of four steps:

GORAZDE II HPP	- with retention water level at 352,00
SADBA HPP	- with retention water level at 362,00
USTIKOLINA HPP	- with retention water level at 373,00
PAUNCI HPP	- with retention water level at 384,00

174

A4 - <u>Variant D</u> is a combination of the following scheme elements:

GORAZDE II HPP   -   with retention water level at 375,00

PAUNCI HPP   -   with retention water level at 384,00

A5 - <u>Variant E</u> is contained of three steps:

GORAZDE II HPP   -   with retention water level at 362,00

USTIKOLINA HPP   -   with retention water level at 373,00

PAUNCI HPP   -   with retention water level at 384,00

A6 - <u>Variant F</u> is also a series of three steps, and this is:

SADBA HPP   -   with retention water level at 362,00

USTIKOLINA HPP   -   with retention water level at 373,00

PAUNCI HPP   -   with retention water level at 384,00

5. CRITERIA FUNCTIONS

In a certain case, having in mind the design task and the necessities of the riparian municipalities development, the following group of eight criteria functions essential for making a final decision, has been formed:

f_1 - present value of the profit - benefit (Din)

f_2 - present value of the costs (Din)

f_3 - total energy output (GWh)

f_4 - total peak energy (GWh)

f_5 - the number of households to be submerged or endangered (pcs.)

f_6 - the number of hectares of agricultural areas to be submerged (ha)

f_7 - the number of settlements or area to be submerged or endangered (pcs.)

f_8 - the impact on the environment (quantified by the grades from 1 to 5)

The households to be submerged or endangered are those households which have to be completely flooded as well as that number of households which are located for the time being in Cvilin Polje, and in the variant wherein the retention water level is at 383,00 for instance, the households will live in a changed way, qualitatively, for it is situated beyond the embankment which retains water level above their roofs.

The most positive impact on the environment is represented by the objective mark 5, while the most negative impact is represented by the mark 1. The points from 1 to 5 are the result of the analysis of the hydroelectric power plants construction impact on the environment, Chapter X(4).

Criteria functions: f_1, f_3, f_4 and f_8 are maximized while with the others only minimum is required.

6. MULTICRITERIA COMPROMISE RANKING

Multicriteria compromise ranking was made using the program package VIKOR (3).

The group of six aforementioned alternative solutions: A, B ,C, D, E and F was ranked, applying different weights of criteria ranging from f_1 to f_8.

Within the Table 1 there are the values of criteria functions. GO mark was entered in addition to the data which had been obtained together with Gorazde expropriation costs and FO with Foca expropriation costs.

TABLE 1.

CRIT		EKST	A1.A	A2.B	A3.C	A4.D	A5.E	A6.F
f_1		max	4184,3	5211,9	5021,3	5566,1	5060,5	4317,9
f_2	GO	min	2733,2	3322,7	3898,7	3762,8	3214,9	2904,3
	FO	min	2914,0	3630,0	3920,5	3957,9	3293,5	2925,9
f_3		max	407,2	501,7	504,0	559,5	514,1	432,8
f_4		max	251	308,3	278,6	335,3	284,2	239,3
f_5		min	195	282	12	167	69	12
f_6		min	244	346	56	268	90	55
f_7		min	15	21	3	16	7	3
f_8		max	2,41	1,41	4,42	3,36	4,04	4,36

Four rankings were made through the data from Table 1, for four groups of criterion weight values, such as instance:

1. RANKING - ID   - identical weights were assigned to all criteria functions
2. RANKING - DIF   - criteria f_2, f_5, f_6, f_7 and f_8 gained the importance in relation to the other at scale 3:1
3. RANKING - SOC   - the group of social, urban, human and ecological criteria (f_4 to f_8) gained greater importance at scale 3:1
4. RANKING - ECON - the group of economic and electricity supply criteria (f_2 to f_4) gained the importance at scale 2:1.

Ranking DIF provided with Gorazde acquisition costs was selected for interpreting the results of the program package VIKOR.

The intention was to select the solution which due to present economic situation in the society required rather small investments (f_2 criterion) and at the same time did not disturb the environment from ecological, perspective urban and social aspect (f_5, f_5, f_7 and f_8 criteria). Therefore, three times greater weights were assigned to these criteria, and thus they where on the list as regards the item No.2 in the first horizontal line RANKING-DIF the values of the criterion W (I) weights are noticeable (normalized, sum = 1) : 0,056; 0,167; 0,056; 0,056; 0,167; 0,167; 0,167; 0,167.

Compromise solution for making a final decision is in this case a group of compromise solutions of alternatives:

A6-F (Sadba 362, Ustikolina 373, Paunci 384) and
A5-E (Gorazde II 362, Ustikolina 373, Paunci 384).

For the given weights of criteria, the advantage of F Variant was printed in relation to E variant. It amounts to 11,2%. Till the advantage is less than 20,0%, the first position is not considered convincing, so E Variant is mentioned at the second position. C Variant is at the third position (it is evident from the Table 2) but it is not proposed as a possible solution for the given weights of the criteria, because the advantage of the second position (E) is in relation to the third one (C) 41,4%.

The results of VIKOR

Rank order according to the QR, Q and Qs measurements.

QR   - min./max strategy

Q   - compromise list

Qs   - strategy of the majority of criteria

R.L.QR	R.L.Q. and	Q(J)	R.L.QS
A6	A6	0.000	A6
A5	A5	0.112	A3
A1	A3	0.526	A5
A4	A1	0.560	A4
A3	A4	0.683	A1
A2	A2	1.000	A2

Compromise solution for making a final decision is:

A group of compromise solutions:

Alternative	Advantage
A 6.F - Sadba 362, Ustikolina 373, Paunci 384	11,2%
A 5.E - Gorazde 362, Ustikolina 373, Paunci 384	41,4%

This is a solution for the given weights of criteria.

The results of multicriteria compromise ranking are presented in the forthcoming table 2 as regards aforementioned combinations of criteria functions weights and FO-Foca and GO-Gorazde acquisition costs.

TABLE 2.

	1. RANKING ID		2. RANKING DIF		3. RANKING SOC		4. RANKING ECON	
	GO	FO	GO	FO	GO	FO	GO	FO
	E	E	F	F	E	E	E	E
	C	C	E	E	C	C	B	C
COMPROMISE	D	F	C	C	F	F	D	B
RANK ORDER	F	D	A	A	D	D	C	D
	A	A	D	D	A	A	F	F
	B	B	B	B	B	B	A	A

The program package VIKOR proposes a compromise solution on the basis of the obtained rank orders for the given values of criteria weights. The obtained compromise solutions are presented in the Table 3.

TABLE 3.

	ID		DIF		SOC		ECON	
	GO	FO	GO	FO	GO	FO	GO	FO
I	A5.E	A5.E	A6.F	A6.F	A5.E	A5.E	A5.E	A5.E
II			A5.E	A5.E	A3.C	A3.C		
III					A6.F	A6.F		

The compromise solutions from Table 3 have got a sufficiently stable position in relation to the change of weight of strategy of decision-making (V in the VIKOR method) and a sufficient advantage to be proposed for selection of final (optimal) solution.

7. SELECTION OF OPTIMAL SOLUTION

It is evident from the table provided with the results that as a compromise solution for making a final decision, the following variants are imposed:

179

1. E Variant (Gorazde 362, Ustikolina 373, Paunci 384), then
2. F Variant (Sadba 362, Ustikolina 373, Paunci 384) and then
3. C Variant (Gorazde 352, Sadba 362, Ustikolina 373, Paunci 384).

Analyzing these solutions one can reach the following conclusion: A decision maker is recommended to adopt the profiles Paunci 384 and Ustikolina 373, for these reservoirs arise in the compromise solutions. For the normal retention water level of 363,00 it is proposed to determine in the next design phase if the dam profile would be at Gorazde II or at Sadba. In case of Sadba option it should be necessary to examine if it would be economically viable to construct also the fourth step of Gorazde II at 352,00. It means a multicriteria optimal solution of hydroelectric power system for the Drina hydropotential exploitation at Foca-Gorazde reach.

8. CONCLUSIONS

To date environmental and legal aspects cannot be neglected when decision making. They change the present scenario of the hydropotential operation of the river. If considered on time, the optimal solution would be most often found in the series of cascades of small scale hydropower plants, sometimes with the same head and uniform equipment. This paper contains a mathematical model helping the Client in making the corresponding decision.

LITERATURE

1. S. Vukotic, - Selection of installed power of the hydroelectric power plants by application of multicriteria optimization, M.A. Examination Work, Faculty of Civil Engineering, Belgrade, 1990.

2. S. Opricovic, Multicriteria optimization, Naucna Knjiga, Belgrade, 1986.

3. S. Opricovic, The Program Package VIKOR for multicriteria compromise ranking Sym Op IS Kupari, 1990.

4. S. Vukotic, Realization of the Preliminary Design of the Drina river hydropotential exploitation at Foca-Gorazde reach, Energoprojekt Co. - Hydroengineering Ltd., Belgrade 1990.

First International Conference on Renewable Energy—Small Hydro
3 – 7 February 1997, Hyderabad, India

OPTIMISATION MODEL FOR DECENTRALISED POWER SYSTEM PLANNING

Hubert Hildebrand

Fichtner GmbH & Co. KG
Sarweystrasse 3, 70191 Stuttgart, Germany

1. SUMMARY

The availability of energy is considered to be the primary prerequisite for any economic development in the countries of the Third World. This applies especially in the rural areas, where the shortage of energy leads to an uncontrolled exploitation of the natural sources of energy. The lack of commercial energy sources not only hinders economic development, but also goes a long way towards destroying the ecological balance of nature and the basic living conditions.

The future development of the electricity supply system is looked upon as being the solution to the energy problems in the rural areas. The utilisation of hydropower as being a proven and environmentally acceptable technology is again becoming increasingly important. The structure of the settlements in the rural areas favours the application of decentralised supply systems with small power station units.

The economic demands made on the design of decentralised electricity supply systems call for the use of a planning instrument which is able to cope with the complex problems. Efficient and field-proven optimisation and simulation models have been developed to optimise the design and operation of supra-regional, thermic-oriented power supply systems. However, the structural pre-requisites of large supply networks differ greatly in their characteristics from decentralised, hydropower-oriented systems which supply electricity in the rural areas of the developing countries.

Basically, two significant structural prerequisites of large, supra-regional power station systems justify simplified assumptions adopted in the afore-mentioned models. Firstly, there is a closed system of power stations and load centres and secondly, the costs for the power generation components are very high, compared to those for the power transmission components. By way of contrast, electricity supply systems, in rural areas of developing countries, are characterised as:

- widely scattered load centres with a comparatively low demand,
- low load factors and short-term high peak loads on the part of the consumers,
- few existing power plants and transmission lines,
- high cost quota for the transmission elements.

A planning model has been developed, which takes into account the above-mentioned structures of the electricity supply in developing countries and avoids assumptions and simplifications often made with regard to large systems.

A verification of the optimisation model was carried out using the power system on the Caribbean island of Grenada, where the diesel based electricity supply system is to be extended by means of constructing small hydropower plants. During the course of the Hydropower Development Plan Grenada, the following basic questions were answered using the optimisation model:

- site selection and optimisation of design capacities of the hydropower plants,
- optimum extension of the transmission system,
- economic comparison of the extended power supply system with the initial state,
- sensitivity of the hydropower development in connection with variable input values such as the price of diesel, interest rate and service life.

With regard to the structure of decentralised hydropower systems, using the optimisation model described in this presentation, the following basic parameters for practical planning can be determined:

- determination of the design capacities of power plants (hydro and/or diesel) and selection of plant sites out of a given catalogue,
- determination of routes for the transmission lines and identification of isolated power supply centres,
- calculation of transmission losses,
- evaluation of measures to reduce transmission losses (e.g. by developing decentralised power stations, by increasing the transmission voltage),
- optimisation of the operation of the power station within the system,
- testing reliability of supply should there be a power station failure or a power cut in the system,
- evaluation of measures to ensure reliability of supply (e.g. by interconnecting power plants and load centres),
- reviewing the sensitivity of system design to changes in input values (e.g. fuel price, interest rate and operating costs).

The optimisation model therefore, represents a suitable instrument for planning small, decentralised and hydro power oriented electricity supply systems.

2. OPTIMIZATION OF MIXED HYDRO-THERMAL ELECTRICITY SUPPLY SYSTEMS

It is obvious that the optimization of a power supply system with several load centres and options for power stations is far more complex than the optimization of single hydropower stations as a component in a supra-regional network or an isolated power station/consumer unit. Power plants within a system can no longer be optimized individually, on the contrary, the most economical design capacity is influenced by the possibility of linking power plants and load centres. From the many possible alternatives the following conceivable solutions have to be compared according to their economic efficiency, e.g.:

- the construction of a few, large power stations,
- the construction of several, small power stations,
- the interconnection of power stations to form a power supply network,
- the construction of isolated sub-systems.

In general the planning of an energy supply system is a task with multiple objectives, which becomes more complex the more the hydropower contribution within the system increases. Due to the natural fluctuating power supply of hydropower oriented systems, the determination of the most economically utilizable hydropower potential plays a central role. In connection with the temporarily fluctuating power generation and power demand, predictions as to the reliability of the supply in the system are required. Moreover, the planning of the power stations and the distribution system has to be combined.

2.1   Description of the Model

In the following a planning model is described, which takes into account the afore-mentioned structural features of power supply systems in developing countries and avoids assumptions and simplifications with regard to large systems.

The use of mathematical methods in order to optimize a system pre-supposes that the real problem which has to be solved can be expressed in mathematical terms in such a way that the loss in reality with regard to the objective is negligible and within the limits of the existing data accuracy.

An iterative operation research algorithm based on the "Branch-and-Bound" principle was selected to solve the problem. The decision variables are represented by the design and operating capacities of power stations and by the capacities of the transmission lines. The constrains include the minimum and maximum capacities of the power stations and the routes of transmission lines to be implemented. Transmission losses, the required reliability of supply and the time dependent availability of hydropower are taken into consideration in the optimisation procedure. Within a given planning period, the model provides the cost optimum design for the power supply system.

Taking into consideration defined secondary conditions, the decision making model supplies the design for a power plant system with minimum investment and operating costs within a given planning period. The objective function of the optimization model can be expressed somewhat simplified as follows:

$$\min z = \sum_{k=1}^{K} (IKW_k + BKW_k) + \sum_{m=1}^{M} (ITL_m + BTL_m)$$

with
z = NPV of system costs, (NPV = net present value),
IKW_k = NPV of investment costs of power station k (diesel or hydropower),
BKW_k = NPV of operating costs of power station k,
ITL_m = NPV of investment costs of transmission line m,
BTL_m = NPV of operating costs of transmission line m.

The main secondary conditions consist of:
- power demand of the load centres,
- capacities of existing power stations and transmission lines,
- site options and maximum design capacities of new power stations,
- transmission losses of power lines,
- required reliability of supply within the system.

3. CASE STUDY "HYDROPOWER DEVELOPMENT PLAN GRENADA"

A verification of the optimisation model was carried out using the power system on the Caribbean island of Grenada, where the diesel based electricity supply system is to be extended by means of constructing small hydropower plants.

The power supply system which was used to verify the optimization model consists of the following elements:
- 6 load centres with a total electricity demand of 50,2 GWh/a,
- 1 existing diesel power plant (Queens Park DPS) with a capacity of 11,6 MW,
- 9 options for hydropower plants with capacities from 100 to 600 kW,
- 13 existing transmission lines,
- 1 option for an additional transmission line (TL 13 = connecting the Castaigne SHP to the St. Georges - Grenville transmission line)

During the course of the Hydropower Development Plan Grenada, the following basic questions were answered using the optimisation model:

♦ selection of sites and optimisation of design capacities of the hydropower plants,
♦ optimum extension of the transmission system,
♦ economic comparison of the extended power supply system with the initial state,
♦ sensitivity of the hydropower development in connection with variable input values such as the price of diesel, interest rate and service life.

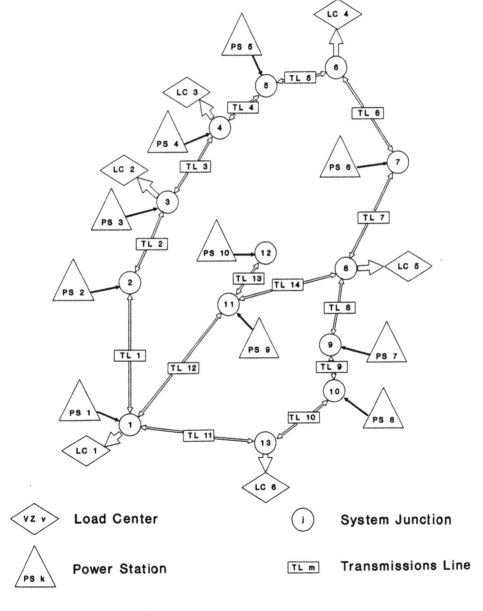

Figure 3.1:   Power Supply System in Grenada

185

The following data and information on the power supply system of Grenada refer back to 1991 when the **"Hydropower Development Plan Grenada Study"** was carried out on behalf of the Government of the Federal Republic of Germany.

The electricity supply on the island is in the hands of the state owned company GRENLEC (Grenada Electricity Services Ltd.). The distribution network consists of an 11 kV circular power network which follows the coastal road all round the island. From the main power line 400 V and 220 V sub-feeders lead to the load centres. The electricity demand is met solely by the Queens Park Diesel Power Station in St. Georges, which has a design capacity of 11,6 MW. The peak demand during business hours and during the evening amounted to 7,0 - 7,5 MW, the base load being 4 MW (see load curves in Figure 3.2).

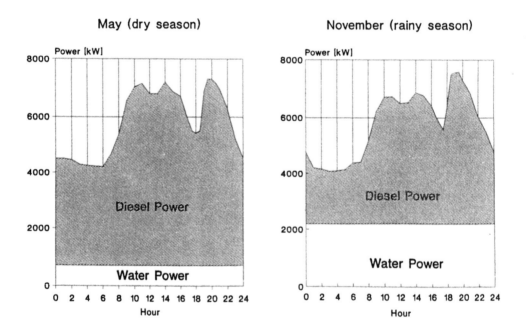

Figure 3.2:   Daily load curves in Grenada [*GRENLEC*]

The sole dependence on diesel oil as a power source induced GRENLEC to look for alternative sources for generating electricity. For this purpose, various studies were carried out to determine the hydropower potential. In total 9 suitable sites for hydropower stations could be identified on Grenada. The design capacity of the power stations amounts to between 100 and 600 kW.

3.1 Results of the System Optimization

System optimizations with differing presuppositions were carried out:

- extension of the existing power supply system by hydropower schemes;
- improvement of the system operation, e.g. by increasing transmission voltage and system load factor;
- sensitivity analysis, with regard to changes in fuel cost, interest rate and planning period;
- development of a new power supply system.

3.1.1 Extension of the Existing Power Supply System

In Table 3.1 the results of the system optimization are given, comparing the initial state and the extension of the system.

	Initial stage (centralized electricity supply): - Queens Park DPS (existent) - distribution (existent)	Extended power supply system (decentralized electricity supply): - Queens Park DPS (existent) - distribution (existent) - optional hydropower plants
Total cost of power supply system	83,9 Mio EC$	75,0 Mio EC$
N_A Queens Park DPS	8045 kW	7197 kW
ΣE Queens Park DPS	58,4 GWh/a	42,8 GWh/a
ΣN_A hydropower plants	/	2806 kW
ΣE hydropower plants	/	13,1 GWh/a
Developed hydropotential	/	85%
Load factor of the power plants	0,83	0,64
Σ Energy production	58,4 GWh/a	55,9 GWh/a
Energy demand	50,2 GWh/a	50,2 GWh/a
Transmission losses	8,2 GWh/a	5,7 GWh/a
Fuel price	0,15 EC$/kWh	
Interest rate	10%	
Planning period	30 years	

Table 3.1: Results of the system optimization with a 11 kV distribution network

A cost comparision between the existing diesel and the extended mixed hydro-thermal power supply system emphasizes the economic efficiency of the hydropower utilization on Grenada. Out of 9 hydropower options 8 plants are to be built. With reference to the total utilizable hydropower potential this corresponds to a utilization factor of 85%. By developing the hydropower plants the system costs can be reduced by 8,9 million EC$ or 10,6%. A large part of the savings arise when diesel oil is directly substituted by hydropower and further savings occur on account of reduced transmission losses. Within a central electricity supply system the transmission losses amount to 14,0% of the energy generated by the Queens Park Diesel Power Station. The energy losses could be reduced to 10,1% if the electricity generation was decentralized and included hydropower plants.

On account of the strictly centralized power supply, when considering the optimum utilization of the transport system the criterion of the transmission losses is the most important. As a result the lines with the least losses are selected to distribute electricity. Should an optimum transmission line fail, inevitably less favourable transmission routes have to be used and greater transmission losses occur. In order to guarantee the reliability of supply within the system, reserve capacities with regard to the dimensioning of the power plant have to be considered accordingly.

	Complete transmission system	Interruption of transmission line No.		
		TL 1 St.Georges ↔ Gouyave	TL 11 St.Georges ↔ St.David	TL 12 St.Georges ↔ Grenville
N_A Queens Park DPS	8.045 kW	8.368 kW	8.242 kW	8.203 kW
ΣE Queens Park DPS	58,4 GWh/a	60,7 GWh/a	59,8 GWh/a	59,5 GWh/a
Energy demand	50,2 GWh/a	50,2 GWh/a	50,2 GWh/a	50,2 GWh/a
Transmission losses	8,2 GWh/a	10,5 GWh/a	9,6 GWh/a	9,3 GWh/a

Table 3.2:   Increase in the required power station capacity in case of interruption of transmission lines

If one of the transmission lines from the Queens Park DPS fails, the transmission losses rise by 13% to 28%. Such a defect has the most serious consequences if the line TL 1 (St. Georges - Gouyave) fails. In this case, in order to maintain the electricity supply to the villages Gouyave and Victoria, the Queens Park DPS has to operate with a 323 kW higher capacity than an optimally utilizable supply network (see Table 3.2).

On the other hand, in a decentralized supply network, the transmission losses are of less consequence than in a central supply network. When all the hydropower plants have been constructed the total transport system apart from line TL 10 will be used (see Figure 3.3).

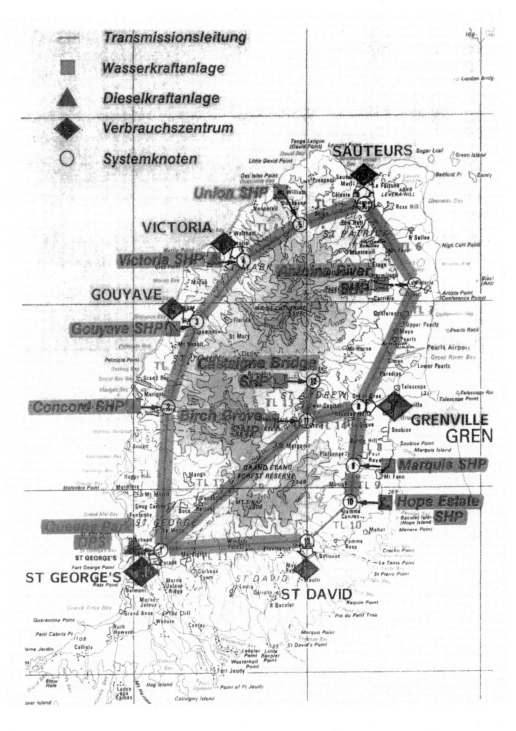

Figure 3.3: Optimization of the system including hydropower options
(Required system elements are marked red)

189

3.1.2 Improving the Economic Efficiency of the Power Station System

Increasing the transmission voltage:

The economic efficiency of the electricity supply system on Grenada is to be substantially improved by the substitution of diesel oil. For this purpose the Grenada Electricity Services Ltd. plans to increase the transmission current of the main supply network from 11 kV to 20 kV, as well as utilizing the hydropower resources. The urgency of this project is confirmed by the results of the system optimization in Section 6.3.1. Due to the relatively low transmission voltage of 11 kV, a considerable amount of the electricity produced is lost during transmission. Admittedly by developing the hydropower plants the transmission losses will drop from 14,0% to 10,1%, but, considering the limited hydropower potential and a rising electricity demand, a significant reduction in transmission losses can only be brought about by further developing the decentralized electricity supply by means of diesel generators or by increasing the transmission voltage.

	Initial stage (centralized electricity supply): - Queens Park DPS (existent) - distribution (existent)	Extended power supply system (decentralized electricity supply): - Queens Park DPS (existent) - distribution (existent) - optional hydropower plants (see Table 6.1)
Total cost of power supply system	75,6 Mio EC$	69,1 Mio EC$
N_A Queens Park DPS	7243 kW	6491 kW
ΣE Queens Park DPS	52,6 GWh/a	38,8 GWh/a
ΣN_A hydropower plants	/	2672 kW
ΣE hydropower plants	/	13,0 GWh/a
Developed hydropotential	/	81%
Load factor of the power plants	0,83	0,65
Σ Energy production	52,6 GWh/a	51,8 GWh/a
Energy demand	50,2 GWh/a	50,2 GWh/a
Transmission losses	2,4 GWh/a	1,6 GWh/a
Fuel price	0,15 EC$/kWh	
Interest rate	10%	
Planning period	30 years	

Table 3.3: Results of the System Optimization with a 20 kV Distribution Network

The result of the system optimization with 20 kV transmission voltage shows a significant reduction in transmission losses with a central electricity supply from 8,2 GWh/a down to 2,4 GWh/a. In this way the system costs drop by 8,4 million EC$ to 75,6 million EC$. The additional development of hydropower plants leads to a further improvement in the economic efficiency of the power supply system.

3.1.3   Sensitivity Analysis

The economical analysis for power stations and power supply system contains uncertainties, due to the fact that in the calculation the input data itself is affected by uncertainties to a greater or lesser extent. Most of the input data is projected far into the future and can with regard to its development during the course of time only be estimated (e.g. operating costs, interest rate, service life).

In view of these problems it is recommendable to check the economic consequences, which result from changing the input data for the investment calculations, with the help of a sensitivity analysis. It must be made clear that the sensitivity analysis does not eliminate the uncertainties when estimating the investment costs. Though, input data whose divergencies from the expected value have a considerable affect on the economy of the project can be identified with the help of the sensitivity analysis. If possible, by means of exact data collection and development controls, uncertainties inherent in the input data, which would considerably influence the success of the project, should be minimised.

The influence of the diesel price, interest rates and service life on the optimum design of a power station system can be seen in Table 3.4.

System elements with high investment costs, such as hydropower plants, are very sensitive to changes in the interest rate. If the interest rate rises from $i = 5\%$ to $i = 15\%$, the economically efficient share of realizable hydropower drops from 90% to 47%.

On the other hand, the rise in the price of diesel leads in consequence to an increased use of hydropower. From this example it is clear however that the present economically developable hydropower resources on Grenada are already approaching the technically utilizable potential. In view of the present energy production costs in a diesel power plant of 0,15 EC$/kWh, 83% of the hydropower potential is already utilizable. By a rise in the energy costs up to 0,25 EC$/kWh the hydropower share increases to 91%.

Changes within the planning period have little affect on the design capacities of diesel and hydropower plants.

Interest rate	5 %	10 %	15 %
Total costs of power supply system	114,7 Mio EC$	75,0 Mio EC$	55,5 Mio EC$
N_A Queens Park DPS	7168 kW	7190 kW	7427 kW
ΣE Queens Park DPS	42,1 GWh/a	42,9 GWh/a	47,8 GWh/a
ΣN_A hydropower plants	2950 kW	2727 kW	1552 kW
ΣE hydropower plants	13,6 GWh/a	13,0 GWh/a	9,0 GWh/a
Developed hydropotential	90%	83%	47%
Fuel price		0,15 EC$/kWh	
Planning period		30 years	

Fuel price	0,15 EC$/kWh	0,20 EC$/kWh	0,25 EC$/kWh
Total costs of power supply system	75,0 Mio EC$	95,2 Mio EC$	114,9 Mio EC$
N_A Queens Park DPS	7190 kW	7190 kW	7168 kW
ΣE Queens Park DPS	42,9 GWh/a	42,9 GWh/a	42,0 GWh/a
ΣN_A hydropower plants	2727 kW	2738 kW	2982 kW
ΣE hydropower plants	13,0 GWh/a	13,0 GWh/a	13,7 GWh/a
Developed hydropotential	83%	83%	91%
Interest rate		10 %	
Planning period		30 years	

Planning period	20 years	30 years	50 years
Total costs of power supply system	68,7 Mio EC$	75,0 Mio EC$	78,4 Mio EC$
N_A Queens Park DPS	7190 kW	7190 kW	7190 kW
ΣE Queens Park DPS	43,1 GWh/a	42,9 GWh/a	43,0 GWh/a
ΣN_A hydropower plants	2596 kW	2727 kW	2702 kW
ΣE hydropower plants	12,9 GWh/a	13,0 GWh/a	12,9 GWh/a
Developed hydropotential	79%	83%	82%
Fuel price		0,15 EC$/kWh	
Interest rate		10 %	

Table 3.4:   Sensitivity analysis for a system extension with hydropower plants

3.1.4 Layout of a New Power Supply System

The aim of this concluding investigation was the optimum design of a completely new power station system to be built on Grenada. The pre-requisites were the known load centres and hydropower options from the preceding system optimization. An 11 kV supply network was to be newly built, i.e., the transmission lines TL 1 up to TL 14 were counted as fixed costs and could be constructed if required. As an alternative to hydropower and for peak load operating, every village was to be provided with an optional diesel power station with a design capacity covering the demand.

As Figure 3.4 shows, the power station system which is to be newly built relies to a great extent on a decentralized electricity supply; the various load centres are not connected to each other by transmission lines. The villages Sauteurs and St. David are to be provided with electricity by thermal means alone. To cover the electricity demand in St. Georges, Gouyave, Victoria and Grenville the hydropower resources nearby are to be used to supplement the diesel power plant. Around Grenville a small supply network will develop to which four hydropower stations are connected. Two hydropower options, Union SHP and Antoine River SHP, are not taken into consideration when developing the system. The results of the system optimization with regard to the initial state, the system extension and the newly designed power supply system are summarized in Table 3.5.

	Initial stage	Extended power supply system	New power supply system
Total cost of power supply system	83,9 Mio EC$	74,7 Mio EC$	70,7 Mio EC$
ΣN_A diesel power plants	8045 kW	5811 kW	6309 kW
ΣE diesel power plants	58,4 GWh/a	42,9 GWh/a	38,9 GWh/a
ΣN_A hydropower plants	/	2727 kW	2343 kW
ΣE hydropower plants	/	13,0 GWh/a	11,9 GWh/a
Developed hydropotential	/	83 %	71 %
Load factor of the power plants	0,83	0,75	0,67
Σ Energy production	58,4 GWh/a	55,9 GWh/a	50,8 GWh/a
Energy demand	50,2 GWh/a	50,2 GWh/a	50,2 GWh/a
Transmission losses	8,2 GWh/a	5,7 GWh/a	0,6 GWh/a
Fuel price	0,15 EC$/kWh		
Interest rate	10 %		
Planning period	30 years		

Table 3.5: Optimization of a New Power Station System

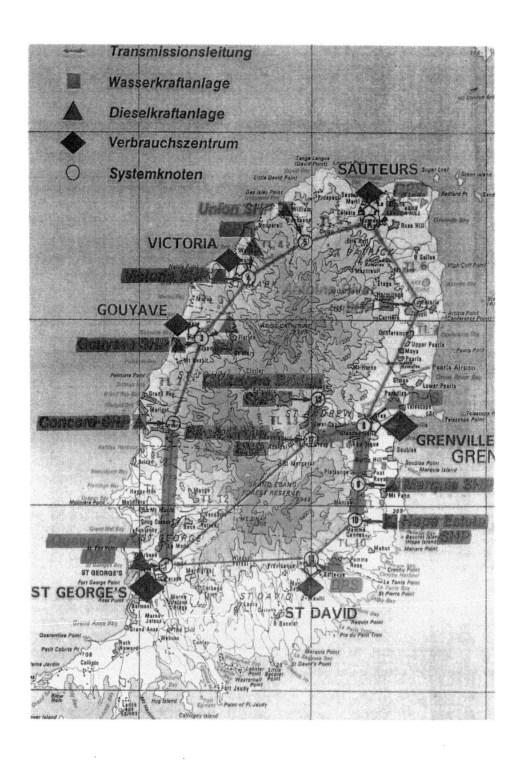

Figure 3.4:   Design of a new power supply system (Required system elements are marked red)

SMALL AND MIDDLE DAMS PARTICULARITIES, CATEGORIZATION, DESIGN FLOOD, SPILLWAYS AND BOTTOM OUTLETS

L.N. Vajda

Energoprojekt-Hidroinzenering, Consulting Engineers Co. Ltd.
11 070 Beogard, Bul. Lenjina 12, Yugoslavia

ABSTRACT

This paper considers characteristics of small and middle eartfhfill dams, as well as their categorization according to dam factor (DF), dam height, reservoir volume and corresponding design flood of certain probability. The failure peak flow (Q_{fpf}) in relationship to dam factor(DF) is determined as a basic physical parameter for each category of dams.

Necessity of using both the hydrologic and nonhydrologic criteria for the determination of the design flood and free board and particularities of small and middle dams are emphasised.

Important issues to be studied (SPS) are pointed out for all categories of dams. Possibilities of new and combined types of inexpensive spillways including one completed over an earthfill dam, as a case study, were indicated for limited conditions.

This issue of the bottom outlets was considered from various points of view: design, material, size, operation, maintenance, reconstruction and definition of dead storage capacity. The time of emptying of storages was emphasised with some limitations. Furthermore, the bottom outlet's role in sediment flushing and flood wave relief are described in this paper.

1. GENERAL

Small and middle dams are built to support water resources management. They have to satisfy requirements regarding stability and safety. Their damage and failure can cause significant effects upon goods and human lives, different economic, cultural, historical, communication and agricultural structures in relation to dam height and size of storage. Within the area affected by flood wave in case of failure of small and middle dams there are very serious consequences due to abrupt releasing of greater water amounts from the storage and natural flood which is overflowing the damaged and partially or completely collapsed dam. For this reason planning, designing, construction and utilisation of all dams as well as of dikes are regulated by law, technical standards and special rules.

Best of all would be if planning, designing, construction and utilisation of all types of dams are conferred to a specialised organisation for such type of structures.

Small dams with height less than or equal to 10 m, or 15 m for middle dams, are designed on the bases of experience and general principles presented in literature from this field.

Number of small and middle dams is steadily and rapidly increasing particularly in poor and undeveloped countries. The reasons for that are: small investments, short construction terms and short term for commencement of usage of created storage, as well as relatively minor influence on the environment of all the elements, except for the deposit of sediments which is specially pointed out.

Experience with respect to the behaviour of small and middle dams differentiates essentially from the experience of the performance of large dams. Failures and errors frequently occur by in their designing as well as when they are constructed and thus adverse events take place during their utilisation. They necessitate additional or remedial works for the purpose of increasing their safety and providing their certain upgrading. Because of that, behaviour, damages, and failures of small and middle dams are of interest for further improvement of their design, construction and utilisation.

2. PARTICULARITIES OF SMALL AND MIDDLE DAMS

Small and middle dams can control relatively small catchment areas. The occurrence of exceptionally heavy precipitation is an incident to a flood characterised by high specific discharges that range even to q=8 to 15 cms/km^2 for small catchment areas in our country.

Impounding of the reservoirs created by small and middle dams can be realised relatively quickly by flood waters. They appear as a rule after long dry periods within which cracks occur in the earthfill body close to its very surface. Under such conditions it is difficult to provide adjustment of body and foundation of dam to water impact and carry out a controlled sealing of cracks. To protect the body and foundation of a dam it is necessary to apply appropriate measures for the control of seepage.

Some factors which are disadvantageous for small and middle earthfill dams point out to the necessity of their studding. They are usually neglected in the course of design and construction and refer to the conditions of construction and foundation and even to the conditions of their utilisation. Adjusting of the basic elements of small dams to the construction equipment represent requirements which have to be taken into account as to ensure proper dam construction by rational slopes of dam body and foundations.

The basic dimensions of dam are: the earthfill dam crest width (not less than 5.0 m), safety freeboard size above the maximum level in the reservoir, inclinations of the external slopes, the width of possibly foreseen shoulders at the downstream slope (not less than 2.0 m), as well as the depth of dam foundation, taking into account geological and hydrogeological conditions in rocky and nonrocky material.

Safety freeboard is adopted on the basis of the standard risk referring to design flood, and the conditions of design flood, wave in the reservoir formed by wind in local conditions, as well as to the possibility of cracks created within the dam body. The cracks in small and middle earthfill dam body can reach the depth of few meters and can cause disadvantageous degradation consequences in the most sensitive parts of the dam, i.e. the vicinity of dam crest and its slope.

Small and middle dams are usually constructed using three basic materials at least (clay, sand, gravel) that can be found in borrow areas in close vicinity of the dam. These borrow areas can be in the reservoir or at the appurtenant structures excavations (spillway, chute, stilling pool, bottom outlet) or at dam foundation.

Clay material for impermeable dam body part has to be of suitable grain-size distribution which is compacted, and is characterized by permeability less than 10^{-5} cm/sec. It is understood that the presence of unfavourable materials (soluble salts) should be less than 2%. Dispersion clays are particularly unfavourable and should be avoided whenever possible.

Semipermeable and permeable materials as sand, gravel and their mixes with clay and silt can be used for filling of body of small and middle dams without special requirements. The only requirement is connected with the rational fitting procedure when minimum energy is used for compacting of earth materials. The materials with uniform and very nonuniform grain-size distribution are relatively unfavourable because of danger of segregation, creation of nonhomogeneous embankment, nonuniform settlement and formation of cracks within dam body.

Special attention shall be paid to foundations due to small dead load which does not change their compacted condition. Impounding of the reservoir has got a substantially greater influence with them, particularly if the foundations are made from macroporous materials, within fissured zone subject to erosion and at medium with greater permeability. It is assumed that materials with small consistency index or of unstable structure have to be removed from the foundation since they can cause serious deformation of foundation and dam body due to disadvantageous geotechnical properties in the course of construction and after impoundment of the reservoir.

Within unstable terrain's and at those which are liable to such a phenomenon, construction of small and middle dams should be avoided.

Reservoirs formed by small (V_t<0.50 hm^3) and middle (0.5≤V_t< 2.0 hm^3) dams are susceptible to seepage and water loss through foundations and dam site abutments. This fact points out to well-grounded experience of them from the aspect of hydrogeological properties, as well as the immediate environment conditions of them and the reservoir, provided that the corresponding sealing up measures have been undertaken.

Small (6.0 m<H_d<10.0 m) and middle (10.0 m≤H_d<15.0 m) dams that are susceptible at the contact with the appurtenant structures (spillway, chute, bottom outlet and the other) due to considerably unfavourable conditions for their connection.

The possibility of privilege paths at the contacts is often present at small and middle dams especially if the corresponding measures are not foreseen to be taken to prevent seepage. Complete drainage protection and prolonging seepage paths are necessary for the purpose of a full seepage control, as well as for the allowable exit gradients. Except for the said characteristics, the impact of external erosion on the slopes of small and middle dams is outstandingly disadvantageous. The following effects are particularly unfavourable: heavy showers, frost, mole-hills, settlement and usage of dam crest as a road at both sides, and at the downstream section there is an additional effect of waves in the zone of their influence that must not be neglected.

3. CATEGORIZATION OF DAMS

Our and world literature do not include more detailed dividing which would indicate the basic characteristics of small and middle dams. ICOLD provisions are governing only the large dams.

In water engineering practice, the dikes are earth structures intended for periodical protection of plane areas (agricultural, industry zones, settlements) from flood. Large dams represent structures which height amounts to H_d≥15.0 m, total storage capacity V_t≥1.0 hm^3, and standard flood discharge (SFD) may be Q≥200 cms. They are classified as large dams although their height can be H_d≥10.0 m on the condition that the dam crest length is B≥500 m, and the total reservoir volume is V_t≥1.0 hm^3.

The dam factor (DF) for large dams can be calculated on the basic of the said criteria (Fig 2). It is not less than DF>12 (in 10^{6}m^4). It is formally so. However, when the competent flood discharge (CFD) is Q≥200 cms the catchment area cannot be less than A=80 km^2 and the corresponding average discharge is Q_{av}=0,70 cms (W_{av}=20 hm^3/year). It is evident that a live storage capacity (LSC) is indispensable for partial water regulation, provided that relatively LSC can be minimum V_{us}/W_{av}=0.10 corresponding to V_{us}=0.1x20.0=2.0 hm^3 for large dams. This means that the dam factor for all large dams has to be from DF≥24.0 (for dam height H_d=15.0 m) to DF>480 for dam height H_d=30 to 60 m, if the competent flood discharge is Q≥200 cms. This means that the SFD does not correspond to V_t and has to be enlarged to 2.0 hm^3.The failure peak flow (FPF) that corresponds to dams height is in relation to DF. The magnitude of a possible failure peak flow (Q_{fpf}) is provided with a relatively modest number of data, specially as regards the height of dams up to 15.0 m. Therefore it shall be necessary to supplement these (Q_{fpf}) values with the new data in the forthcoming period Q_{fpf}=f(DF) and Q_{fpf}=f(H_d) and define them in a better way. On the occasion of determining the magnitudes of possible flood wave in case of dam failure (Q_{fpf}) referring to their categorization and thus their values have been given within the limits (Fig. 2).

In both our and world practice there is no clear definition and criterion for small and middle dams. Taking into consideration such circumstances and above mentioned definition of dikes and large dams we consider that small and middle dams should be of the following characteristics (Table 1):

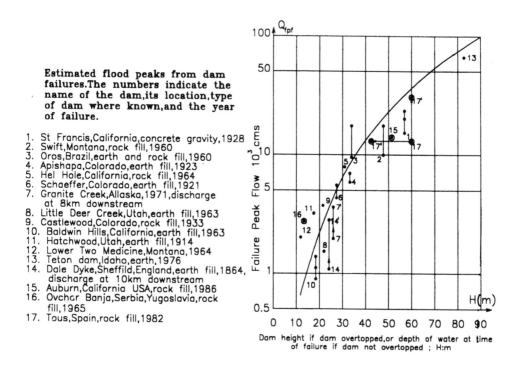

Estimated flood peaks from dam
failures.The numbers indicate the
name of the dam,its location,type
of dam where known,and the year
of failure.

1. St Francis,California,concrete gravity,1928
2. Swift,Montana,rock fill,1960
3. Oros,Brazil,earth and rock fill,1960
4. Apishapa,Colorado,earth fill,1923
5. Hel Hole,California,rock fill,1964
6. Schaeffer,Colorado,earth fill,1921
7. Granite Creek,Allaska,1971,discharge
   at 8km downstream
8. Little Deer Creek,Utah,earth fill,1963
9. Castlewood,Colorado,rock fill,1933
10. Baldwin Hills,California,earth fill,1963
11. Hatchwood,Utah,earth fill,1914
12. Lower Two Medicine,Montana,1964
13. Teton dam,Idaho,earth fill,1976
14. Dale Dyke,Sheffild,England,earth fill,1864,
    discharge at 10km downstream
15. Auburn,California USA,rock fill,1986
16. Ovchar Banja,Serbia,Yugoslavia,rock
    fill,1965
17. Tous,Spain,rock fill,1982

Dam height if dam overtopped,or depth of water at time
of failure if dam not overtopped ; H:m

Fig.1.Dam failure flood v.dam height (R.3)

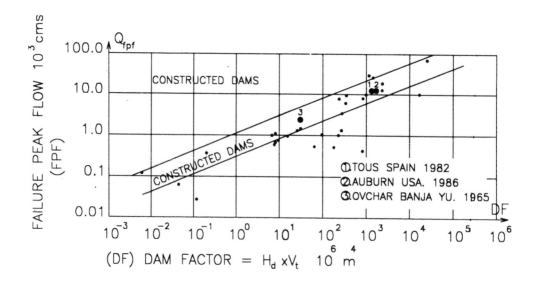

Fig.2.Dam failure flood v.DF (R.1)

The **small dams** are considered with $6.0 < H_d < 10.0$ m; $V_t < 0.50$ hm^3; $B < 300.0$ m; $DF < 4.0$ (10^6m^4); $100.0 \leq SFD < 150.0$ cms and $300.0 < Q_{fpf} < 1000.0$ cms.

The **middle dams** are considered with $10 \leq H_d < 15$ m; $0.50 \leq V_t < 2.0$ hm^3; $4.0 \leq DF < 24$ (10^6m^4); 300 $m \leq B < 500$ m; $1000 < Q_{fpf} < 3000$ cms and $150 \leq SFD < 200$ cms.

The **large dams** are considered with $H_d \geq 15.0$ m; $V_t \geq 2.0$ hm^3; B - no limitation; $24 \leq DF < 480$ i.e. $DF \geq 480$, $Q_{fpf} \geq 3,000.0$ i.e. $Q_{fpf} > 10,000.0$ cms.

The author decided to classify the dams (small, middle and large) using the dam factor (DF) because it represents a basic physical parameter. Dam factor is composed by two major elements: dam height (H_d) and total storage capacity (V_t).The values of peak flow (Q_{fpf}) determined by DF characterize approximately each category of earthfill dams (See table 1 and fig. 2).

It is obvious that the determination of failure peak flow is neither easy nor simple. The mentioned Q_{fpf} depends both on dam height H_d (Fig. 1) and DF (Fig. 2), but it was decided to use the relation to DF as more logic and more precise, but with upper and normal limits.

Dam height for DF estimation has to be reduced by a factor 0.8 due to the fact that in case of earthfill dams, its failure and forming of breach is neither instantaneous nor total and does not appear during the maximum water level in the reservoir (Auburn cofferdam, Cal. USA; Tous dam, SP etc.). Certainly, this is not a case when concrete dams are concerned (Vajont, I; Malpasset, FR; Gleno, I; St. Francis, USA; Vega de Tera, SP; etc.). As a rule, the failures of concrete dams are less frequent than of earthfill dams, but such a failure is instantaneous and mostly total. Also it is not usual that the failure of concrete dams is caused by its overtopping (Vajont, I).

According to the proposed categorization which primarily refers to earthill dams, if any of conditional elements is exceeded or human lives are endangered, the considered dam passes to the higher category with a corresponding design flood. For all kinds of concrete dams (except buttress) applied category has to be one degree less in comparison to earthfill dams.

All dividing in dams categorization with subcategorises, basic data and corresponding design floods are presented in table 1 as a useful and practical proposal.

Design flood and freeboard which directly depend on the dam category, numerous of non-hydrological and hydrological conditions are considered in the following chapter.

4. DESIGN FLOOD AND FREEBOARD

Design flood is one of the basic elements in dam design practice from the hydrological point of view. Design flood depends on the category and type of dam. Dikes are designed and built only in order to pass safely the design flood.

In former hydrotechnical experience design floods were defined for two boundary cases i.e. for dikes and large dams. According to the introduction of small and middle dams criteria, the corresponding design floods which are between the design floods for dikes and large dams are proposed. It is also proposed the confidence interval for the design flood in the potential earthquake areas with the value of MCS8°. The reason for such proposal is due to the high sensitivity of all types of dams (especially small and middle dams) to the earthquake impact.

Design floods determined according to DF and criteria known in hydrotechnical practice for dikes small, middle and large dams are presented in table 1.

Examples of earthfill dam failures have proved that the danger of their failure are significantly increased in two cases: during the floods and due to earthquakes. Meanwhile, the dams failure can be prevented in these two cases from dam operation service if the function of hydromechanical equipment and the water level drawdown are ensured on time. Also, the regularly observation of steep slopes and seepage, especially at the contact with appurtenant structures and with earthfill dam concrete elements as well as with its foundation,

CATEGORISATION FROM DIKE TO LARGE EARTH FILL DAMS

CATEGORY OF DIKE/DAMS	DAM FACTOR (DF) 10^6 m⁴ ①	HEIGHT OF DIKE/DAM (HD) m ②	TOTAL STORAGE CAPACITY (TSC) hm³ ③	DAM CREST LENGTH (DCL) m ④	STANDARD FLOOD DISHARGE (SFD) $Q_{o,f}$ cms ⑤	FAILURE PEAK FLOW (FPF) Q cms ⑥
DIKE	—	⩽6.0 à 7.0	—	—	—	—
	Dam/Dike Design flood ⑦ (optimisation of protection towns,agriculture and industry) (DDF) p=1% to 0.2%	SEISMICITY (S-MCS>8) ⑧ CONFIDENCE INTERVAL (CI) p%	FREE BOARD (FB) m ⑨ 0.5<FB<1.0	SPECIAL PROBLEMS STUDED (SPS) ⑩ -Sedimentation;-Flood wave duration -Wind and waves;-Filtration		
① **SMALL**	(DF) 10^6 m⁴ ① ①.1 DF<2.4 ①.2 2.4<DF<4.0	(HD) m ② 8.0<H_d<10.0	(TSC) hm³ ③ 0.2<Vt<0.50	(DCL) m ④ <300.0	(SFD) Q cms ⑤ 100<Q<150	(FPF) Q cms ⑥ 300<FPF<1000
	DDF p% ⑦ ①.1 – 0.1% ①.2 – 0.05%	(S-MCS>8) ⑧ (CI) p=95%	FB (m) ⑨ 1.0	(SPS) ⑩ -Lifes paradize; -Protection Down Stream section; -Drow Down velocity;-Sedimentation;-Wind & Waves;		
② **MIDDLE**	(DF) 10^6 m⁴ ① ②.1 4.0<DF<12 ②.2 12.0<DF<24	(HD) m ② 10.0<Hd<15.0	(TSC) hm³ ③ 0.5<Vt<1.0 1.0<Vt<2.0	(DCL) m ④ 300<B<500	(SFD) Q cms ⑤ 150<Q<200	(FPF) Q cms ⑥ 1000<FPF<3000
	DDF p% ⑦ ②.1 p=0.05% ②.2 p=0.02%	(S-MCS>8) ⑧ (CI)–p=95%	FB m ⑨ 1.0<FB<1.5	(SPS) ⑩ -Lifes paradize; -Protection Down stream section; -Drow Down velocity;-Slides in reservoir; -Sedimentation;-Wind & Waves;		
③ **LARGE**	(DF) 10^6 m⁴ ① ③.1 24<DF<480 ③.2 DF>480	(HD) m ② 15.0< H_d >30.0	(TSC) hm³ ③ Vt>2.0	(DCL) m ④ no limitation	(SFD) Q cms ⑤ Q>200.0	(FPF) Q cms ⑥ 3,000<DF<10,000 DF>10,000
	DDF p% ⑦ ③.1 – 0.01% ③.2 – MPF	(S-MCS>8) ⑧ (CI)–p=95%	FB m ⑨ For each case separately	(SPS) ⑩ -Lifes paradize;-Protection Down Stream section; -Flood volume probability;-Slides in reservoir; -Sedimentation;-Wind & Waves;-Drow Down velocity;		

REMARK:For all kind of concrete dam (except butress) category one degree less.
If human lifes are endangered for all categories one degree higher

is necessary. Both above mentioned facts are taken into account considering design flood for small and middle dams.

Small and middle earth fill dams are more sensitive than other types as far as freeboard is concerned (drying, cracking, frost, seepage, instantaneous flood occurrence, seismicity, smaller compactness of the dam body, drawdown, showers, communication over the dam crest, etc.). According to the hydroengineering practice, freeboard for small and middle earth dams is usually, but not less than 1-1.5 m, and has to be checked by corresponding wind and wave.

5. SOME NEW AND ECONOMICAL TYPES OF SPILLWAYS

It is evident that the type and construction of spillways depend on local, topographical, geological, hydrological, operational and construction conditions. Properties of the small and middle earthfill dams spillways should adapt to the first place to the dam body behaviour.

Dividing of spillways according to its location is possible. They may be defined in three main groups: a) spillways incorporated in the dam body; b) spillways out of dam body located on river banks; and c) spillways overflowing the earthfill dam crest.

Here below are stated five types of spillways applied with small and middle dams as possible economical hydroengineering solutions.

First type

The structure of this type of overflowing spillway depends on the type of the fill dam, and requires strengthening and flattening of the upstream and downstream dam slopes. The structure of the overflowing earthfill dam body depends on the environment conditions. Downstream part of the overflowing crest and dam body must properly be drained in order to meet winter conditions as well as to provide better stability and resistance against erosion.

For this type of spillway the dam height is limited from 10 to 15 m. It may be applied to fill dams, of all rockfill types with clay core or earthfill dams regardless whether being permanent and temporary structures. The overflowing dam slopes are flattened, and the overflowing surfaces have to be strengthened.

In practice, the strengthened downstream slope with stepped concrete slabs is one of the best alternatives. They may be designed for any discharge capacity.

Here below is stated, as an example, completed uncontrolled spillway over earthfill part (H_d=10.50 m) of the Tisza River Concrete dam with the following main data: overflowing velocity V≤2.0 m/s; specific discharge q=6.30 cms/m'; denivelation Δz≤8 cm; nappe z≈4.50 m and Q_{max}=465.0 cms. This additional spillway strengthened with concrete slabs and drainage has been in operation since 1977. It compensates the part of discharge over inundation (which was closed by excavated material), without any significant problem (R.4).

Second type

This type represents application of prestressed geotextile tubes for spillway protection of small earthfill dams (J.1). This technical solution hydraulically tested in laboratory is based on the geotextile tubes filled by sand or gravel, with adequate grain-size distribution and without loss of fine material. The hydraulic tests showed that if the fill material is compressed then the surfaces of the tubes become very resistant to hydrodynamic forces. The proposed spillway protection is applicable for all hydrostructures based on deformable foundation and as a protection of the earthfill downstream dam slopes during their controlled overflowing. At present the application of such type of spillway is limited to small and middle dams (H_d<15.0 m).

Third type

This type of spillway (HYDROPLUS) has been initially designed only for concrete dams in order to increase the useful storage capacity (J.2.). However according on our opinion it may be used together with the second type (prestressed geotextile tubes) for earthfill dams. Its application is possible for small and middle earthfill dams with limited height ($H_d<15.0$ m). This type of spillway enables increasing of the existing storage capacity with this type of fuse gates which are spilling by moderate floods, but with releasing singly for larger floods ($p\geq1\%$).

This type of spillway has the following features: metallic labyrinth crest shape, inlet wells admitting uplift pressure when the water reaches a designed water level, and standardised design to keep low costs. It is usually possible to recover approximately 70% of currently wasted storage capacity.

Fourth type

This type represents, in fact, the application, with a real reason, of well known rubber gates. It is not so frequently applied due to its possible exposure to damaging during overflowing, by floating matters and bed load sediments.

It may be applied with small and middle fill dams ($H_d<15.0$ m), since it does not require any power and lifting devices. Also, it may be applied together with the second type of spillway.

Fifth type

This type represents an emergency spillway. It is a dike adapted for controlled breach to meet the extremely large flood (or the design flood). It enables safe using of total live storage capacity.

The choice of appropriate material for its construction and its placement as well as results of laboratory breach procedure for this type of spillway, are tested in hydraulic laboratory, taking into account routing through the reservoir. Practical recommendation are elaborated in the PH dissertation (R.5).

6. BOTTOM OUTLET STRUCTURES

6.1. General

The purpose of bottom outlets as regards the dams is wellknown. Their role essentially in small and middle dams does not differentiate in relation to large dams. It can be expanded in the form of supply pipe extending to the pumping station or small hydroelectric power plant. In critical conditions it can be used for partial releasing or for preliminary emptying of the storage. The bottom outlets can be used as an additional capacity during floods as well as for the needs of meeting minimum biologic discharge.

In case of flushing deposit of sediments, bottom outlets can be of certain use. Their role is vital during the total emptying in frequent cases when these storages are predicted for fishery. The types of bottom outlets depend on the local and construction conditions as well as on river diversion during the dam construction period. Their layout together with the other appurtenant structures can be within or outside the dam body. However selection of structure are available and could be applied offers possibilities of utilisation of various materials.

6.2. Design Requirements

As regards the bottom outlets, the following data have not been defined:

a) **criteria and conditions for defining their diameter** from the standpoint of their maintenance, inspection, repairs, replacement and restrictions in respect to maximum realised discharge of the sector downstream of the dam. Limited size of diameter in relation to backwater level in the reservoir and Q_{max} which must not exceed artificial flood with p=20%-10%. The second condition as regards diameter refers to the possibility of maintenance, and possible replacement. If we want to fulfil this last two conditions, min. diameter of bottom outlet should be not less than 1000 mm

b) **selection of the material for their completion** is of essential importance. Steel or iron pipes should be avoided because of corrosion with time regardless of anti-corrosion or cathodic protection. If, for any reason steel or iron must be applied, outside reinforced concrete and cathod protection us well as inside anti-corrosion protection are necessary to be checked after each use of bottom outlet. Therefore, application of reinforced concrete is recommended for the bottom outlets due to its possible construction at the dam site, and provided with a casing in the gate or valve zone. In all cases, it is necessary to extend the seepage paths along the bottom outlet, at its contact with earthfill dam body.

Special attention should be paid to the grain size distribution of deposit sediments that must be flushed from the storage with maximum velocity if it serves for partial releasing of floods or during its total emptying. The attention should also be paid to its settlement along the whole length of the bottom outlet.

c) **elevation of the bottom outlet** must meet a few conditions: <u>first,</u> to provide river diversion during dam construction; <u>second</u>, to enable efficient flushing of sediments from the storage; <u>third,</u> to provide complete emptying of the storage either for the needs of fishing or inspection of dam body and bottom outlet; <u>fourth,</u> the upstream end of the bottom outlet must be beyond the zone of possible landslide and sedimentation silting for 100 years period; <u>fifth,</u> the top of joint-use reservoir's capacity if it is assigned to flood control; <u>sixth,</u> the top of the dead reservoir capacity when the reservoir has not got a joint use; <u>seventh,</u> a bypass outlet works for small HPP and PS may be used if they are isolated from the turbine(s) or pump(s) by gates or valves.

d) **location, selection of the type and the size of bottom outlet gate or valve** depends on several conditions (grain-size distribution of sediments, max. velocity, cavitation etc.) especially with earthfill dams. As a rule its location should be at the upstream section. In this way, the bottom outlet is not subject to constant pressure that is exceptionally disadvantageous for earthfill dams. Technical solution which implies upstream location of the gate is more expensive than downstream location of it, but condition of operation, maintenance and possible replacement of bottom outlet comparing to the location of the gate at the downstream section are more favourable. Certainly, technical solution with the gates at both ends of the bottom outlet is possible, on condition that the upstream one is an auxiliary or emergency one. The solution with two gates is technically possible and conforms to the conditions of bottom outlets utilisation as a water supply, pipeline leading to the small hydroelectric power plant or as a water supply pipeline provided with a branch pipe extending to downstream pumping station.

e) **determining of the emergency emptying time** of the small ($V_t<0.50$ hm^3) and middle ($0.50V_t<2.0$ hm^3) dams is limited by our analyses and proposals from to 2.0 to 10.0 days. Velocity of drawdown reservoir levels can be adopted between 2.50 to 4.00 m/day but this value must be checked by stability upstream slope of the dam due to porous pressures. At the same time the max. discharge released from reservoir through the bottom outlet has to be examined analysing the hazard and risk assessments for downstream river section.

f) **the possibility of emptying the reservoir** in emergency conditions or for inspection, maintenance and repair of the dam and appurtenant structures that are normally submerged. Releasing the part of flow during the floods has to be used. We point out to four limitations and recommendations concerning above mentioned activities: <u>first,</u> the river section downstream of the dam must not be endangered from any reason; <u>second,</u> if possible, at the same time to make the flushing of settled sediments out of storage; <u>third,</u> the turbidity of flushing current saturated with sediment is not allowed to exceed max. natural values; and <u>fourth,</u> at the end of the bottom outlet adequate stilling basin has to be built for Q_{max} (p=20% to 10%) which can be released from the reservoir.

g) **the releases which are to satisfy downstream rights** as min. discharge during dry periods shall be regulated by the gate of the bottom outlet. Criteria for this Q_{min} means that it is equal to $0.10xQ_{av}$, to probability of Q_{av} (p≤90%) and to yearly min. Q_{min} (p=75%).

Operational requirements are dictated by irrigation and water supply demands, for recreation, fishery and for water quality improvement.

h) **dead storage** as a part of the reservoir must be reserved for the predicted 100 years sediment inflow. For the time being in our country it is required that the ensured dead storage capacity be provided for 50 years sediment inflow, which is not sufficient.

7. CONCLUSIONS

In addition to small dams, at our proposal, the category of middle dams should be introduced. They are classified into the group of sensitive and specific structures as regards to series of parameters. Stability, safety in operation and economy are required to be reached by them. They should serve the initial development of agriculture, water supply and energy supply first of all of poor and economically underdeveloped countries.

Many aspects of small and middle dams which are of interest of above mentioned countries as well as of the countries which are recovering after abolition of sanctions, have been studied to a great extent and solved.

However, regardless of the limited heights of small dams (up to 10.0 m) and of middle dams (up to 15 m), sizes of total storage capacity ($V_t<1.00$; i.e. to our proposal 2.0 hm^3) dam crest length (B=300.0-500.0 m), the SFD. (Q<100, 150 i.e. 200 cms) they are charakterized with corresponding failure peak flows. They can cause significant damages on the occasion of failure and endanger human lives at the downstream dam section. Therefore, besides the hydrological, nonhydrological and seismic conditions, the dam factor (DF) is the basic parameter for the appropriate categorization of small, middle and large dams. Categorization of dams depends also on the HD, TSC, SFD and possible failure peak flow (Q_{fpf}) that would appear.

Particularities of small and middle earthfill dams are numbered and explained in the paper.

In this paper are used the **criteria for categorization** and interducing of small and middle earthfill dams between dikes and large dams (by ICOLD). Only three modifications are suggested: for total volume of large dam reservoirs (V_t2.0 hm^3) to harmonise SFD (Q$\geq$200.0 cms), with omission of all dams with H<15 m as large dams and using the DF>480 value as a criterion of MPF application for large dams, and for SFD to be determined with p=0.1%.

Design flood (DDF) is given within the limits since all basic sizes of small, middle and large dams are given within the limits. If any of the sizes would be exceeded, the design flood shall be adopted from the following higher category. Categorization includes also the **confidence interval** (CI) of the design flood if earthquake is MCS>8°.

The paper gives presentation some new, economical and combined **types of spillways** for small and middle dams. Some of new types of overflowing spillways including one completed many years before over the earthfill dams are described.

As regards **freeboard** values for small and middle dams which are accustomed the same must be separately checked for each case.

Bottom outlets were considered from the various points of view, as well as of emptying the total volume of reservoir in case of emergency and for any other purposes.

The sizing of **dead storage** is also emphasised in relation to the life time of dams, which must be predicted for 100 years sediment inflow.

REFERENCES

I Manuals

M.1   Manual for Hydraulic Engineering, Moscow 1955, USSR

M.2   Design of Small Dams, Denver, Colorado 1987, USA

II Journals

J.1   Boreli M., Tanackovic V.: Application of prestressed geotextile tubes for Spillways protection of small earthfill dams. WDI J. Cerni 1985, Communications 19, Belgrade, Yugoslavia

J.2   HYDROPLUS, GTM, Nanterre Codex, France, First presentation on XVI Congress on Large Dams, 1991, Vienna, Austria

III Reports

R.1   N.Y.-ASCE. 1986 Schuster R. and J.E. Costa, Geotechnical Special Publication No.3 - USA

R.2   Vajda L.N., Energoprojekt - Hidroinzenjering, Belgrade, Serbia - Yugoslavia: Design flood of Large Dams - Non Hydrological Aspects. II Intern. Symposium on Design of Hydraulic Structures. Fort Colins 1989, USA

R.3   Kirckpatrik G.W.:Guidelines for Evaluation Spillway Capacity. Intern WATER POWER and DAM CONSTRUCTION No.8 1977.

R.4   Vajda L.N., Energoprojekt - Hidroinzenjering, Belgrade, Serbia - Yugoslavia: Part of Free Spillway over Earth fill Tisza River Dam. ICOLD XVI, Contributions, Vienna - Austria, 1991.

R.5   Fabian Djula: Emergency Spillway - dike adapted for controlled breach to meet extremely large floods. PHD dissertation, Civil engineering faculty, Subotica 1992, Yugoslavia

SOME ASPECTS OF SELECTION OF SMALL HYDRO PROJECT SITES USING REMOTE SENSING AND GIS

C.J. Jagadeesha and S. Adiga

Regional Remote Sensing Service Centre
Indian Space Research Organisation
Banashankri, Bangalore 560 070, India

ABSTRACT

The evaluation of potential dam sites is usually performed in three phases. During the first phase all the potential sites and their comparable characteristics are determined. The second phase comprises of evaluation of these potential sites for prioritization. In the third phase the detailed design is taken up for the selected ones and these are taken up for development. Thus during the first two phases satellite data and the Geographic Information System (GIS) can be of immense value. The map scales for the entire area under construction should be 1:250,000 for Phase I study and 1:25,000 to 1:50,000 for Phase II study. The third phase has very large map scale requirements wherein aerial remote sensing and field surveys are needed. Using the satellite data and collateral data from SOI toposheets the geologic maps of surficial deposits, slope map, slope stability map, lithology map could be prepared. These are used as vector/raster layers in a GIS environment for selection of suitable hydroelectric sites. The information on construction materials, foundation conditions, internal drainage conditions, weak planes and mass-wasting process which facilitate good planning and safer/cheaper construction could also be derived using satellite and ancillary data. Geologic maps on the topographic base, together with accompanying hydrologic information, provide just such knowledge, which are basic to any rational understanding of the ground conditions that the planner must have.

Digital Terrain Models (DTM) may be created by either digitising contours and spot heights from existing topographic maps, collecting elevations with field survey or as a product of photogrammetric stereo-compilation. The spatial frequency of the sample elevations and the precision of the data are two other factors that must be considered with respect to DTM. The ability of IRS-1C to "point" on-board sensor upon command at anywhere upto ±26 degrees off-nadir provides the possibility for acquiring high resolution (5.8m) stereoscopic data in panchromatic mode over large geographic areas (swath of 70km) and revisit frequency of five days. Thus the DTM which can be generated using either topomaps or satellite data can be used to get computer derived slope, exposition and height or relief for each pixel. Further there is the methodology of DTM analysis in generating three types of data sets, i.e., original DEM with depressions filled, a data set indicating the flow direction for each cell, and a flow accumulation data set for each cell which define the topographic structure. Thus the digital representation of topographic surfaces after suitable accuracy assessments help in estimating peak discharge values in map units chosen for small hydroelectric sites. Hence suitable sites for hydroelectric power can be arrived in a reconnaissance way.

Seismotectonic analysis and predictive earthwork cost estimates can be carried out using satellite data or aero-space data at a targeting level in the selection of site for small-hydro. Knowledge based selection of geotechnically favourable attributes for hydel site can be carried out using GIS package.

The ERDAS and EASI/PACE image processing software, ARC/INFO GIS software and micro photogrammetry equipment with Geocomp software are available at the Regional Remote Sensing Service Centre (RRSSC) of ISRO in Bangalore. Geocomp Version-8 of DTM can handle data sets upto 32670-3D points and a modeling speed of over 30,000 points/minute. These can be put to use in arriving at such predictive models for peak flow, estimation of earthwork, seismicity and landslides near a small hydro site.

1.   INTRODUCTION

Small hydro-electric projects have become imperative as there is a great deal of opposition for large dam construction on grounds of environmental degradation and unsustainable development. The role of remote sensing in earth resources and terrain evaluation has been well documented and accepted. Remote sensing has proved to be a useful tool in the assessment of large areas in addition to locating and monitoring more regional and local phenomena. The synoptic view provided by the Indian Remote Sensing Satellite (IRS) imagery permit extraction of regional applied geoscientific information on land forms, drainage, weak planes in rock masses and landslides. Such information can be utilised for preliminary planning of civil engineering works. Before undertaking any study with satellite data several decisions have to be made, viz., which sensor to use? (for example best spatial resolution and large area coverage); which type of data is required? (whether digital or photoprints or transparencies); and what is the most suitable type of methodology (visual/digital interpretation, DTM/GIS use or hybrid interpretation, etc.). Automated methods of classification are available to eliminate the subjectivity associated with visual interpretation.

Every terrain must be evaluated thoroughly before taking up any construction work related to engineering projects. The information about terrain relief plays an important role and is one of the inputs to a geo-information system. This demand could be easily fullfilled by generating DTM either by topomap contours or satellite stereo-pair derived contours.

Using satellite data, different thematic maps, such as lithological, geomorphological, lineament density and forest cover density can be prepared by visual/digital interpretation. These thematic maps are to be scanned and loaded to GIS system for analysis. A qualitative selection of favourable attributes is to be done for each thematic map and then integrated using GIS software to get an area which fulfils conditions favourable for constructing a small hydro reservoir. An attempt is made to cover these aspects in this paper.

2.   PHASES OF SELECTION OF SMALL HYDRO-ELECTRIC SITES

Storage may be small or large, simple or complex, serving one purpose or several, the general approach for selection of a site is similar over the entire range of dams. One of the key elements of selection of small-hydro projects is the requirement to examine feasible and prudent alternatives to a given proposal [01]. If likely discharges based on Horton's law of morphometry are given, the head required to generate a certain intended horse power of a small-hydro can be arrived at using turbine efficiency equation. For 1MW hydro power generation in a stream carrying 50 cusecs approximately 80m height of water fall is required. For 100 cusecs only a fall of 40m height is enough. The digital elevation or digital terrain models generated using topomaps or stereopairs of satellite photos can be used to have various alternatives of potential sites ranging from 40m to 80m height. These alternatives can be compared with the morphometry and geology as occurring and delineable using satellite data.

During second phase while prioritising the features like lesser drainage density sites, higher discharge sites (corresponding to low height of water falls) at least a few parameters of environmental impacts (on landuse and sedimentation) can be studied using satellite remote sensing and geographic information system. The models like WARA/USLE which uses remote sensing and field inputs can be utilised for qualitative assessment of small-hydro sites [07].

During the third phase either aerial photographs or satellite data equivalents of aerial photos (stereopair data sets obtained from satellite data) can be used to delineate the outlines of landslide scars within short vicinity of the selected hydroelectric sites.

3.   INTERPRETATION FOR GEOLOGY

The quantity of geologic information that can be obtained from remote sensing data depends on terrain, climatic environment and stage of the geomorphic cycle. Sedimentary areas may yield more information on lithology and structure than those formed by igneous and metamorphic rocks. Intrusive igneous rocks tend to be homogeneous over large areas and hence have massive topography, lack bedding but frequently exhibit pronounced joint or fracture pattern. In case of metasediments the stratification is less pronounced while in case of metamorphosed intrusive the foliation trend give a diffused striped pattern on remote sensing data. Lineaments are numerous, short and parallel to one another, and never form long continuous ridges and valleys. Lineaments/fractures are nicely picked up on remotely sensed data products. These help in knowing structured relationships of different formations [09].

The geologic maps of surficial deposits like late glacial and post glacial deposits, outwash and associated deposits etc., which comprise most of the materials critical for development can be prepared. The bed rock units which are relatively uniform over wide areas can be mapped in a reconnaissance way. The bed rock units of mesozoic and tertiary age, pliestocene age and recent age are to be mapped using both satellite images and other geologic maps available for the region [04].

4.   INTERPRETATION AND USE OF SLOPE AND SLOPE STABILITY

The generalised slope maps using topomaps showing categories of slopes chosen to meet small hydro location needs can be considerably enhanced by field observation and use of aerial photographs (or satellite equivalents). Digital elevation models (DEMs) give percent slope maps in the region. For example the areas having greater than 45% slope and randomly distributed slopes are to be checked for knowing the limits of manouvering tracked vehicles.

The slope stability maps can be derived from DTM generated slope maps or toposheet generated slope maps. Landslides occur mainly on steep slopes and along sharp topographic discontinuities such as bluffs. They do not occur in areas of low relief that are far from breaks in topography. By combining the elements of slope and geology a slope stability map can be generated. For example, low to moderate slope areas have high stability, etc. The following types of slope stability zones can be identified: (a)the areas which are unsuited for most development because of slope stability problems that cannot feasibly be surmounted except in unusual circumstances, (b) those areas in which slope stability presents problems that must be considered in planning and design but for which a solution can be worked out, and (c) those areas which are relatively free of slope stability problems.

5.   INTERPRETATION OF SURFACE LITHOLOGY TO ASSESS CONSTRUCTION MATERIALS

A map of construction materials can be prepared from a geologic map. It is essentially a lithologic map, in which geologic units of various ages and origins that have similar lithologies, and therefore similar utility as construction materials, are grouped together in a single map unit. The satellite images are capable of identifying surface lithology which are to be interpreted along with geologic units for arriving at map units for unconsolidated materials, bed rock, interbedded gravel and sand, silt, till and swamp deposits, etc., on images. The intent of the map here is not to serve as a resource map in the sense of estimating gravel reserves or detailing the uses for which a particular deposit is most suited. The map does serve primarily as a guide so that the planner can know which areas are to be selected for further investigation and which have no construction materials and can be considered for other types of development. The tone and colour of land surface on images help identifying construction aggregates, i.e., as in piedmont fans and river terraces. The steeper slopes, rugged outlines and lack of land use in the area provide material for rip-rap or buildings [04].

6. INTERPRETATION FOR FOUNDATION STRENGTH, WEAK ZONES, LAND SLIDES, ETC., FOR ASSESSING SUITABILITY OF SITES

Foundation strength and stability are based on the structure of the unconsolidated material, porosity, permeability and strength. The hydrogeomorphology and lithology maps generated from satellite data, slope maps, field and laboratory criteria like pedological, geophysical, groundwater level data are to be used in assessing foundation condition and reclamation. For example, high porosity and permeability in flat to gently sloping terrain of piedmont fans and river terraces means that though these provide good foundation to buildings and power houses, the canals will have to be lined since unlined canals would cause loss of very huge quantity of water through infiltration. The bridges must provide sufficient free board for free movement of sediment underneath due to high sediment transport in the geological unit.

The zones of weakness in the earth's crust like major faults/folds/thrusts/joints and fractures can be interpreted on satellite imagery by tone, texture and change in topography or land form of the region. The dam and other appertinent buildings should never be placed on these weak zones since they may not only cause excessive leakage from the reservoir but endanger the stability of structure itself. For example, the information on joint spacings across the region of the imagery shows relative rock mass strength in the area. Close joint spacings are low strength areas than widely spaced ones. Landslides and slope instabilities are caused by adverse combination of water, soil and rocks in the area. The orientation and altitude of joints, fractures and faults also facilitate landslides in adverse conditions. The light tone and crescent shape on imagery point major landslide(s) area. The landslide areas and deforested areas in the thickly forested hill regions contribute silt to the reservoirs downstream.

7. DIGITAL TERRAIN MODELING (DTM)

Digital terrain model can be defined as a digital representation of the terrain relief (earth surface) suitable for computer processing. DTM may be defined aptly as the numerical/digital and mathematical representation of a terrain surface in terms of adequate planimetric and height measurements which are compatible in number (i.e., sample size) and distribution (sampling pattern) with terrain and which are suitable for computer processing. There are three techniques to generate DTM:

(a)   Field survey Techniques · The input information is extracted directly from the terrain. The technique is accurate but is labour intensive and expensive.

(b)   Cartographic Technique - The contour lines on the existing topographical map are digitzed manually or by automatic digitisation. Manual digitisation is laborious and time consuming. It is least expensive but may not be very accurate.

(c)   Photogrammetric Technique - This is the most commonly applied technique using aerial photographs although the accuracy of this data is restricted by the scale and quality of aerial photographs. IRS-1C satellite is capable of producing stereo pairs as the panchromatic camera (sensors) are steerable when satellite is in orbit.

The development of a Digital Terrain Model involves, (i) preparation of an adequate contour map in a format necessary for converting analog data to digital data, (ii) digitisation of the contour data, (iii) development of appropriate methodology for interpolation of height values, (vi) conversion of vector data to raster format to represent elevation, slope and exposition as gray scale images and (v) development of methodology for computation of shaded relief image after fixing the direction of sun illumination [10].

There are grid based terrain modeling and triangulation based terrain modeling based on type of interpolation of random spaced elevation data. With the formation of the DTM of 3D data, either as a grid model or a triangular model, contours can now be interpreted through the area. This data base can now be used for surface modeling and volumetric analysis. The ability to display the DTM as a 3D surface model is very useful for gaining insight into area structures, illustrating spatial relationships, deriving relationships of trends and communicating vast amounts of information at a glance. The combination of surface modeling and volumetric analysis is extremely useful for visual impact studies, e.g., land fill or waste disposal studies, borrow-pits, reservoir construction, the flooding of lowlying land or landscaping. For e.g., a DTM constructed from topomap which gives variables such as elevation, slope gradient, slope aspect, and configuration are to be incorporated in a raster GIS model. A linear polynomial equation can be developed which incorporates the four variables and their weights to model the landslide potential of each gird cell in the study area [02]. By integrating thematic maps, viz., lithological map, forest cover map, geomorphological map, lineament density map, using GIS geotechnically favourable attributes for hydel site selection can be arrived.

7.1   Preliminary Reconnaissance Level Cost Models Using GIS

Landsliding zones, route planning and a knowledge based selection of geotechnically favourable attributes for a small-hydro site can be carried out using GIS and image analysis packages capable of generating digital elevation/terrain modeling. Landslide zones can be arrived at by suitably weighing the variables like (a) topographic and hill slope characteristics, (b) geology and structure of the underlying bedrock and (c) rainfall intensity, earth quakes and human interference. A DEM is generated to obtain elevation, slope gradient, slope aspect and configuration and these variables are weighted suitably to model landslide stability. For highway route planning near hydroelectric sites the pavement construction cost model, earthwork cost model and drainage crossings cost model are to be developed. The drainage lines, catchment boundaries, land use, etc., which define that catchment characteristics can be interpreted from small scale images obtained from IRS-1C or SPOT. The mean valley widths, maximum and minimum terrain reliefs, lithologic details in vector and attribute data base, etc., can be put in GIS to obtain cost models [05].   Knowledge based selection of geotechnically favourable attributes for hydroelectric site can be carried out from thematic maps like lithological map, geomorphological map, lineament density map, forest cover / vegetation cover map and their integration by overlaying, boolean operations on layers, etc., [08].

However while using GIS packages great care is to be exercised in overlaying. Various combination of two and three layers are used to determine the amount of identification type operational error which might be present in output products [11]. Spatial resolution data of 100 sq.m or 200 sq.m for the land cover, terrain and soils were analysed. Output product accuracy were determined by counting sample points which are correctly labeled throughout each of the data layers analysed. The most accurate output product was possible using the thematic data bases in a two layer map created by compositing the 100 sq.m spatial resolution land cover map and the 100 sq.m cell dominate soil type map. Only 10 of the 35 sample points on the 2.5 acre (100 sq.m) resolution output maps had both land cover and soils correctly labeled (a combined accuracy of 29 percent only). The least accurate output map was a 3 layer product derived from a combination of the 2.5 acre resolution map of land cover, slope aspect, and soil type (dominant method). In this instance only 11 percent are correctly classified through all three map overlays. Operational error reduces the accuracy of a GIS output from its theoretical best.

The terrain data base generated earlier on the basis of geologic, geomorphic and geotechnical characteristics can be used in evaluating mean valley widths, maximum and minimum terrain reliefs for each land system mapping units for the principal directions to evaluate possible variations in earth work volumes. The computation of earth work volume uses typical symmetrical v-shaped valley concept in any GIS. This can be adjusted from the valley density by one or two standard deviation points to compensate for the natural terrain [03].

Similarly, drainage structure cost model can be estimated from terrain and watershed hydrology data using simple hydrologic estimation techniques to determine peak flows. The boundaries of selected watersheds can be delineated, digitised and stored in the GIS data base. Using standard hydrologic estimation procedures (rational methods for small watersheds and SCS runoff curve model for larger watersheds) relationships between peak discharge and catchment area are developed for each map unit. The basic data such as runoff co-efficient, lengths of channels, and corresponding elevation differences for the investigated catchments, area and rainfall intensity data can be integrated in GIS data base. Peak discharge values can be used to develop predictive relationships between stream order (a measure of size) and anticipated peak runoff for major map units. The drainage net can then be digitised into the graphics data base according to the peak discharge values. Intersection of a road alignment with the drainage net gives

213

a value of peak discharge at that crossing site. Assessment of suitable floodways is also possible using terrain data base.

7.2   Extracting Topographic Structure from DTM

For extracting topographic structure from digital elevation data conditioning is to be done. Conditioning phase generates three data sets: the original DEM with depressions (that hinder flow routing) filled, a data set indicating the flow direction for each cell, and a flow accumulation data set in which each cell recites a value equal to the number of cells that drain to it. The original DEM and these three derivative data sets can then be processed in variety of ways to optionally delineate drainage networks, overland paths, watersheds for user specified locations, sub-watersheds for the major tributaries of a drainage network, or pour point linkages between watersheds. These can be checked with topomap derived drainage network, watersheds, information using suitable sampling procedure, etc [03].

In satellite stereopairs the image of same area from two different view angles is obtained (such as SPOT stereo pair and IRS 1C PAN stereo pair). DEM can be created from this stereo pair using an image analysis package. As the resolution of the satellite image itself is low (5-10 meters), the elevation accuracies will be lesser than the stereopair resolution. These procedures will eliminate the inherent errors in the digitising process and subjectivity of using topographic map for DEM generation. To go directly from a stereo photo model to either a triangulated network model (TIN), a series of profiles, or strings of digitised contours requires that the points and lines to be digitised be determined by a skilled operator. Such photogrammetric procedures are employed to produce most of the very large scale topographic data bases developed for municipal mapping activities and engineering site analysis. Accuracy assessments need to be carried out for a stereopair generated DEMs with reference to map generated DEMs. A balanced combination of both the processes of DEM/DTM generation helps in best reconnaissance surveys for earthwork estimation near selected hydroelectric sites. The so called fourth dimension in digital photogrammetry can also be made use of by using expert systems based image interpretation on DEM/DTM data.

A new method to represent natural surface using DTM and fractals is gaining popularity now-a-days. For a given material of elevation data, the conventional spatial interpolation techniques so far used (like patchwise interpolation, pointwise interpolation, or the spline interpolation) do not incorporate the randomness of the surface at the interpolating points. They just try to connect the existing elevations by smooth curves or planes, which is in reality not the case. Most of all natural shapes are characterised by a unique parameter called fractal dimension, for example mountains, coastlines, etc. The fractal approximation of the fractional brownion motion gives us an impression of a mountain profile. This has driven scientists to experiment for generating a surface by making a number of random cuts on it and displacing the two cut portions by a random amount. All these ideas lead to think of using fractals for Digital Terrain Modeling in getting a more realistic image. This helps in more accurate estimation of cost models needed for hydroelectric sites.

8. SIESMOTECTONIC ANALYSIS

Seismotectonic analysis has to be carried out while selecting small hydro sites in hilly terrains. The analysis is aimed at knowing the probable location of future earthquakes, frequencies of these occurrences, the extent of energy release and their impact at a construction site. Seismological, geological, geophysical and man-induced data base forms basic data for such analysis. Morphotectonic analysis involving interpretation of geological structures can be best attempted using the remote sensing data. Identification of faults and active faults can be done using aerospace (or its equivalents) data in conjunction with seismological data and supporting ground truth. Geologically active faults can be identified on the basis of certain features like fault scarp, rift valley, oversteepened base of mountain fronts, faceted triangular spurs or ridges, drainage offsets entrenched meanders, sag ponds, multiple river terraces, alignment/association of hot springs, stratigraphic offsets in younger unconsolidated sediments and many other elements [06].

9. GEOCOMP SOFTWARE WITH A MICRO-PHOTOGRAMMETRIC WORKSTATION TO AID SELECTION OF SMALL HYDRO-ELECTRIC SITES

The Geocomp software with digital photogrammetric equipment at RRSSC/ISRO, Bangalore has the following capabilities. This can handle data sets of upto 32670 3D points to generate a single DTM provided 2MB of extended or expanded memory is available. Modeling speeds of over 30000 points/minute can be achieved on fast PCs/workstations. The gridded data can be used or the height of each grid intersection can be interpolated in the spatial data systems (SDS) to form a DTM. Gridded volumes can be shown between two irregular surfaces or between a plane and one irregular surface. The output can be displayed at a larger grid interval than that of the original grid. The grid volumes can be output as an ASCII file suitable for importing into a spread sheet package. The cut volumes, fill volumes and the net volumes can be displayed. The dynamic contour display come as an aid for designers where contours can be drawn on the screen based on the current Z values (heights of DTM triangle points) not the values which existed when the DTM was formed. This helps in getting intersection layout and the effects of changes in plans of a hydroelectric site. Geocomp has the facility to interpolate heights from the DTM of a surface onto a set of points. This can be used to place design heights on a series of strings. If the points already have heights then they are given a height equal to the vertical difference between their original height and that interpolated from the triangle fill. These height differences can be used to create isopachs ("contours" of cutfill depths).

All volumes can be obtained from either using cross sections or DTM surfaces. DTM volumes can be between certain surfaces within a number of volume boundaries. Bulking factors can be applied on the basis of the volumes being in either cut or fill. A separate volume, with a separate bulking factor, can be determined for a uniform capping layer of nominated depth. The cross section volumes can be from either one combined cross section file or as a net volume from two files. For example if one cross section file represents the total fill for an approach road to hydroelectric dam site and the other cross section file represents the fill upto a certain RL (Reduced Level), the fill above that RL can be obtained and tabulated under the headings - chainage, volume, progressive volume and percentage completed of the total computation. Thus existing software and work stations are having high ability to facilitate selection of a hydro-electric site.

10. CONCLUSIONS

At the reconnaissance phase of dam site selection, aerial photographs have been routinely used for many years. The proven utility of satellite data derived maps on geology, lithology, hydrogeomorphology, land use, etc., in arriving at a suitable alternatives of small hydroelectric sites needs to be emphasised in all concerned institutes or agencies. As satellite imageries are now capable of producing stereopairs they can also be used at the reconnaissance level. IRS-1C satellite has stereo-capability to permit production of maps with contour intervals of about 6 to 10 meters in panchromatic mode. The efforts are to be put in exploiting the utility of data from IRS series satellites for selecting small-hydro electric sites. This has to be done by generating Digital Elevation Models (DEMs)/Digital Terrain Models (DTMs) and selecting favourable attributes (parameters) from different thematic maps and integrate them using GIS technique. Utmost care is to be exercised in using the new techniques by knowing their limitations for practical use. Detailed project investigations covering various other aspects like fixing levels for dam site, approach road, quarry sites, etc., should concentrate on targetted alternatives produced by remote sensing data.

11. Acknowledgements

The authors would like to acknowledge the continued encouragement for such studies by Sri K Radhakrishnan, Director, NNRMS-RRSSC and Dr. M G Chandrasekhar, Scientific Secretary, ISRO. The secretarial support and neat error-free rendering of the text is provided by Ms. K B Savithri Radhika Devi, RRSSC, Bangalore.

12. REFERENCES

01 ESCAP Report, 1986, "Report of the workshop on the Applications of Remote Sensing Techniques to Water Resources Development", Seoul, 29 October - 3 November 1986.

02. Jay Gao, Lo C.P., 1995, "Micro-scale Modeling of Terrain Susceptibility to Landsliding from a DEM: A GIS approach", Geocarto International, Vol.10, No.4, December 1995, Hongkong.

03. Jenson S.K., and Domingue. J.O., 1988, "Extracting Topographic Structure from Digital Elevation Data for Geographic Information System Analysis", Photogrammetric Engineering and Remote Sensing, Vol. 54, No.ll, USA.

04. Jha, V.K., 1994-95, "Utility of Remote Sensing in Preliminary Terrain Evaluation for Engineering Geological Projects", Proc. ISRS Silver Jubilee Symposium, Dehradun.

05. Joseph O Akinyede, 1993, "A geotechnical GIS concept for highway route planning", ITC Journal 1993.

06. Katti, V.J., Satya Saradhi Y.R., Kak S.N., Banerjee D.C., 1994, "Application of Remote Sensing Techniques for Nuclear Power Project Site Selection in India", Proce. of ICORG-94, Hyderabad.

07. Manavalan P., Krishnamurthy J., Manikiam B., Adiga S., Radhakrishnan K., and Chandrasekhar M.G., 1993, "Watershed Response Analysis using Digital Data Integration Techniques", Advanced Space Research, Vol.13, No.5, UK.

08. Manoj Dangwal, and Asis Bhattacharya, 1994, "Site evaluation of a storage reservoir using Remote Sensing and GIS techniques - A case study in Mahadayi basin, Karnataka", Proc. of ICORG-94, Hyderabad.

09. Narain. A., 1989, "Remote Sensing in Water Resources", Lecture notes for training course on "Scientific Source Finding", Space Applications Centre, Ahmedabad.

10. Reuben Paul P., 1988, "Documentation of the work done during the period 1-2-87 to 30-8-87 at Zurich, Switzerland", An internal report of Hydrology Division, NRSA, Hyderabad.

11. Stephen J.Walsh., Dale R.Lightfoot., and David R Butler, 1987 "Recognition and Assessment of Error in Geographic Information Systems", Photogrammetric Engineering and Remote Sensing (PE&RS), Vol.53, No.10. USA.

***
**
*

First International Conference on Renewable Energy–Small Hydro
3–7 February 1997, Hyderabad, India

TYPES OF LAYOUTS AND MECHANICAL STRUCTURES FOR SMALL HYDRO POWER STATION ON EXISTING CANALS DROPS

R.K. Sharma, N.K. Bhalla and D.S. Sahni

Punjab Irrigation Department
S.C.O. 142-43, Sector 34-A
Chandigarh 160 001, India

1.0 INTRODUCTION

With the increasing costs of Medium and High Head Hydro Power Stations, necessity has been felt to utilize the available drops on the existing canals so that the investment of the project vis-a-vis its completion period is comparatively less and the benefit is accrued in shorter span of time. Ministry on Non-Conventional Energy Sources, Govt. of India, has notified a number of incentives for small Hydro Power Stations and it will be quite useful for the various states to tap their resources for these Hydro Stations as early as possible so that gap between the overall requirement of energy and the available power in the country may not increase to a non-manageable limit. In this context, it has been felt that the drops available on the existing canals in various states, which are of the order of 2.5 mt. to 3.0 mt, should be utilized for power generation.

2.0 TYPE OF RECOMMENDED LAYOUTS

2.1 Considering the various factors, such as health of the existing drop structure, availability/cost of land and positioning of existing water distribution system from the branches canals, it has been experienced that following layouts are possible, keeping in view, the very fact that the generation cost should also be on a competitive rate. The type of mechanical structures such as Intake Gates, By Pass channel Gates and draft tube gates and their lifting arrangements are also shown in the various sketches annexed herewith:

2.2 Where the cost of land is high and there is less width of land available on the canal Banks:-

    2.2.1 Every small hydro power station will have the following structures.

        (i) Power House Building and switch yard

        (ii) Intake/forebay structure

        (iii) By pass channels to take care of the tripping of the units, due to electrical/mechanical failure.

        (iv) Tail Race structure.

In case where the width of the land along the canal banks is a restriction, the layout in plate-I has been adopted and found to be cost effective.

    2.2.2 In this layout, the by pass channel has been made underneath the power house building. When any of the generating units trip, due to some electrical/mechanical fault the water from the penstocks travels in this channel and the tail race gates (which act as draft tube gates also) operate automatically and allow the water to pass to the main channel. In this case, the in-take forebay is constructed on an off-take from the main/branch canal.

2.3   These type of power houses have been built along the Upper Ganga canal in U.P., from app. 20 K.M. from Roorkee towards Delhi and Seven power houses are constructed upto Bulend-Shehar. The canal drops have been utilized and the capacity of the power houses is varying from 2 MW to 4 MW and available head is from 4 mt. to 6 mt.

2.4   The distribution off-take are atleast, 500 mt. away from the intake forebay and the quantum of irrigation water in the adjoining distribution canals is not affected when any of the generating unit trips.

2.5 Type of Mechanical Structures:

(i)   Intake Gates: Fixed wheel gates with electrically operated double drum wire rope hoist with emergency lowering provisions.

(ii)   By pass gates and draft tube gates: Flap gates, bottom hinged suitable for automatic opening with hydrostatic force and closing with counter balance force provided with counter weight.

These are clearly explained in Plate I.

2.6 Merits and Demerits

2.6.1 *Merits*

(i)   Land required is less

(ii)   Tail race structure and by pass gates are combined in one structure.

2.6.2 *Demerits*

(i)   Tail race structure requires frequent maintenance as there is a likelihood of deposit of silt on the u/s of the crest of the gates which do not allow the gates to open. This has been experienced on Nirgajni power house on the upper ganga canal system.

(ii)   More deep foundations of the power house building are required to accommodate the By pass conduit.

3.0 WHERE THE EXISTING FALLS ARE HEALTHY, THE BRANCH CANALS TO BE UTILIZED AS BY PASS CHANNELS

3.1   On sites, where the existing canal drop structures are of adequate strength and their condition is healthy, then the branch canal itself can be utilized as a By pass channel. Flap gates/Radial gates installed just u/s the drop will open automatically as the water level in the intake forebay rises. The intake forebay shall be built on an off-take channel to be constructed on one side of the branch canal. This type of layout is shown in Plate II.

3.2   This layout will save the construction of the By pass channel provided there is no distribution outlet at about 500 ft. from the drop structure upstream the branch canal.

3.3 Merit and Demerits

3.3.1 *Merits*

(i)   There is a net saving in the cost of construction of By pass channel and the drop structure.

(ii)   There is no involvement of deeper foundation of the power house building as is required for the layout at Plate I.

3.3.2 *Demerits*

(i)   Dependence on branch canal as By-Pass channel involves, disturbing the irrigation distribution on the u/s distributory off-takes, during the tripping of the generating unit. Similarly when the power house starts the water from the tail race channel disturbs the irrigation pattern of a distributory off-take d/s the tail race structure. If these off-takes are at a distance of 500 ft., then this phenomenon is not significant.

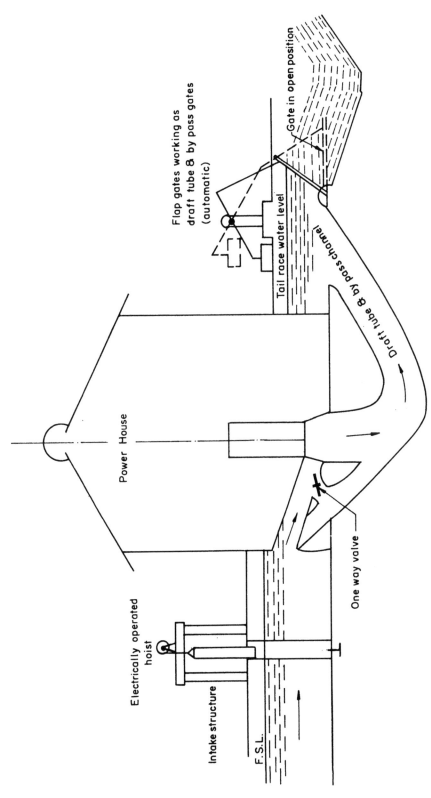

Plate-I : Layout - By Pass & Draft Tube Gates are Common Structure

221

3.5   This type of layout has been used in Rohti and THUI power houses on Punjab Canal System by PSEB. In one of the Micro Hydro Power Station at Kakroi near Panipat town on the branch canal of western Yamuna canal, the bypass gates have been installed on the branch canal itself and the float chambers for the automatic operation of the gates have been constructed on the d/s of the fall utilizing the space in between the d/s glacis and top of the canal banks as per Plate V.

3.6 Types of Mechanical Structure

3.6.1 Intake Gates and Hoisting

Fixed wheel gates with electrically operated wire rope drum hoist. The span of the gates may not be sufficient to accommodate two separate drums. A single drum with its two segments of wire rope grooves is the possible answer as shown in Figure A below:

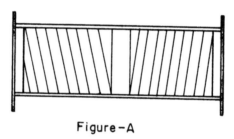

Figure–A

3.6.2 Draft Tube Gates

These gates are to operate under balance head conditions. So fixed wheel gate of self lowering characteristics and a, A-frame with a monorail hoist, lifting beam arrangement is the possible hoisting system instead of a gantry crane.

3.6.3 By Pass Gates

As the existing branch canals are having shallow water depth, in these cases, flap gates with counterweights are best suited gates for the bypass channels for automatic operation.

These types of gates are successfully working at:

(i)   Nidampur Micro Power station 2x500 KW installed on Gogger Branch of Punjab Canal System. Project executed by BHEL on turnkey basis.

(ii)   Devi Ghat Project in Nepal. Gates installed by Haryana State Minor Irrigation.

(iii) Dumber H.E. Project Agartala, Tripura State. Gates installed by T.S.L.

4.   SITES WHERE SUB SOIL WATER LEVEL IS VERY HIGH AND WIDTH OF LAND ON THE SIDES OF CANAL BANK IS A RESTRICTION, EXISTING DROP IS NOT IN HEALTHY CONDITION.

4.1   The layout of the intake, power house and tail race is as per conventional practice but the Bypass channel in having its common bank as the power house wall. This layout is explained in Plate III.

4.2   The Bypass canal will have Fish Belley gate/flap gates for automatic operation.

4.3   Intake Gates and draft tube gates will be conventional fixed wheel gates with hoisting arrangements as in layout at Plate II.

222

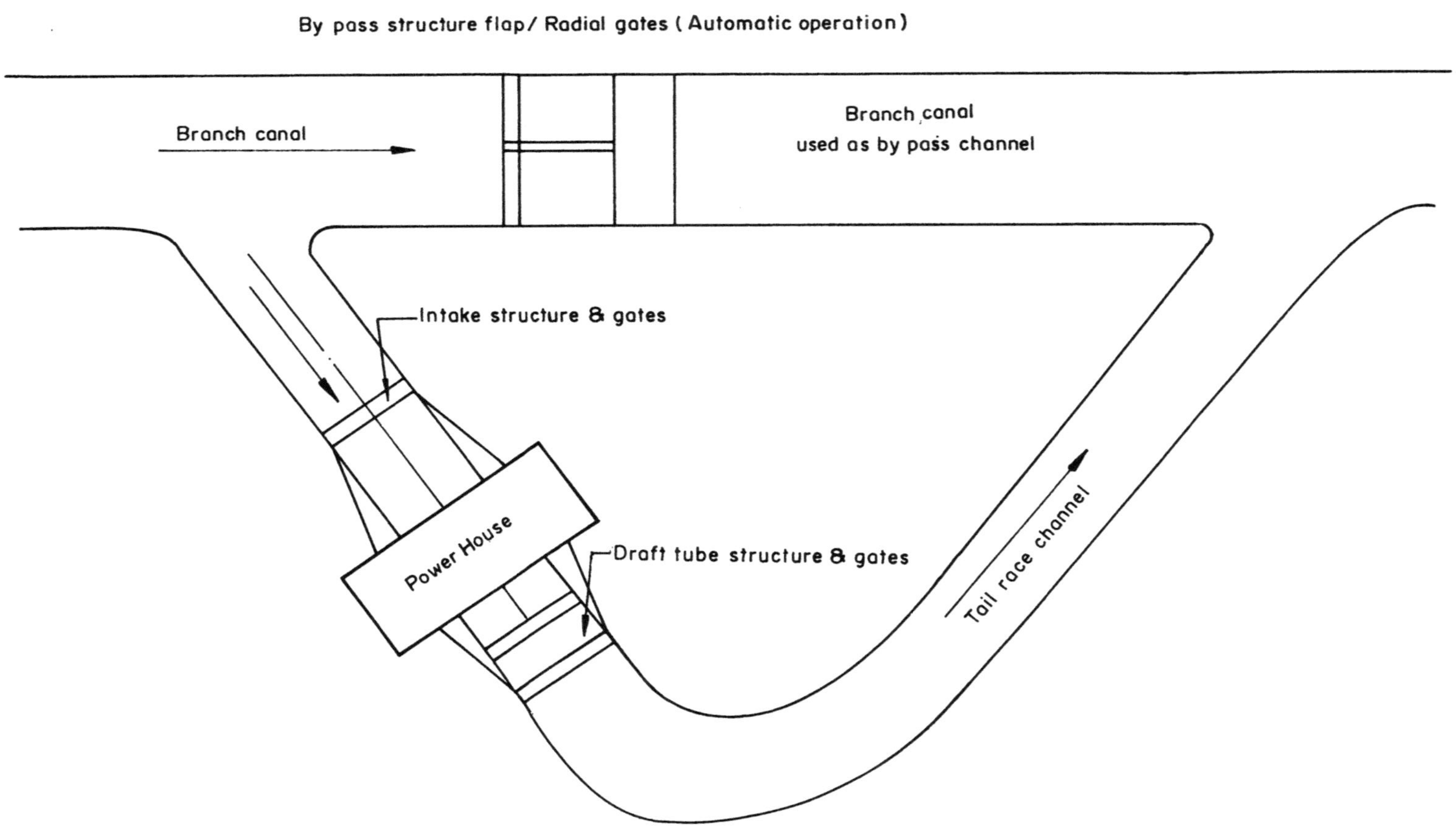

By pass structure flap/ Radial gates (Automatic operation)

Branch canal
used as by pass channel

Branch canal

Intake structure & gates

Power House

Draft tube structure & gates

Tail race channel

Plate-II : Layout- Existing Branch Canal Being Used as by Pass Channel

223

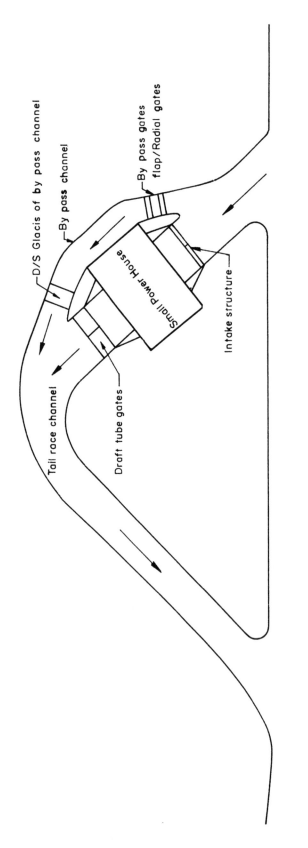

Plate-III : Layout- Existing Canal Not Being Utilized as by Pass Channel

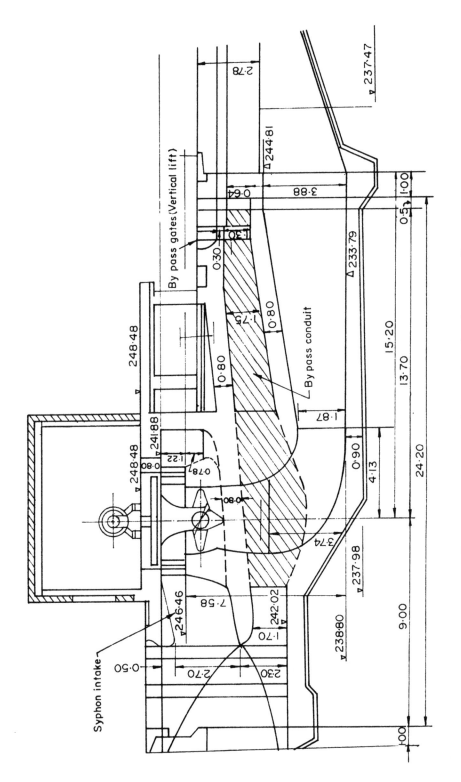

Plate-IV : Layout - Syphon Intake, by pass Conduit

5.0 WHERE MAIN BRANCH CANAL CAN'T BE USED AS BY PASS CHANNEL AND NATURAL LAND ADJOINING THE POWER HOUSE TO BE USED FOR BY PASS CHANNEL

5.1 These layouts are generally used for syphon Intakes. There is no necessity of installing in-take gates.

The water with syphon action enters the turbine for which an actuating device is kept. With the tripping of the units, water goes into a conduit/open bypass channel shown by hatching in the enclosed Plate IV.

5.2 The by pass gates are vertical lift gates, to be operated electrically being actuated automatically by sensors installed in the bypass conduit/channel.
This type of layout has been installed in some Micro power houses in Russia.

5.3 Merits and Demerits

5.3.1 *Merits*

    (i)   Saving in the cost of intake gates/Hoists.

    (ii) Saving in the cost of by pass channel.

5.3.2 *Demerits*

The system is totally dependent on sensors which is not having any backup. So dependence on the main canal has to be done in case of failure of the operation of the by pass gates; where the installation of Flap Gates has necessitated even to divert water into the Off-take Canal.

Courtesy

    1.   Punjab Energy Development Agency.

    2.   Irrigation Design Organisation Roorkee (U.P.)

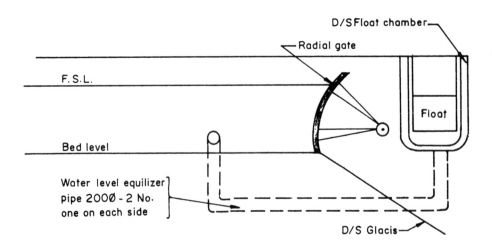

Plate-Ⅴ : Installation of Float Chamber on D/S Glacis

DESIGN AND CONSTRUCTION

First International Conference on Renewable Energy—Small Hydro
3 – 7 February 1997, Hyderabad, India

SELECTION OF NUMBER AND UNIT SIZE OF A SMALL HYDRO PROJECT

S.P. Singh, Arun Kumar and J.D. Sharma

Alternate Hydro Energy Centre
University of Roorkee
Roorkee 247 667, India

ABSTRACT

This paper present the method for selecting the number of units for a small hydro power plant. A new technique is developed to calculate the annual energy generation by simulating flow duration. The method may be used to calculate the energy generated by more than one units or a single unit. The criterion for selecting the number of units have been discussed in detail.

INTRODUCTION

It is well known that the Small Hydro Power Plant (SHP) generates electrical energy at practically no fuel cost. and almost free from inflation. The accurate annual energy calculation is important for projecting annual income from the SHP installation. In the storage schemes where out flow is constant throughout the year, the annual energy generation is the function of size of the unit and plant load factor. Normally the SHP schemes are run off the river. The energy calculation based on unit size shall not give the true picture if the variation of in flow is not accounted for (1,2,4). This will lead to unrealistic projections of cach flow from the project. It is therefore necessary to account for the variation of in flow over a period of one year (one hydrological cycle) while computing the annual energy.

The increase in unit size result in output of the plant for a storage scheme. But in case of run off river schemes the increased output is confined for the particular duration for which desired flow for power generation is available. The increase in unit size would result in loss of generation during lean discharge. On the other hand there will be loss of energy if the unit size is small as the discharge during high flows shall spill over. The annual energy generation can be increased if more than one unit of smaller size are selected to utilise overflow of water during high discharge.

Therefore, the selection of number of units is important as the financial success of project (3) shall depend on the annual energy generation (annual income) of the project. The single unit or less number of units would result in loss of energy. The project cost will increase with increase in number of units. Which results increase in annual energy. Thus the number of units should be selected in such a way so as to extract the maximum energy generation from the available flow. However, the maximum energy generation can not be sole criteria for selecting number of units. There must be balance in increase in energy generation with the increase in project cost. More over, it must be ensured that return from the project is achieved within stipulated time to meet debt service if any.

2. SELECTING THE NUMBER OF UNITS IN PROJECT

The first step is to select appropriate installed capacity. There may be two kind of installation ie grid connected or isolated. It is usually recommended that installed capacity of grid connected installation may be based on 50% dependable discharge. and for the isolated plants it may be based on 90 % dependable flow. Once the installed capacity is finalised, the next logical step is the selection of number of units. In the same way it is also necessary to carry out energy economic analysis for the different number of units.

2.2 SELECTION CRITERIA FOR NUMBER OF UNITS

The selection of unit should be based on following.

(A)   Flow Duration Characteristic

(B)   Category of Plants

2.2.1 (A)   SELECTION BASED ON FLOW DURATION CHARACTERISTIC

Single Unit :   If the variation of in flow over a period of year is not appreciable,the single units is considered most appropriate from the economy point of view. The addition of another unit in such situation will not justify the increase in energy generation vis a vis increase in cost of the project.This type of installation are most suited for grid connected projects. SHP on Return channel of thermal projects may fall in this category due to almost constant discharge our a large period.

If the project is non grid connected , the two unit must be considered even its flow is constant so as to serve the utility (consumer) in case of failure of either of the one unit.

Multiple Units :   In case the variation of flow is appreciable the optimum energy generation can not be obtained from single unit. During lean discharge it may be operated at full or derated capacity and during high discharge the flow will spill over.

The over flow can be utilised by installation of an additional unit. The installation of additional units would result in increase in annual energy generation but shall need additional investment. The energy economic analysis is to be made arrive at final decision. This criteria for selecting number unit will hold good irrespective of category of plant.

In some cases the variation in flow is too much, one should consider to have more than two number of units in a power plant. The energy economic analysis shall be the final resort to achieve at optimum number of units .

More than four numbers of units in a small hydro plant is a costly affair. There may be exception to this statement only in the cases where size of unit restricts installation of larger size units. Therefore energy generation by single unit , two unit, three and four units should be considered for comparative energy economic analysis.

2.2.   SELECTION BASED ON CATEGORY OF THE PLANT

Non grid connected plants

If plant is non grid connected at least two number of units are recommended irrespective consistency in flow. If the variation in flow is appreciable more than two units may also be considered.

Grid Connected Plants

In case of grid connected plant one unit is preferred in order to save the overall cost. The addition of second or third unit should justify the economic viability of the project.

The aim of the paper is to arrive at technique for computing energy generation by a single unit and different combination of units of same size.

3. TECHNIQUE FOR COMPUTING ENERGY FROM FLOW DURATION

Normally flow duration represents the variation in flow which shall be available over a period of one year. For energy economic analysis flow duration for a year ie. for one complete hydrological cycle should be available. The accuracy of result will depend on the number hydrological cycles or years for which data is available or considered in energy calculations. The average flow duration can the obtained from historical data or past hydrology of a site for computing annual energy .

The annual energy generation can be obtained from flow duration. The area under the flow duration curve represents the energy in system. If flow duration is multiplied by suitable multiplying factor gives out the annual energy in the system.

The following data shall be required for annual energy calculations.

(1)   Flow duration ie variation of flow with time (% of time availability vs discharge). The flow may be taken daily or 10 daily average flow.

(2)   Head available

3.1 COMPUTER ALGORITHM

Step required for Computing Energy Generation by Single Unit

The following steps shall be required for computing energy generation by single unit.

1.   Read flow duration data in terms of percentage of time P(I) Vs available Discharge Q(I) for I=1, N, head, Multiplying factor etc.

2.   Select the size of unit equal to installed capacity.

3. Calculate design discharge for power generation (QD)corresponding to installed capacity.

4. Initialise Area = 0.0, and Q2 = 0 , I = 0

5. Calculate the minimum discharge available Q(I) corresponding for P (I) = 100- I (100% time) from flow duration data, using linear interpolation technique.

6. Check if Q1 > QD If yes go to step 8

7 I = I + 1 go to 5

8. Calculate area of the under the flow duration. Using relation Area = P1 (Q1 - Q2)

9. Calculate Energy generated by base unit by multiplying the Area by multiplying factor.

For the calculation of energy generation by different units of same size calculate the design discharge (QD) for each unit based on installed capacity and number of units and modify the above algorithm as follows.

For annual energy generation from second unit

Replace Q1 by Q2, assign QD = 2QD, I = I+1 and go to step 5

For annual energy generated from third unit

Replace Q1 by Q2, assign QD = 3QD and I = I + 1 and go to step 5

For annual energy generation by fourth unit

Replace Q1 by Q2, assign QD = 4 QD and I = I+1 and go to step 5

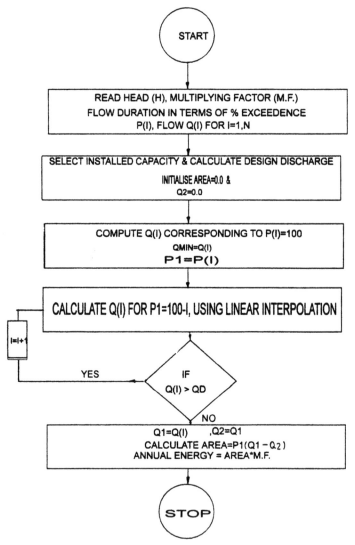

FIG 1. FLOW CHART FOR COMPUTING ANNUAL ENERGY GENERATION
(FOR SINGLE UNIT)

4.0. CASE STUDY

The nett head considered for power estimation is 3.0 meter. The flow duration of a typical canal is shown in Fig. 2.0. The installation capacity corresponding to various dependable flows is given in Table 1.0.

Fig 4.2 flow duration

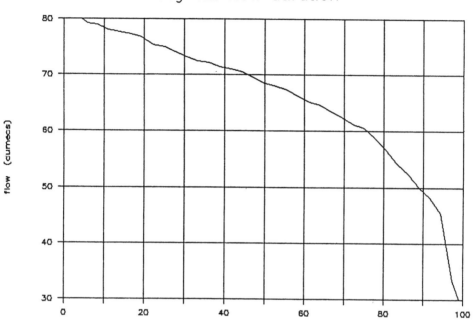

% of the time flow equalled or exceeded

Table 1.0

% of Time	Discharge (Cumecs)	Installed capacity (kW)
90	42	846
80	54	1088
60	68	1370
50	70	1411
40	71	1431

The computer programme as explained in preceeding section has been used to calculate the annual energy generation.

The following assumptions are considered for calculating annual energy.

i.   The Head remains constant irrespective of variation in tail water level.

ii. The machine is expected to generate only when the available flow is equal or more than the QD rated flow for the particular machine size.

iii. The conversion efficiency of the system as considered is 75% (turbine, generator, gearbox)

iv. All machines of equal size are considered in order to reduce inventory for spares.

v.   The efficiency of turbine is considered to be constant with variation in flow.

235

4.1 Size and number of units for grid connected scheme

In case of grid connected schemes the installed capacity may be based on 50% dependable discharge. The corresponding capacity of the plant will be about 1400 kW. The energy generated by single unit of different capacity is given in Table 2.0

Table 2.0

Installed Capacity (kW)	Unit Size (kW)	Energy generated in M.U.	No of days machine in operation
1500	1500	3.59	124
1400	1400	5.44	200
1200	1200	6.40	277
1250	1250	6.24	259
1100	1100	6.17	292
1000	1000	5.93	306
750	750	5.23	343

The Table 2.0 shows that a single unit of 1200 kW will generate maximum energy and operate for 277 days in a year.

The Table 3.0 shows the energy generated by the multiple unit of same size with the installed capacity limited to 1500 kW

Table 3.0

Installed Capacity (kW)	No of Unit and Size kW	Energy generated in million units (M.U.)		No of days machine in operation
		Each Unit	Total	
1500	I - 750	5.23		343
	II - 750	1.40	6.63	102
1500	I - 500	4.37		346
	II- 500	2.49		292
	II- 500	.75	7.61	92
1500	I - 425	2.96		350
	II - 425	2.71		328
	III- 425	2.04	7.75	351
1500	I - 400	2.96		350
	II - 400	2.45		339
	III- 400	2.10	7.51	262

From the Table 3.0 and 2.0 it is observed that the maximum annual energy would be 7.75 MU from 3 x 425 kW and 6.40 MU from a single unit of 1200 kW. In grid connected installation more than two numbers of generating may be rejected due to economic considerations. The comparable option in terms of annual energy production by single and two units is given Table 4.0

Table 4.0

Installed Capacity	No of Units	Annual Energy (MU)
1500	2 x 750	6.63
1200	1 x 1200	6.40

The additional energy generated by 2 x 750 option shall be more than 0.23 MU as compared to single unit installation of 1200 kW.

The additional revenue will be to the tune of Rs 0.5175 million per annum considering the unit selling price @ of Rs. 2.25 per unit. It is to be ensured that additional revenue is sufficient to meet out the annual expenses incurred on account of increase in installation capacity and to repay the loan within stipulated period, if any. The final decision should be based on revenue accrued over the life of the project.

4.2   Size and No of Units of Non-grid Connected Schemes

The non-grid connected (Isolated) schemes are expected to generate throughout the year. Therefore, the installed capacity should be based on minimum discharge available over a period of one year. However, the installed capacity should be based on 90% or 80% dependable is normally recommended.

The installed capacity at the 90% dependable discharge is 846 kW. At least two number of unit of same size should be selected to meet out in event of failure of one unit. Therefore two units of 425 kW may be adopted. The energy generated by two number of units of of standard size and the same rating is computed and shown in Table 5.0.

Table 5.0

Installed Capacity -(kW)	Unit Size (kW)	Annual energy generated (M.U.)		No of days machine in operation
		Each Unit	Total	
850	850	5.46		332
850	I - 425 II - 425	2.96 2.71	5.67	350 328
1000	I - 500 II - 500	4.37 2.49	6.86	346 292
1200	I-600 I-600	4.37 2.91	7.28	346 270
1500	I - 725 II - 725	5.23 1.46	6.69	343 102

Table 5.0 shows that the energy generated by 2 x 600 kW installation is the maximum. The installed capacity of 2 x 600 kW is most suited in this case. The energy economic analysis is necessary to justify the unit size.

Limitations of the Algorithm

1. The proposed algorithm does not include variation in head with flow.

2. The efficiency of turbine varies with flow and may be accounted for realistic analysis .

3. The machine can also be operated at derated capacity. The minimum discharge required for operating the machine without any detrimental effect is to be specified by manufacture. Thus the machine can also be operated even when the flow is below the rated discharge of the machine. This may be included in annual energy calculation to make the analysis more effective.

5.CONCLUSIONS

The proper selection of a size and number of units in a SHP Project can ensure its economic viability. The selection of number and unit size be based on annual energy generation. The variation in flow should be accounted for while estimating the annual energy generation to project actual return on the investment . Finally the energy production cost (unit cost) must satisfy energy-economic criteria. The technique presented in the paper is for computing annual energy generation by single and multiple units. This technique simulates flow duration curve and accounts for variation in flow. The technique is capable of computing annual energy with certain accuracy even if sufficient discharge data are not available.

6. REFERENCES

1. Diamant Y.Y, Harpeth R.G., 1983 Computer Optimization for run of river enenrgy, Water Power Dam Construction, Nov 1983, pp. 17-20.

2. Da deppo L., Dateio C., Fioretto V. and Rinaldo A., Capacity and Type of Units for a Small run off River Plants, Water Power and Dam Construction, pp. 33-38.

3. Bahamondre r., 1981, Economic Criteria for Selecting the number of Units for a Hydro plants, Water Power and Dam Construction.

4. Fahlusch F., 1983, Optimum Capacity of a run of river Plant, Water Power Dam Construction, pp. 45-48.

First International Conference on Renewable Energy—Small Hydro
3 – 7 February 1997, Hyderabad, India

NEW SOLUTIONS FOR ECONOMIC RETROFITTING OF MICRO HYDRO POWER GENERATION UNITS IN EXISTING IRRIGATION WEIR GATES

Valentin Schnitzer

Pumpenanlagen Kleinwaserkraftanlagen Neue Energiesysteme
IndustriestraiBe 100, 69245 Bammental, Germany

Generally Hydro Power Plants are designed and optimized to their sites, individually. There are however many existing plants in irrigation canal systems, river control for flood and navigation, dams and reservoirs, for storage and protection, that are not utilizing their hydro power potential yet, as they were built for another prime purpose.

When trying to utilize power potentials of these plants by fitting typical turbines with their conventional arrangements, the idea of retrofitting is often given up for technical unsuitability or price reasons for the extensive measures to be taken to match the plant:
Usually complicated by–pass systems have to be built within irrigation canals, gates, weirs and locks have to be redesigned totally to fit any conventional turbine plant. Extensive additional works for major modification cause sometimes even more cost than a new plant would cost.

A system has been developed and successfully applied in several plants in Germany:
A horizontal Kaplan turbine with integrated generator is prepared as a unit to fit in the existing gap of a weir gate. The existing or reinforced guide is prepared to absorb the axial thrust, fitting the premanufactured unit by crane into the weir gap. Inlet with integrated weed screen cleaning facilities and draft tube are designed to fit the site conditions and joint as premanufactured components. The great advantage of the system in saving of civil works and erection time is leading to economic solutions for ultra–low–head turbines.

After the approval of this type of plant in Europe it is proposed to apply it in irrigation plants worldwide.

INTRODUCTION

The small hydro power potential in Germany has been developped to a high degree along with the industrialization, the river development for navigation, flood control and power utilisation especially. Although most sites are exploited, a substantial number of ultra–low–head sites can still be developed where old mills had been given up or power generation was left out for economic reasons. The construction of ultra–low–head water power plants in conventional design is very expensive and generally not economic, – in spite of the standardization efforts of turbine manufacturers and drastic increase of tariffs for production of renewable energy. Substantial additional cost occurs nowadays by environmental requirements and conditions set forth by the Authorities for approval of the project.

The catalogue of requirements for development comprises of:

- minimal use of land, retaining the natural shape of surrounding,
- minimal volume of civil work, (often civil work super structures not permitted,)
- no negative or minimum effect in case of floods
- assurance of fish ways for free passage of aquatic life
- cost and time–effective implementation of construction

MAIN FEATURES OF THE SOLUTION

The solutions discussed here refer to existing structures with retrofitted micro hydro generation equipment; newly installed in various applications in European plants. The techniques demonstrated may also suit many other plants worldwide, especially in irrigation drops where similar conditions prevail.

- The turbine generator unit is installed at the place of a weir gate within the free rectangular water passage.

- There are no civil works superstructures and civil works are limited to modifications for adaption of the unit.

- The unit is installed in the open space as no superstructure or concrete covering is submersible and exposed to the weather.

- The unit is premanufactured and mounted using the existing weir gate guide.

- The equipment is fitted with the help of a mobile crane having no permanent crane or lifting tackle installations.

- Switchboard control, hydraulic power pack are accommodated adjoining the plant, preferable in a premanufactured container.

- The unit consists of three main elements: Inlet, turbine–generator and tail pipe.

GENERAL ARRANGEMENT

Layout

In general the existing gates with their rectangular passages may be located within the weir, adjacent to the weir, or in a by-pass (Fig.1 a,b).

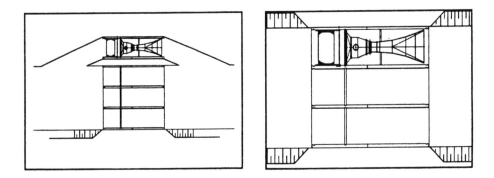

Fig. 1: Location of water power plant in existing weir

After first applications in existing plants, new plants were also designed on the base of the same arrangement. They are then located within the weir or only adjacent to it, not in a diversion (by-pass). Fig.2

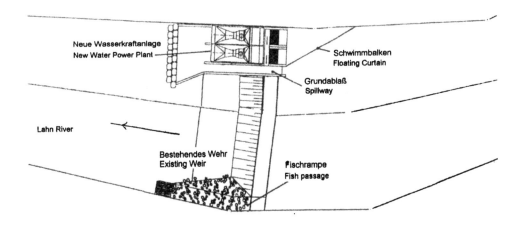

Fig. 2: New power plant in weir of river Lahn

Plant Section in Comparison to Conventional Solutions

Conventional micro hydro design refers mostly to the "downscaling" of "classical" designs. Even with the standardization of turbine units for small hydro power, the simplification of buildings and underground structures the benefits are obvious of an open rectangular passage without the need for superstructures. Fig. 3 a,b

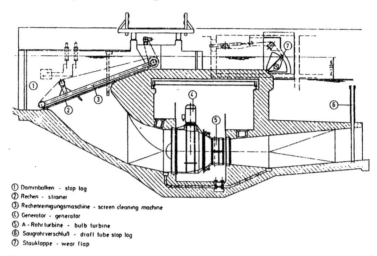

① Dammbalken - stop log
② Rechen - strainer
③ Rechenreinigungsmaschine - screen cleaning machine
④ Generator - generator
⑤ A - Rohrturbine - bulb turbine
⑥ Saugrohrverschluß - draft tube stop log
⑦ Stauklappe - wear flap

Fig. 3 a: Classical design for plant without civil works above ground [1]

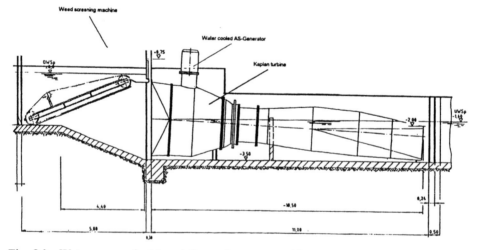

Fig. 3 b: Water power plant in existing weir–passage with
minor civil works modifications [2]

COMPONENTS OF PLANT

The *Kaplan turbine*, double regulated with its ideal characteristics for various heads and flow, is delivered pre-assembled. The adaption to the gate requires a solid steel welded construction to match into the existing gap with the u-shaped guide. In this guide the hydraulics thrust from the head water is absorbed. For installation the unit is lowered into the guide of the gate by a mobile crane. The connections to the land-based switch-board, control and hydraulic hydraulic power pack is by means of quick couplings for speedy installation.

Inlet and outlet are shaped for optimal flow and energy recovery requirements. The premanufactured tail pipe is resting on the bottom; a sub-wall is separating the tail water to permit a full drainage of the installation pit (service area). Inlet and outlet are only integrated for extremely small units with concessions to efficiency.

The *weed screen* is directly mounted to the inlet front frame. The cleaning may be by conventional facility or by automatically working, submerged chain-operated rack for the models to be totally flooded.

Submersibility

The whole unit. is exposed to the weather and may be flooded, it has therefore to be of submersible design. Moving parts are of stainless steel and sealed appropriately. Additionally they are equipped with the protective cover against floating matter and dirt.

The submersible motors are water cooled if mounted on top, driven by rectangular gear.

Integrated in the bulb it may be a standard submersible asynchronous motor, if working in parallel with the grid.

Special Submersible Solutions

Twenty years ago a French motor and generator manufacturer had introduced a submersible hydro generating unit of a simple design; it disappeared from the market after the company was taken over and the product emerged.

SUBMERSIBLE PUMP TECHNOLOGY FOR TURBINES

Operating centrifugal pumps as turbines is nowadays an approved technology for energy recovery. Limitations do exist for propeller pumps. A Swedish pump manufacturer has solved the problems by his own turbine design on the base of his pump components. The design with its simplicity has spread fast over the last fifteen years. Due to the vertical installation the design is however limited to the lowest fall of approx. 3m; horizontal installation is not possible.

A system of a self-starting syphon with minimum requirement in vertical position is an optimal solution for flooded stations without superstructure.

Reversed Pumps

Due to flow conditions the standard propeller pump cannot be reversed and still maintain the good efficiency as experienced in the centrifugal pump's reverse. Leading pump manufacturers are now researching to modify their pumps with the target to use their pumps modified at good efficiencies as turbines. Units so far applied in the field of ultra-low-head are not yet optimized in the efficiencies.

POTENTIAL FOR FURTHER DEVELOPMENT AND OUTLOOK

Innovative solutions have to leave the path of the "downscaling" and look out for original new designs or combinations. Much research has been done at hydraulic research institutions recently and results have still to be applied in the products and plants. Our research initiated now is aiming at model tests for future plants, i.e. for the design of minimum tail pipe dimensions the following solutions are to be investigated:

- using the spin of flow when leaving the turbine, the diffusor angle, limited normally by 8°-10° may be extended to above 15°, thus shortening the tail pipe.

- A multi-cone can further shorten the diffusor, recovering the high speed energy in a shorter distance.

- special shaping of tail pipe to match various existing structures with their typical shallow tail water.

With the private investors in the field of micro hydro power the "least cost" developments will gain momentum and unconventional solutions as outlined here will be taken serious in the future.

References

[1]   KW Bamberg, Kössler GmbH Austria

[2]   KW Ruttershausen, SFL Wasserkraftanlagen GmbH Germany

First International Conference on Renewable Energy–Small Hydro
3–7 February 1997, Hyderabad, India

DESIGN ASPECTS OF HYDRAULIC STRUCTURES FOR HYDRO-POWER DEVELOPMENT AT CANAL DROPS

J.S.R. Mohan Rao

Water Resources Engineering Department
Chaitanya Bharati Institute of Technology
2-2-18/41/1/A, Meher Nilayam, Bagh Ambarpet, Hyderabad 500 013, India

ABSTRACT
A canal drop structure is primarily meant to negotiate the fall in water levels in the canal from upstream to downstream of the drop, without endangering the safety of the structure. Normally, no gate control arrangement is provided at these drops, unless cross regulators are clubbed with the drops, which is not of common occurrence. A drop structure without gate control arrangement is designed for safety of the floor against uplift forces due to unbalanced head when the canal is flowing full or when there is no flow in the canal. With the subsequent installation of a hydro-power plant at the canal drop, the drop structure itself needs to be provided with gate control arrangement to regulate the canal flows over the drop and divert them partially or fully through the newly excavated by-pass channel on which the power plant is installed, thereby converting a mere canal drop into a drop-cum-regulator.
In this new setup, the worst condition for the safety of a drop structure aganist uplift is when water stands upto top of closed gates on upstream with or without flow in the bypass channel and, in any case, water not backing up to the rear of the structure. In such a situation, the unbalanced head causing uplift will be enormously increased to a value equal to the difference between upstream full supply level and downstream bed level of the canal. Strengthening the floor and deepening the end cutoff of the existing drop structure for safety against such unduly high head, will be neither feasible nor desirable, necessitating dismantlement of the entire downstream floor with end cutoff and reconstruction of a new floor to the required length and the required thickness with a new end cutoff taken to the required depth, which adds to the cost of civil works of the mini-hydel project.

If the drop structure were initially designed and constructed as a drop-cum-regualtor, the overall cost of the combined structure would have been much less, with the added advantage of there being no necessity for a costly by-pass channel, but for a widened channel section for a short length at the drop site to accommodate the open regulator vents and power units side by side.

As regards the returns from such mini-hydel plants at canal drops, the total annual power generation possible at a canal drop is very much less, compared to the installed capacity. While the installed capacity corresponds to the full supply discharge in the canal and the corresponding fall in water levels in the canal at the drop, the actual power generation is for partial flows in the canal during greater part of the crop periods with no generation at all during the canal closure periods. All these aspects need to be borne in mind, while assessing the cost of installation including civil works and the power generation potential at canal drops and working out the returns.

INTRODUCTION

A large number of canal systems of major and medium irrigation projects are laid in most parts of the country. The distributory channel networks of such of those canal systems which are spread in submountainous and other upland areas and excavated in hard strata, are generally aligned as ridge canals with a series of drop structures provided all along their lengths to negotiate the steep falls of the terrains normally obtaining in such areas. When these canal systems were planned in the post-independent era, mini-hydel power generation possible at these canal drops was not considered at all and the drop structures designed and constructed without provision for mini-hydel projects even at future dates. Perhaps, such sort of design and construction of canal drops is continuing even on the ongoing canal systems under execution, even though construction of quite a few mini-hydel projects at some of the existing canal drops on the existing canal systems was taken up in the eighties and is continuing these days.

CANAL DROP STRUCTURES

A drop structure on a canal is primarily meant to negotiate the fall in water levels in the canal from upstream to downstream of the drop site without damaging the canal section. When once a rigid structure is introduced for the safety of the flexible canal section at the drop site, the drop structure itself will be exposed to the differential head created due to fall in water levels. And to safeguard the structure against this differential head, a sufficiently long and sufficiently thick floor is required at the base of the drop wall, with sufficiently deep cutoffs underneath at the upstream and downstream ends of this floor. The purpose of the floor is for safety against uplift pressures exerted by seepage flow occurring in the sub soil below the floor and that of the cutoffs for safety against scours caused in the canal bed at upstream and downstream ends of the floor due to surface flow in the canal above bed. In. addition, the downstream end cutoff serves to keep the exit gradient of seepage flow at the downstream end of the floor within safe limits of the bed soils. The differential head considered in the floor design is for either of the two flow conditions in the canal(viz) (i) when there is full flow in the canal from upstream to downstream side of the drop and (ii) when there is no flow in the canal but with still water standing upto crest level.of drop on upstream side and no water on downstream side, whichever in more. The maximum differential head for full flow condition is the difference between the upstream and downstream full supply levels in the canal plus the average unbalanced head in the region of formation of hydraulic jump or standing wave in the cistern forming part of the downstream portion of the floor, while that for no flow condition is the difference between the upstream and downstream canal bed levels plus the height of crest of drop wall above upstream canal bed level or, in other words, the difference between the crest level of drop wall and downstream canal bed level. The crest of drop wall is raised over the upstream canal bed level to an extent of not more than 40%of full supply depth in the upstream canal in order to maintain the depth-discharge relationship of the normal canal section upstream of the drop at lower discharges in the canal.

DROP-CUM-REGULATORS

Normally no gate control arrangement is provided at the drops, unless cross regulators are clubbed with the drops, which is not of common occurrence. In case a drop is clubbed with a regulator, the combined structure becomes a drop-cum-regulator. Cross regulator on a canal serves several purposes such as (i) to raise the water levels in the upstream reach of the canal when it is carrying low discharges and (ii) to reduce the flows in the downstream reach of the canal; both by regulating the gates and throttling the vent openings, as well as (iii) to shut out the flows altogether in the downstream reach of the canal by fully lowering the gates and closing the vents completely and divert the available flows into the various offtake channels taking off from the canal in the upstream reach. The worst condition for the safety of the floor of this combined structure is when all the regulator vents are completely closed by fully lowering the gates, with water level in the upstream canal standing upto top of the closed gates at full supply level and no flow and no water in the downstream canal. The maximum differntial head in this case works out to full supply depth in the upstream canal plus the differnece between the upstream and downstream canal bed levels or, in other words, the difference between the upstream full supply level and downstream canal bed level.

COMPARATIVE PICTURE OF THE TWO TYPES OF STRUCTURES

Normally the discharge in the canal is not reduced and the section of the canal not altered at drop site. In some exceptional cases, even if the canal discharge is not reduced, the canal section may be altered,(ie) bed width to depth of flow ratio, for the same discharge to suit the varying substrata in the reaches above and below the drop. But, in the case of a combined structure of drop-cum-regulator, it is more likely that the canal discharge is reduced below the regulating arrangement resulting in reduction in canal section from upstream to downstream of the structure. Even with reduction in canal section, such reduction may be effected by reducing the bed width only to the required extent, with little or no reduction in the depth of flow in many a case. Hence, assuming no change in depth of flow from upstream to downstream of the structure in either case (ie) whether it is a mere drop or a drop-cum-regulator, the magnitude of differential head to be considered in the floor design will be considerably more for a drop-cum-regulator than for a mere drop. While the maximum differntial head in the case of a mere drop can be as high as the fall in full supply levels or bed levels plus 0.4 times the full supply depth, the same in the case of a drop-cum-regulator is even higher at the fall in full supply levels or bed levels plus complete full supply depth. In case the fall in full supply levels at drop site is nearly equal to full supply depth in the canal, then the differential head will substabtially increase from about a maximum of 1.4 times the full supply depth in case of a mere drop (usually it is 1.20 to 1.25 times) to about 2.0 times the full supply depth in case of a drop-cum-regulator (ie) the differntial head acting on a combined structure is more than 1 1/2 times that on a mere drop structure.

INSTALLATION OF MINI-HYDEL PLANTS AT CANAL DROPS

For the subsequent installation of a mini-hydro-power plant at an existing canal drop, a few more additional civil works are required at the drop site for the effective functioning of the canal and the hydel plant. It is neither desirable nor feasible to widen the existing canal section and locate the turbines in continuation of the drop wall on one side, involving dismantlement of the wing walls and return walls of the drop structure and removal of the canal bank for the required lengths on upstream and downstream sides of the

structure, all on the widened side of the existing canal
section, without interfering with the canal flows during
the period of construction. Hence a bypass channel is to
be excavated taking off from and merging back with the
existing canal on one side at reasonable distances
upstream and downstream of the drop structure for smooth
flow into the bypass channel, whenever required, and
hydel plant located on this bypass channel, preferably
in line with the drop structure. Next, with the
installation of the hydel plant on the newly excavated
bypass channel, the drop structure on the existing canal
itself needs to be provided with gate control
arrangement in order to regulate the canal flows over
the drop structure and divert the flows fully or partly
through the bypass channel for hydro-power generation at
the hydel plant, thereby converting the existing mere
canal drop into a drop-cum-regulator. In this new setup,
a situation may arise when the regulator gates installed
over the drop structure are fully lowered to divert the
entire canal flows through the power plant with the rear
water in the bypass channel not backing right upto the
rear of the drop structure. Or, occasionally, it is
quite likely that the canal below the drop is closed by
closing both the regulator vents over the drop and the
power vents adjacent completely, yet maintaining full
supply levels above, though at lower discharges in the
canal upstream. Further, the provision of regulating
arrangement over the drop structure will always tempt
both the irrigation engineer and the power engineer ·at
site to maintain higher water levels upstream of the
drop, whenever the canal runs at low flows, the purpose
being to push the required flows through the various off
take sluices located on the canal close to the drop
structure on upstream side in the case of irrigation
engineer and to produce more power with greater heads on
the power plant thus created with lower discharges drawn
through the power plant in the case of power engineer.
In such situations, the unbalanced head causing uplift
pressures on the floor will increase enormously to a
value equal to full supply depth plus drop in full
supply levels from upstream to downstream of the drop,
being the difference between the upstream full supply
level in the canal and downstream canal bed level.
Strengthening the downstream floor, both in length and
thickness, and deepening the downstream end cutoff for
safety aganist such unduly high heads, which may be of
the order of 1 1/2 times the original design heads,
becomes neither feasible nor desirable, the only
practical solution being dismantling the entire
donwstream floor, downstream of the drop wall, together
with the downstream end cutoff and relaying the floor to
the required length and required thickness, with the new
downstream end cutoff taken to the required depth. All
this remodelling of the drop structure for such
increased unbalanced heads, apart from the normally

contemplated installation of gates and hoists with elevated hoist platform etc, over the drop wall, adds to the cost of civil works component and pushes up the overall cost of the mini-hydel project to a considerable extent.

Alternatively, a separate regulator structure will have to be constructed immediately upstream of the drop structure, segregating the old and new structures with a proper pressure relief arrangement introduced in between, in order to relieve the old drop structure completely from the additional water heads created upstream of the new regulator structure and regulate and divert the canal flows into the bypass channel. In this case, apart from the higher cost of an additional independent regulating structure, the length of the bypass channel will increase, increasing this cost also, all of which adds still further to the cost of civil works of the mini-hydel project.

Yet another possible alternative arrangement to avoid either part dismantlement of the existing drop structure or construction of an independent regulating structure, is to raise the existing end sill or deflector wall, as the case may be, at the end of the downstream cistern floor of the existing drop structure and convert it into a second drop wall, thus inducing the total drop in water levels to form in two stages and construct a new second cistern in rear of this second drop wall, with a final deep end cutoff at the end of this second cistern, in continuation of the existing floor of the existing drop structure for further energy dissipation to take place below this newly formed second drop wall. In this revised setup, the existing side walls in the existing first cistern portion of the drop structure need to be raised to cover the higher water levels created in the zone due to the raised crest level of the end sill or deflector wall-turned newly formed second drop wall. The deep water cushion, thus created in the first cistern in rear of the existing first drop wall, will serve to counterbalance the higher uplift pressures caused due to installation of gates over this drop wall, provided the water cushion is maintained in the cistern by filling up the cistern and keeping it full before lowering the gates and cutting off flows downstream. While construction of a new additional independent structure will be costly, remodelling of the existing structure, either with part dismantlement and reconstruction or with additional constructions to serve as supporting structures, will be both costly and cumbersome in construction and may not be economical, after all.

In any case, it is not mere installation of gates and hoists with an elevated hoist platform over the existing drop wall, but also substantial remodelling of the entire drop structure as such, for structural safety and hydraulic efficiency, involving substantial costs on

civil works component of the mini-hydel project. If the
drop structure were initially designed and constructed
as a drop-cum-regulator, the overall cost of the
combined structure would have been much less, with the
added advantage of there being no necessity at all for a
costly bypass channel,but for a widened channel section
for a short length at the drop site to accommodate the
open regulator vents and power vents side by side. For
the existing drop structures, there is every need to
carry out hydraulic model studies and comparative cost
analyses for all the possible alternative setups that
can be thought of, to determine the relative efficiency
in hydraulic performance and relative economy in cost,
in view of a great many number of canal drops available
on a good number of existing canal systems all over the
country, awaiting installation of mini-hydel plants,
provided such mini-hydel plants prove to be effective in
catering to the local power needs at reasonable initial
capital costs and subsequent operational costs.

RETURNS FROM MINI-HYDEL PLANTS

As regards the financial returns from such mini-hydel
plants at canal drops, the installed capacities may be,
no doubt, fairly large, corresponding to the maximum
design discharges or full supply discharges of the
canals and the corresponding falls in full supply levels
in the canals at the drop structures. But, the fact
remains that the canals flow with full supply discharges
only for a few spells during the crop periods and,at all
other times, the discharges in the canals are very much
reduced with, perhaps, higher water levels maintained in
the canals for commandability of the off taking
channels, apart from canal closure periods when there
will be no flows at all in the canals. These closure
periods also are longer in case of single seasonal
canals (ie) canals with irrigation only in one season,
either khariff season or rabi season, when compared to
two seasonal canals or perennial canals with irrigation
in both khariff and rabi seasons. So much so, the total
annual power generations possible at canal drops are
very much less, compared to the installed capacities.
This aspect also needs to be borne in mind while
assessing the power potentials at the canal drops and
working out the returns with lower outputs at higher
installation costs due to additional civil works. Hence,
in all future constructions, it is desirable to design
and construct all drop structures, wherever hydro-power
generation is possible, at the initial stages itself as
drop-cum-regulators with power vents also by the side,
accommodated in widened channel sections, for
installation of power units at later dates, as and when
the canals are completed and commissioned, in order to
reduce the costs of civil works chargeable to the power
plants.

PLANT AND EQUIPMENT

WATER CURRENT TURBINES FOR PUMPING AND ELECTRICITY GENERATION

P.G. Garman and B.A. Sexon

Thropton Energy Services
Physic Lane, Thropton, Morpeth, Northumberland NE65 7HU, UK

ABSTRACT

The development of river and canal systems for hydro power has so far focused on the construction of civil engineering works to create a head to drive a turbine. The potential for decentralised power production using "zero head" turbines requiring no dams or barrages has largely been ignored. A water current speed of 1m/s represents an energy density of 500Watts/m^2 in which a turbine of a few square metres swept area can produce a useful output.

The Water Current Turbine is a prime mover which extracts kinetic energy from flowing water in a river or canal. The energy extracted by the turbine provides a power source for remote areas and is used to drive a water pump or electricity generator. The paper describes the principle of operation of the turbine, its performance over a range of site conditions and the lessons learned during extensive field testing and commercial operation in several countries.

Used as a water pump, the turbine is a replacement for a three inch diesel pump set. Its numerous advantages include:
  It uses renewable energy and needs no fuel or oil,
  It is easy to operate and maintain,
  It has proved highly suitable for use in isolated locations,
  It is non polluting and almost silent when running,
  It needs no large scale civil engineering work and can be moved,
  It is designed for local manufacture and maintenance,
  It can operate 24 hours per day without a full time attendant.
  The capital cost of the turbine is low compared to other renewable energy systems. Its very low running cost means that it becomes competitive with a diesel pump set after between one and three years depending on local price of fuel and materials.

Case studies from Sudan illustrate the turbine's use for both irrigation pumping and water supply applications.

The potential for the introduction of water current turbine technology to the canal and river systems of India is discussed.

Manufacture of the turbine has been established in U.K. and Sudan and the paper describes how the necessary infrastructure to install and maintain the machines has been set up in Sudan. The paper gives an indication of the likely manufacturing cost of the turbine in India and discusses its economics compared to a diesel pump over its design life of twelve years.

1) THE POTENTIAL FOR ENERGY EXTRACTION FROM WATER CURRENTS

Many large rivers and canals flow for hundreds of kilometres through terrain where the installation of conventional hydro plants is impossible because of lack of head. In these areas there is often a need for power for irrigation pumping or domestic use but the demand is not concentrated enough to warrant connection to any centralised energy supply such as a grid. Where these conditions are present, the moving water in the canal or river can be used as an energy resource. Like most other renewable energies, this resource is diffuse but it is predictable, reliable and available 24 hours a day. A water current speed of 1m/s represents an kinetic energy density of 500Watts/m^2 of river cross section. Speeds of this order are commonly found rivers or canals throughout the world (for example, see Table:1). Water current turbines (WCTs) extract kinetic energy from flowing water to provide a power source for water pumping or electricity generation.

In the four countries (Sudan, Egypt, Somalia and Peru) where Thropton Energy Services have installed water current turbines and/or carried out market surveys, the majority of the rural population lives along river or canal banks. In Sudan and Egypt there is only desert away from the rivers and in the Peruvian rainforest there are no roads and all transport is by the river. Thus the power users live next to the resource.

The scattered nature of the power demand lends itself to small (500Watt-5kW), simple, easily installed units. Such power producing units should be as cheap as possible due to the limited purchasing power of rural people. Local manufacture is a large help in achieving a low capital cost.

2) WATER CURRENT TURBINES

Water current turbines make use of the kinetic energy in a flowing river or canal. The GARMAN turbine (Fig:1) can be thought of as an underwater windmill which floats on the surface of a river with the rotor completely submerged. It requires neither dams nor diversion of a portion of the water flow. It is tethered in free stream in a river or canal and extracts a proportion of the kinetic energy from the moving water (see equation below). Mooring is from one bank only and so navigation in the river or canal is not affected. This design of water current turbine can be installed cheaply and quickly without the use of concrete or machinery.

The different types of WCT rotor are reviewed in (Ref:1). The inclined axis propeller rotor used in the GARMAN turbine has proved to be the most cost effective and appropriate for local manufacture for machines of under 3kW shaft power. Rotor efficiencies of up to 30% are achievable with locally fabricated blades of constant hydrofoil section and fixed pitch. The arrangement of the Sudan built water pumping turbine is shown in Fig:1 below. The three bladed turbine rotor of up to 4m diameter (swept area up to 10m^2) is coupled to a standard centrifugal pump via a two stage belt transmission. This arrangement gives an excellent match between rotor and pump, enabling the rotor to operate within its most efficient speed range over a wide range of water current speeds without any adjustment to the system. Given sufficient current speed, pumping heads of up to 25m can be generated with one machine. Higher heads can be generated by installing machines side by side with their pumps in series. This design is capable of discharges of up to 13l/s.

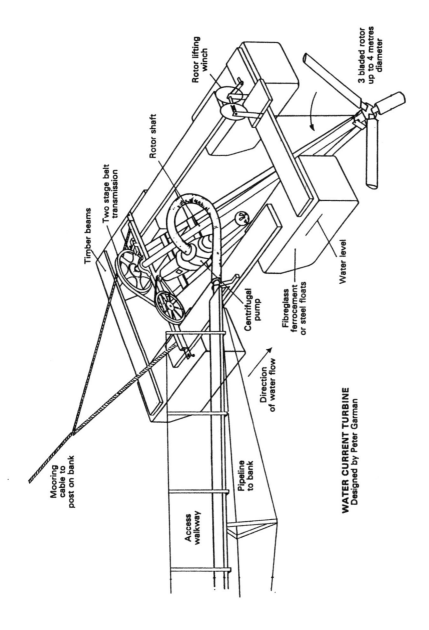

Rotor lifting winch

Rotor shaft

3 bladed rotor up to 4 metres diameter

Two stage belt transmission

Timber beams

Centrifugal pump

Fibreglass ferrocement or steel floats

Water level

Direction of water flow

Mooring cable to post on bank

Access walkway

Pipeline to bank

WATER CURRENT TURBINE
Designed by Peter Garman

Figure:1 Sketch of GARMAN Water Current Turbine

The centrifugal pump can be replaced by a generator. In the present 240volt system operating in Sudan both generator and pump are fitted to the machine, allowing the farmer to pump during the day and have electricity at night by simply moving the belt. For the 240volt system a three phase induction motor is used as a generator with an electronic controller and ballast load to keep voltage and frequency constant.

The useable power produced (i.e. after all system losses such as pump losses and pipe friction or generator losses have been deducted) by a water current turbine is given by the formula:

Useable Power Output = $1/2 . \eta . \rho . A_s . V^3$

where η is the system efficiency (0.15 for a well engineered water pumping or electricity generating system)

ρ is the density of water $(1000 kg/m^3)$

A_s is the rotor swept area (m^2)

V is the free stream water velocity at the rotor hub depth two diams. upstream (m/s)

Water current turbines with swept areas of up to $10m^2$ and useable power outputs of up to 2kW are available commercially and are manufactured in an engineering workshop in Atbara, Sudan and by Thropton Energy Services in U.K.

The range of performance of the turbine can be illustrated as follows:

a) 0.6m/s is considered to be the minimum viable water current speed. At this speed a 4metre rotor diameter water pumping turbine gives a discharge of 2.6l/s at 4metre static head and the output from a 240volt electricity generating machine is about 100Watts.

b) At 1.2m/s a 3.4metre rotor diameter machine would give an electrical output of 820Watts. At 1.9m/s the corresponding figures are 2.2metres and 1750Watts.

c) At speeds above 2.0m/s the drag forces become very high and special mooring and pontoon arrangements are necessary to ensure safe operation.

3) WATER CURRENT TURBINE CASE STUDIES

Typical applications include the replacement of a diesel pump by a water current turbine as in the case study below:

3.1) Supply of Irrigation Water For 12 Acre Date Farm

A GARMAN water current turbine is providing round the clock irrigation for a twelve acre date farm beside the Nile. The farm is situated in a very remote area twenty kilometres north west of Abu Hamed in northern Sudan.

The turbine was manufactured in Atbara, Sudan under license from Thropton Energy Services. It was installed in January 1995 by the Nile State farmers Co-op Union mobile workshop technicians.

System Details:
Output from the turbine flows through a 40m long 3" diameter polythene pipe up the bank to the start of the earth irrigation

canal which is 14 metres above the river level during the summer. From the main canal the water flows through a series of earth channels to small basins about 10m x 10m constructed so that when the first basin is filled the water automatically flows into the second and so on. This system allows the irrigation to go on without a full time attendant and means that the turbine can be run 24 hours per day. This farmer is the first irrigation user to make use of the full potential of the machine in this way.

Delivery:
    At the lowest river level the turbine delivers between 4.8 and 6.3l/s to the start of the channel and basin system and in its first three months of operation delivered 28,000m^3 of water (approximately 57m^3/ha/day). The entire farm is planted with dates underplanted with vegetables and animal fodder crops.

Advantages:
    The turbine replaces an earlier diesel engine powered system which required continuous supplies of fuel and oil. Due to the isolation of the area, obtaining fuel supplies was a constant problem for the farmer. The turbine also operates without a full time attendant and normally only has to be cleaned and checked every week or ten days. Operation is much more straight forward than a diesel pump and maintenance is only required once a year to check the belt tensions etc. Running costs are extremely low. Inspection of the machine in June 1996 after more than 11,000 running hours confirmed that no spare parts had been required.

Costs:
    The capital cost of the Sudan built water pumping turbine is £2,600, about four times the cost of the conventional technology, an Indian made diesel pump and about one tenth of the cost of a solar powered pumping system with similar daily output. When fuel and maintenance costs are taken into account the turbine becomes cheaper than the diesel option over a period of two years or more. Looked at from another angle, the capital cost of the turbine is equivalent to half of the farmer's annual profit from dates.

Back Up:
    The farmer has already made arrangements with the Co-op Unions mobile workshops to carry out the annual maintenance when the turbine is moved into the river bank during the first few days of the flood. In the event of a breakdown the mobile workshop can be summoned by sending a radio message from Abu Hamed.

3.2) Water Supply For 5,000 Displaced People, Juba, Sudan

    The first of four GARMAN turbines purchased by the French N.G.O., A.I.C.F., has been commissioned on the Nile in Juba, southern Sudan to pump water for 5,000-6,000 people displaced by the civil war.
    The turbine was manufactured in Atbara, Sudan under license from Thropton Energy Services. It was installed by Mr. Hugo Molenaar, an engineer seconded to A.I.C.F. by Red R.

System Details:
    Output from the turbine flows through a 750m long pipeline to an Oxfam 45,000 litre tank ten metres above the river level. From the Oxfam tank the water passes through a slow sand filter

constructed in an 1,800 litre tank and from there into another 1,800 litre holding tank to cope with fluctuations in demand. Flow through the sand filter is upwards and a gate valve at the base of the filter tank provides a means of backwashing the filter. The holding tank feeds a pipe system comprising three tap stands with four taps per stand and an additional three taps in the A.I.C.F. dispensary (doctor, lab and patients).

Delivery:
    At the lowest river level the turbine delivers 2l/s to the Oxfam tank (ie 172,800litres/day) which is more than adequate for the population being served. It is planned to use the excess water for irrigated vegetable gardens.

Advantages:
    The turbine replaces an earlier diesel engine powered system which was damaged during the attack on Juba in 1992. Using the river powered turbine has eradicated the previously intractable problem of fuel shortages. The turbine's ability to function 24 hours per day without an attendant is an added bonus and the continuous delivery means that a smaller diameter pipe can be used, partly compensating for the machine's higher capital cost. From the point of view of the funding agency the capital cost of the turbine (£2600) is only a small part of the cost of the whole water scheme and is not high when the freedom from recurrent expenditure on fuel and oil is taken into account.

Back Up:
    During manufacture of the turbine a one week training course in installation and maintenance of turbines was provided in Atbara for Mr. Molenaar and his two Sudanese colleagues (one from the Juba Water Corporation and one employed by A.I.C.F.). One of the Atbara turbine technicians travelled to Juba to assist with the installation and commissioning. Technical support and spares are readily available from Atbara or from Thropton Energy Services in U.K.

4) TECHNICAL CONSTRAINTS

    Detailed design considerations will not be discussed in this paper (Refs:1 & 2). However there are design constraints applying to WCTs operating in rivers which should be mentioned. These are seasonal water current speed change and debris in the water.

4.1) Seasonal Changes in Water Speed

    Where there is a seasonal speed change of more than about 40% it is necessary to change rotor diameter to keep the power levels within the design specifications. This can be done by changing the blades and altering the transmission ratio by moving the second stage belt. Both the intermediate shaft large pulley and the pump pulley are stepped so that the belt can be moved across without the need to retension. This procedure can be done by the operator after additional training. A simple current speed indicator is now fitted to all machines supplied to this type of site to assist the operator. The two sets of blades and stepped pulleys add to the cost of the machine but allow year round operation.

4.2) Debris

Experience has shown that because debris either sinks and moves along the river bottom or floats on the surface it does not damage the turbine rotor. However at some sites floating debris can collect on any part of the machine which cuts the water surface and will eventually block the flow to the rotor. Design improvements resulting from extensive field testing have greatly reduced debris problems but all WCTs must have components which cut the water surface and so some rubbish will be collected. Careful siting (just out of the main stream) will help but there may be some rivers where cleaning of debris from the machines is a daily or weekly operation. Water hyacinth is a problem on parts of the White Nile in southern Sudan and machines there need regular cleaning but there is no debris problem on the main Nile in northern Sudan.

4.3) Short Term Speed Fluctuations

A third constraint which applies only to the 240volt electricity generating machine is the steadiness of the water flow. The generator must run at constant speed so the system will only give good efficiency over a relatively narrow range of river speed(+/-10%). At installation the transmission ratio must be set to give the optimum rotor tip speed at the lowest river speed encountered during the year. Also the second by second variations in water current speed at many river sites are at least 10% of the average velocity measured over 100seconds. This gives a variation in power output of about 30% (due to V^3) and means that the load which can be applied to the user circuit is equivalent to slightly less than the minimum power produced during the 100second cycle. Because of these two factors a lot of power is being wasted at most sites in the ballast and so the system efficiency of the 240 volt machine is similar to the water pumping machine in spite of the higher efficiency of the induction motor compared to the centrifugal pump and pipe line. Canals have been found by experience to have steadier flow and are therefore likely to give better system efficiency.

The option of using a variable speed permanent magnet alternator, transformer, rectifier, battery and inverter is being investigated as an alternative. The battery would only be there to smooth out second by second water speed variations and does not need to be of large capacity.

5) THE AVAILABLE RESOURCE

5.1) Sudan

In 1993 the Sudan WCT manufacturer carried out surveys at 132 sites randomly selected along the Nile between Khartoum and Wadi Halfa (Ref:2). Each site was surveyed twice, once during the flood season and once at the lowest river level. Amongst the measurements taken at each site were the static pumping head plus current speed and river depth 12m from the bank. The proportion of sites which were suitable throughout the year and which would give a minimum discharge of more than 4l/s (the minimum found to be acceptable to the farmers) was 10.6%. Operating experience has shown that turbines can be installed every 50 metres along the river bank without the wake of one affecting the performance of the next one downstream. Using this figure, there is the potential for over

6000 machines(equivalent to about 3.5MW) along the 3000km of river bank between Khartoum and Wadi Halfa. This ignores island sites, side by side installations (with the pumps in parallel to increase discharge or in series for very high heads) and sites between 12m and 24m from the bank which are now practicable. Thus using presently available, mainly locally manufactured technology a realistic estimate of the potential for this stretch of river is between 6 and 8MW. It is worth noting that this power potential, although modest, is sufficient to irrigate approximately 100,000 hectares of land beside the Nile.

5.2) Peru

In 1996 IT Peru and Thropton Energy Services carried out 26 surveys in both the high and low jungle regions of Peru. Over 70% of the sites surveyed would give at least 500Watts electrical output and over 40% would give at least 1.5kW. Most of the rivers surveyed were at their high level and the tests have not yet been repeated at the low river levels. However data obtained from the Ministry of Energy indicate that velocities do not decrease markedly as the river levels drop. Even if half of the sites were unusable at low river levels the potential, using existing technology, along the approximately 8000km of river bank of the Amazon and its main tributaries in Peru is at least 80MW or 10kW/km. Peru is of course only a small corner of the Amazon basin.

5.3) India

In India there is a significant energy resource in both the river and canal systems. An indication of the available resource on some of the canals is given in Table:1.

The values in Table:1 are calculated from the data given in (Ref:3). Outputs are given in terms of electrical power but the hydraulic output from a water pumping machine would be similar ([Discharge in l/s] x [pumping height] x [g]) for pumping heights between 5 and 10m. Water pumping for irrigation of farms beside canal cuttings is a particulary appropriate power use because the maximum farm water requirement will coincide with the period when the canal is running at or near full discharge. It should be noted that mean canal velocities have been used whereas WCTs can be positioned so that they exploit the maximum velocity at any cross section. Thus the output per machine can be expected to be higher in practice.

Two limits have been applied to the possible total power output and the output/km. Firstly, taking the minimum distance between turbines to be 50m as discussed above, the maximum number of machines is 40/km if both banks of the canal are used.

Secondly, WCTs remove energy from the canal and therefore will have a similar effect to weed growth, increasing the friction coefficient and hence the hydraulic gradient. This will reduce the freeboard at the top end of the canal reach and reduce the drop at any control structures. Problems are only likely to be caused in canals where there is a shortage of gradient (i.e. the control structure drops are very small) and only then when these canals are running at or above their design capacity. Thus it is expected that there will be some canals which can cope with large numbers of machines and some in which only relatively few may be installed. Given information about the length of each canal reach, the freeboard and the control structure drop it is possible to

Table:1 Assessment of Power Potential

Canal Name	Speed in main sect.	Depth	Length	Output per machine	Total output	
	m/s	m	km	kW	kW	kW/km
Godavari Central	0.97	2.6	13	0.38	60	05 %
Godavari East	0.73	2.8	6	0.19	40	06 %
Godavari West	1.1	3.2	9	0.65	70	07 %
Jawahar	1.1	3.6	203	0.75	5000	25 %
Nizamsagar	0.9	3.1	155	0.35	2170	14 N
Vamsadhara Left	1.1	1.9	107	0.35*	670	06 %
Dhansiri	1.95	1.6	20	0.85*	680	34 N
Western Kosi	1.37	4.2	76	1.00	2200	29 %
Salauli	0.94	1.6	25	0.22*	70	03 %
Banas	1.48	2.4	46	0.85*	560	12 %
Damanganga Right	1.37	2.6	45	1.00	610	14 %
Kakrapar Left	0.81	3.3	64	0.25	620	10 %
Karjan Left	1.31	2.7	52	0.85	290	06 %
Mahi Right	1.66	5.0	74	1.60	2050	28 %
Narmada	1.66	7.6	443	1.60	28350	64 N
Panam	1.33	1.9	105	0.60*	440	04 %
Ukai Left	1.23	3.0	73	0.70	310	04 %
Ukai Right	0.89	2.9	48	0.22	350	07 %
Bhakra	1.74	5.4	278	1.65	6540	24 %
Augmentation	1.12	4.0	75	0.75	1290	17 %
Jawahar Lal Nehru	1.43	2.6	226	1.00	4020	18 %
Gurgaon	1.06	3.5	94	0.70	460	05 %
Jui Lift	0.93	1.7	80	0.21*	270	03 %
Loharu Lift	0.97	1.7	68	0.24*	360	05 %
W. Jamuna	1.53	4.2	82	1.30	4280	52 N
	TOTALS		2467		61760	25

NOTES:
 * indicates the use of battery charging (12 or 24 volt) machines
   due to the limited depth.
 % indicates that limit of 10% of canal's potential power is used
 N indicates that limit of 40 machines/km is used.
 Typical speed range for unlined canals is 0.75-1.1m/s and for
   lined canals 1.3-1.7m/s.
 Discharges of canals in above table vary from 14m^3/s to 1132m^3/s.
 The output per machine is based on the useful electrical output
   of the machine in Thropton Energy Services range which most
   nearly suits the available speed and depth.

calculate the power which may be extracted from that reach without causing problems. Assuming one large turbine filling the canal rather than many small ones along the edges we can write;

Useable Power Output (Watts) = $\eta \cdot \rho \cdot g \cdot H \cdot Q$

where η is the turbine system efficiency (0.15)

ρ is the density of water (1000kg/m^3)

g is the gravitational constant (m/sec^2)

H is the allowable level increase at upstream end of reach (m)

Q is the canal design discharge (m^3/sec)

In the absence of detailed canal data at the time of writing and purely for the purposes of illustrating the potential, the average permissable power output from the WCTs on a canal has been taken as 10% of the canal's potential power ([Total fall along canal bed in m] x [design discharge in m^3/s] x g x ρ). That is to say 10% of the power normally dissipated in friction on the canal bed and sides. Notwithstanding the uncertainty of the 10% assumption, the figures in Table:1 show that canal flows in India represent a substantial energy resource and that there is a very large potential for providing small, decentralised power supplies using presently available technology.

6) MANUFACTURE

The WCT shown in (Fig:1) has been designed to be made and maintained in regions where it will be used. Much of the manufacture is straightforward fabrication which is labour intensive but uses the minimum of materials. In the design every effort has been made to use standard components and materials easily available in most countries. This can help to reduce the foreign exchange component of the cost. A proportion of local manufacture also provides local employment.

All cost calculations have therefore been made on the basis of as much local manufacture as is consistent with achieving the required quality and reliability. For example, the ex works cost of a Sudan built water pumping WCT is £2,600 of which £650 is parts supplied by Thropton Energy Services. All other materials are bought in the local market for local currency (although they may have been imported originally). Sudan has the highest requirement for specially imported parts of any country where manufacture has so far been considered. In India it is not expected that any specially imported parts will be required.

Manufacture requires cutting, welding and drilling equipment plus a milling machine, large lathe (to machine the 0.5m diam pulleys) and aluminium casting facilities. The Sudanese manufacturer has additionally purchased grit blasting and zinc spraying equipment for corrosion protection of the under water parts. Thropton Energy Services has also supplied items such as hardened drilling bushes for jig and fixture making, form tools for machining pulleys and quality control equipment such as plug and 'C' gauges.

7) COST

As mentioned above, the ex works cost of a Sudan built turbine is £2,600 at a production rate of between ten and twenty per year. This is about one tenth of the cost of a solar powered pump of similar daily output and one third the cost an equivalent wind pump. In Egypt where about 80% of the parts and materials could be locally manufactured the ex works price in 1992 was calculated to be £1,470 at 50 units per year and £1,200 at 200 units per year (Ref:4).

In India it is expected that almost 100% indigenisation will be possible (Ref:5) and costs are expected to be similar or lower than Egypt. It would therefore seem reasonable to expect an ex works cost of around £1,200 (Rs 60,00)at 200 units per year in India.

Table:2 gives an indication of the expected economics of water pumping WCTs in India as compared to diesel pumps. In this table the following assumptions are made:

1) Diesel costs based on Sudan Ministry of Agriculture figures;
   Fuel consumption of river/canal bank pumps averages 21litres/hectare/irrigation. Our tests on two Indian made diesel pumps agreed with this to within 9%.
   Average of one irrigation every 20 days.
   Oil consumption is one gal/drum of diesel.
   One kg of grease/pump/year.
   Typical water requirement is for growing vegetables during dry season is $64m^3$/hectare/day.
2) A growing season of 260 days (i.e. 13 irrigations/year)
3) Static head is 6m.
4) Both pumps are costed on the basis of irrigating an area of five hectares. To do this the diesel pump would be run for an average of 9 hours per day during the season. A turbine delivering 6l/s would irrigate the five hectares if run for 15 hours/day.
5) The diesel pump set has a life of 15,000 hours after which the engine is replaced and the pump overhauled. Spares and maintenance costs are estimated from the expenditure of Sudanese farmers calculated as a percentage of capital cost.
6) The turbine has a life of 12 years (limited by corrosion not wear) with major maintenance every three years (12,000 hours).
7) The cost of diesel fuel in India is Rs8 per litre (Ref:5)

As can be seen from the table, the turbine is cheaper over a period of three years or more. The figures also show the high running cost of the diesel compared to the WCT. This tends to discourage the farmer with a diesel pump from irrigating areas not previously cultivated. However a farmer who has purchased a WCT will try to develop enough land to make use of the full output of the machine and gain the maximum benefit from the low running cost. The WCT therefore tends to encourage the increase of the cultivated area.

Year	1	2	3	4	5	6	7	8	9	10	11	12	Total
						Cost in Rs(000s)							
Ex works cost	60												60
Installation	05												05
3 year spares				05			05			05			15
Annual maint.	01	01	01	01	01	01	01	01	01	01	01	01	12
3 year maint.				01			01			01			03
Annual Cost	66	01	01	07	01	01	07	01	01	07	01	01	95
Accumulated Cost	66	67	68	75	76	77	84	85	86	93	94	95	
Diesel Pump Cost	25						15						40
Installation	02						01						03
Fuel	11	11	11	11	11	11	11	11	11	11	11	11	132
Oil & Grease	01	01	01	01	01	01	01	01	01	01	01	01	12
Spares & maint.		03	05	05	05	05	00	03	05	05	05	05	46
Annual Cost	39	15	17	17	17	17	28	15	17	17	17	17	233
Accumulated Cost	39	54	71	88	105	122	150	165	182	199	216	233	
WCT-Diesel Pump	27	13	(03)	(13)	(29)	(55)	(66)	(80)	(96)	(106)	(122)	(138)	

Table:2 Comparative Costs of the GARMAN Turbine and a Diesel Pump

8) INFRASTRUCTURE FOR DISSEMINATION

8.1 Surveying, Installation and Maintenance

All micro hydro systems are site specific and thus accurate
site surveying is essential to ensure appropriate system design.
Due to the long service life and high reliability required from
hydro plants to compensate for their high capital cost, a very high
standard of installation with meticulous attention to detail is
required. When something does go wrong there must be back up
readily available in the form of qualified technicians and spare
parts. Without doubt the hardest task in the dissemination of any
micro hydro is to put in place the infrastructure required for
surveying, installation and maintenance without adding unacceptable
amounts to the cost of systems. The provision of these services for
such small and of necessity low cost systems is not an attractive
commercial proposition for the large companies in capital cities
which are capable of manufacturing complete sets (including
generators). Thus if a significant fraction of the available micro
hydro resource is to be exploited investment in training of local
technicians will be required (e.g the staff of small agricultural
engineering firms which are already providing services to the
farmers). It is at this point that assistance by aid donors or
government can play an important role.

In the case of WCTs in northern Sudan the farmers' Co-op
Unions have been appointed as agents by the manufacturer. The
Unions already operate mobile workshop services for diesel pump
repair from three centres along the Nile. The mobile workshop
technicians have been trained in site surveying, installation and
maintenance. The training was funded by the German NGO, German
Development Service. Surveying equipment (accurate current meter,
inflatable boat, outboard engine etc) was provided by grant from
the Sudan government) and a spare parts stock have been supplied
to each centre. With the aid of suitable standard forms, the
technicians process the site survey data to specify to the
manufacturer the required rotor diameter(s), pump size and
transmission ratio for each new machine.

8.2) Finance

The relatively high capital cost of the WCT is a major
stumbling block for most prospective purchasers. Unless the
turbines are manufactured in reasonable quantity their price will
remain high. To stimulate the market government help is needed.
This could take the form of a grant equivalent to the government
subsidy on the 16,200litres of fuel saved during the turbine's 12
year life. Alternatively low interest loans from government banks
could serve the same purpose. In Sudan the Agricultural Bank of
Sudan gives the farmers three year loans at half the normal
commercial rate.

9) CONCLUSIONS

1) The extensive canal systems of India can provide significant
   amounts of power for water pumping and electricity generation.
2) Small fractions of this power can be extracted anywhere along
   the canal bank by water current turbines.
3) Water current turbine technology at rotor shaft powers of up to
   3.5kW for pumping and 12, 24 and 240volt electricity generation

is well tested, available and suitable for transfer to India.
4) The ex works cost of a water pumping turbine in India is expected to be approximately £1,200. At this price it can pump water more cheaply than a diesel pump set over any period of three years or more.
5) To introduce WCT technology into India it is now necessary to install demonstration turbines at suitable sites in canals so that their effect on canal flow can be studied and so that potential users and manufacturers can observe them. At the same time a more detailed study of the canal and river resource in India should be carried out.
6) To acheive widespread dissemination of low cost micro hydro technology, investment in training, infrastructure and credit schemes by government and international organisations is required.

REFERENCES

1) Garman P.G. 1986. Water Current Turbines: A Fieldworker's Guide. Intermediate Technology Publications, London, 114pp. ISBN 0 946688 27 3
2) Garman P., Sexon B., Ahmed F.R. and Fahmi M.V. 1993. University of Wadi el Nil, Atbara, Sudan. Water Current turbine Development in northern Sudan, June 1990 to May 1993.
3) Central Board of Irrigation and Power 1986. Major Canals in India. Publication No.187
4) Jones G. 1992. Establishment of Manufacture of Water Current Turbines in Arab Republic of Egypt. Energy for Sustainable Development Ltd., Report to the European Commission.
5) British High Commission, New Delhi 1996. Market Information Report for Thropton Energy Services.

First International Conference on Renewable Energy–Small Hydro
3–7 February 1997, Hyderabad, India

PROJECT OF A PICO HYDROGENERATOR FOR RURAL USE

Geraldo L.Tiago F.,[1] *Osvaldo J. Nogueira,*[1] *Costa Dimas*[2] *and Maria S.C.C. Mendonca*[2]

[1] Federal School of Engineering of Itajuba, Small Hydro Power Lab
   Av. BPS, 1303 - Bairro Pinheirinho, Itajuba - M.G., Brazil 37500-000
[2] Minas Gerais Energy Company, Av. BPS, 1303-Bairro Pinheirinho
   Itajuba - M.G., Brazil

Introduction

Given that the Minas Gerais State has a wide hydro potential, the Minas Gerais Energy Company - CEMIG is developing to evaluate the micro hidroelectric power stations (MHS). The CEMIG has been receiving assistance from the Federal School of Engineering of Itajubá - EFEI through the Fundação de Pesquisa e Assessoramento a Indústria - FUPAI. The MHS are alternative sources of energy for rural properties.

In Minas Gerais State almost 55% of the rural properties, about 300,000 of the 500,000don't have electrical energy. The rural electrical energy brings comfort, information, education and improvement of agricultural production. On the other hand, rural electrification has very small profitability because it requires large amount investments in the distribution network, even for small loads.

Most of the rural properties without electrical energy don't have enough flow or head for a conventional MHS (1 to 50KVA), thus forcing the development of a technology to take the advantage of the energy sites with less than 1KVA as an alternative to the rural electrification. A turbine/generator set, to produce energy in DC mode has been developed, which is called Picohydrogenerator (PHG-DC). The PHG-DC is intenteded to provide energy for illumination, radio, TV,small water pump and battery recharging.

Description of the PHG-DC

The PHG-DC is composed of a DC generator, a hydraulic turbine and a system ti transfer energy from the turbine generator, in this case, belt, as presented in the following figure, where:

The operation of the PHG-DC consists of striking a jet of water over the buckets of the turbine, causing the wheel to spin. The hydraulic energy of the jet of water is then converted into mechanical energy of the shaft, that is transferred to the generator by the belt and is converted in electrical energy.

The Turbine

The turbine developed had to be simple, cheap and adaptable to lower heads and low flows of water. Based on those restrictions, the most suitable is the Turgo action turbine.

The Turgo turbine is made of many buckets attached to the runner in a axle. The buckets are, basically, made of inlaided steel cups, and the runner is a steel wheel with a border at its edges. Although the Turgo is classified as free jet, like the Pelton, it has the jet striking the buckets laterally, according to the diagram shown bellow - see Figure 2.

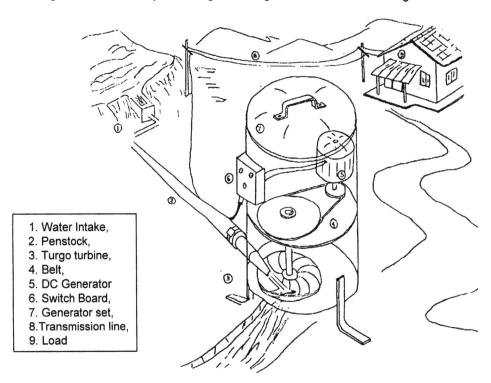

1. Water Intake,
2. Penstock,
3. Turgo turbine,
4. Belt,
5. DC Generator
6. Switch Board,
7. Generator set,
8. Transmission line,
9. Load

Figure 1 Seheme of the picohydrogenerator with Turgo tutbine

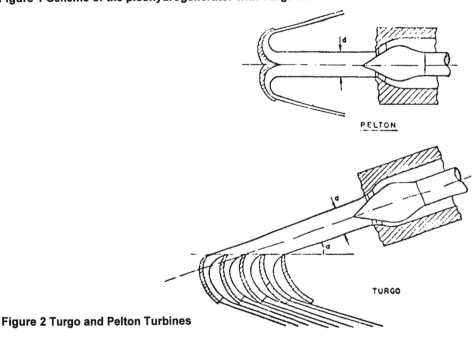

PELTON

TURGO

Figure 2 Turgo and Pelton Turbines

The operation of this turbine is based on striking one, or many, jets of water over the sides of a rotor, and leaving it on the opposite sides, and from the impact of the water over the buckets, the rotor spins. This turbine can have up to three jets of water, thus obtaining more energy without the need to increase the size of the turbine. Usually, the turbine has only one jet of water with the diameter smaller than the maximum allowable. The Turgo turbine offers large power for its size. additionally, the Turgo turbine can reach high speed of rotation, so that it can be linked directly to the generator, without the need od multipliers, and thus requiring simple, economic, reduced and durable installations.

Advantages of the Turgo Turbine

In the Pelton Turbine the jet of water can reach up to 1/8 of the diameter of the wheel, while in the Turgo, the jet can reach up ro ¼ of the diameter of the wheel. Based on that, for the same jet of water, the Turgo can have the half of the diameter of the Pelton, besides requiring lower, or no multiplication os speed to run the generator.

For flows between 0.15 1/s, the tangential hit of the jet of water in the Pelton works better, although theoretically, the Turgo could also be indicated. For higher flows, the lateral hit in the Turgo is more efficient and has lower noise.

Comparing to the reaction turbines, such as Francis, Kaplan and Propeler, the Turgo presents as the main advantage, the lack of cavitation, which can cause damage to the turbine. Another advantage is its lower cost of operation and construction of the Turgo compared to the other turbines.

Conclusions

The performance curves of the Turgo turbine show high efficiency, between 80% and 90%, at different flows and loads. This is important if the turbine is going to be used with changing loads, as the case rural areas.

Base on the chart given below, it can be verified taht the Turgo turbine is very suitable to a wide range of heads, producing enough power to supply low loads, such as illumination, radio, TV and a small water pump.

Heads	(m)	2	3	4	5
Flow	(l/s)	9,86	12,08	13,95	15,60
Hydraulic Power	(W)	193,50	355,56	547,40	765,18
Electric Power	(W)	55,15	80,00	156,00	218,08

By utilizing the components found in the market, it is possible to built a set of less than 1 kW, at low cost to operate at different heads. This work has been developed on that premise.

Bibliography

Alen, C.A. "Funcionamento, dimensionamento básico e aplicação de uma roda Turgo" - monografia, EFEI, Brasil, 1988

Cheng, L. - J. - "A new Turgo Turbine in China" in water power, vol.2 - pp. 934 - 943, 1985.

Ferreira, P.C.G. "Turbina Turgo", rel. pesquisa, EFEI, Brasil, 03/1989.

Harvey A. et all - "A Guide to Small-Scale Water Power Schemes" - Ed. It Publications, Great Britain, 1993, ISBN, 853391034.

First International Conference on Renewable Energy–Small Hydro
3 – 7 February 1997, Hyderabad, India

THE MODULAR DESIGN OF SMALL WATER TURBINES

David A. Williams

Gilbert Gilkes & Gordon Limited
Kendal, Cumbria LA9 7BZ, UK

INTRODUCTION

The scope of this paper is to show how Gilkes, as a manufacturer of water turbines for small hydro-electric schemes (0 to 10 MW), solves the problem of producing a machine which is uniquely designed for each site but keeps the costs to a minimum by using a modular design concept.

THE PROBLEM

Turbines are selected and designed to suit the two basic parameters of a site :

1.   The effective head of water (pressure).

2.   The flow of water available.

No two sites have the same head and flow.

Head is the vertical distance from head water level to the turbine or tail water level. The head water level may be the level of water behind a dam or the intake to the pipeline feeding a turbine if water is abstracted from a river.

The effective head at the turbine will be the static head less any losses in the pipeline due to friction.

Flow depends upon the rainfall on, and the catchment area at, the intake to the pipeline and, of course, it varies with seasonal differences in rainfall.

For large hydro-electric schemes above 20MW, a model is built of the turbine runner and it is tested so that the actual size, efficiency and performance of the full size unit when subjected to the site conditions can be accurately predicted.

Unfortunately this is impractical with small hydro since turbine runner size is often the same as the model and this procedure virtually doubles the cost of the mechanical/electrical equipment. Most schemes would not be viable using this approach.

However, the customer still wants high efficiencies from the equipment at low cost and it is to this end that a modular design concept has been adopted at Gilkes.

Before describing how modular design works, it is necessary to see how turbine selection is achieved.

TURBINE SELECTION

Power available from water in a hydro-electric scheme is the product of head (pressure) and the rate of flow of water (discharge) through the turbine.

Water power available   =   Head x Flow

P   =   H x Q   (1)

Final output from the hydro site must also take into account :

> Head loss in the pipeline due to friction (effective head or net head equals gross (static) head minus friction loss).
>
> Turbine efficiency
>
> Generator efficiency
>
> Drive loss (if generator drive is via belt or gearbox)

Five main types of water turbine are available; Pelton, Turgo Impulse, Francis, Kaplan, Crossflow.

How is one type actually chosen and not another?

Turbines are classified by specific speed.

The definition of specific speed is "the speed (in RPM) of a geometrically similar (to the one in question) turbine capable of developing 1 metric BHP under a 1 M head".

In mathematical terms :

$$\text{Specific speed (Ns)} = \frac{N \sqrt{BHP}}{H^{1.25}} \quad (2)$$

Each specific speed turbine runner design has its own model characteristics. Basically, the specific speed should be thought of as a type number and fig. 1 shows the relationship between runner geometry and specific speed, as well as the limiting head under which they normally operate.

We can now see from equation (2) that for a given head and output (i.e. a given site) to increase the rotational speed of the turbine/generator set one must increase the specific speed.

This is what the turbine manufacturer attempts when he carries out his turbine selection, to enable him to reduce the size of machine/generator and keep costs down.

The starting point for the selection of a turbine as stated earlier is the net head and design flow, and using the specific speed band, one or two likely specific speed runner designs are chosen. This in turn leads us to our model curves (see fig. 2 and 3 for a typical Pelton and Francis performance curve). For very large hydro applications the machine will be designed so that the duty speed coincides with the actual best efficiency point (BEP) machine speed. For small hydro applications this would entail excessive design costs, so we have a range of machine sizes to cover a certain application area and the selected speed will normally be within +/- 5% of (BEP) speed, which keeps one within the (BEP) envelope. This speed is usually a synchronous speed for the relative frequency in question (i.e. 50 Hertz in India etc.). The design flow is that given by the site specification.

Affinity laws relate the variation in size, head , flow and speed between model and prototype :

$$\frac{\text{Actual Flow}}{\text{Model Flow}} = \frac{\text{Actual Head}}{\text{Model Head}} \times \frac{\text{Actual Runner Diameter}}{\text{Model Runner Diameter}^2}$$

Or in symbols :

$$\frac{Qa}{Qm} = \frac{Hna}{Hnm} \times \frac{Da^2}{Dm^2} \qquad (3)$$

$$\frac{\text{Actual Speed}}{\text{Model Speed}} = \frac{\text{Actual Head}}{\text{Model Head}} \times \frac{\text{Actual Runner Diameter}}{\text{Model Runner Diameter}}$$

Or in symbols :

$$\frac{Na}{Nm} = \frac{Hna}{Hnm} \times \frac{Dm}{Da} \qquad (4)$$

From this we can locate a duty point on the model curve and by judiciously altering turbine size and speed we can position the duty point at or near (BEP).

The model efficiency is now "majorated" to take into account effect as we go from model to prototype conditions.

Basically this means one can normally gain 2-3 points of efficiency when majorating from a model to a full size machine.

The Moody formula for majorating Francis turbines is commonly used :

$$\frac{1 - \eta t}{1 - \eta m} = \frac{Dm}{Dt}^{0.5}$$

To summarise, from a basic set of site data; head and flow, we have :-

A. Selected a type of turbine from consideration of specific speed.

B. Selected a size (i.e. runner diameter) from a range of standard diameters using the model curve for the selected specific speed. This has given us an operating speed for the turbine.

C. Obtained a value for efficiency of the turbine by majorating from the model curve.

D. Discovered the potential output of the site using formula 1 and taking into account the generator efficiency and any indirect drive system losses.

Now we know the turbine we want to make, we use modular design to provide a machine at the least possible cost but maintaining the efficiencies and design quality.

MODULAR DESIGN

For the purposes of illustrating the modular design concept, we shall use the Gilkes Turgo Impulse Turbine.

The Turgo Impulse Turbine is a side jet impulse water turbine having a higher specific speed than the Pelton turbine.

Basically the unit has a runner onto which one or two jets are directed to provide the rotation which is transmitted to a generator. The jets are formed in needle valves which have movable tips regulating the flow of water to the runner Jet deflectors are used to direct water from the runner to provide speed control (governing) or to shut the unit down quickly.

279

The turbine is split down into a number of critical components or assemblies which are selected to suit particular site conditions.

Components and assemblies - though in a number of standard sizes and materials (and in the case of electric items, varying in voltage etc.) - are standard designs or pre-selected proprietary items which will all assemble together to give a unit unique for the site conditions specified by the customer.

We now look at the build-up of the turbine from its components pieces:

Runner   (Fig. 4)

The heart of the turbine, the runner size is taken from the range of standard runner sizes and has been selected as described in the previous section by consideration of head, flow, specific speed, optimum efficiency from model curve and best speed to suit local supply frequency.

Runner material depends upon the head of water and operational speed. The higher the head and speed, the higher the internal stresses in the runner and a higher tensile strength material is employed.

Case   (Fig. 4)

The case, of fabricated steel construction, is sized to suit the runner and the maximum flow of water which this diameter of runner is likely to pass.

The front of the case is designed to carry the needle valves and, where applicable, a pedestal bearing. The back of the case is merely a splash cover directing water into the tail race.

The runner and case are the two major basic components onto which all other items are designed to fit.

Shaft and Bearing   (Fig. 5)

Three designs are possible :

1. Turbine shaft with sleeve type thrust and journal bearings.

    This assembly is used when high loads are transmitted to the bearing because of high jet loads, as a result of high heads and flows.

    The bearing is mounted onto a pad which is part of the case fabricated and the shaft is flange mounted onto the runner. This design also facilitates the "sandwiching" of a flywheel between the turbine and generator shafts should extra inertia be required. This may be necessary for applications in which large instantaneous loads are applied to the generator.

    The bearing itself has four possible configurations depending upon the loads applied to it :

    1a.   Bearing with internal loose oil ring for lubrication during start-up and shutdown.

    1b.   Water cooling of the oil by a coil in the bearing sump. Pipework is required with pressure reducer and filter to provide water directly from the penstock upstream of the turbine.

    1c.   An external circulating pump (AC or DC) which pumps oil from the bearing sump to oil pockets on the thrust and journal faces of the bearing.

    1d.   The separate lubrication console with pumps (AC or DC or both if oil is required for start-up and shutdown) and a heat exchanger which obtains its water in the same manner as item 1b).

2. Turbine shaft with anti-friction type (roller) bearings.

    This method is used for lower loads than above and comprises a bearing cartridge having two

280

grease or oil lubricated spherical roller bearings.

The cartridge is mounted onto the front plate of the turbine case and the runner is keyed to the turbine shaft. A flexible coupling is used to transmit the drive from turbine to generator.

3.   Generator shaft extended into the turbine case.

In this design the runner is flange mounted onto the generator shaft end. It has price advantages over design 1. above but only when the generator can accommodate bearings which can tolerate the side loads and thrusts owing to jet forces. It is also only applicable when no extra inertia is required over the inherent generator inertia or when any extra can be built into a flywheel at the non-drive end of the generator.

Shaft Seals   (Fig. 5)

A labyrinth seal is used on all Turgo Impulse applications.

For shaft and bearing designs 1. and 3. above, the seal is mounted on a plate bolted to the case front plate.

For shaft design 2. above, the seal is attached to the bearing housing between the drive end bearing and the runner.

Variation in shaft seal designs are only dependant upon the variation of shaft diameter which is a function of power output of the turbine and speed.

Needle Valve (Fig. 6)

The Turgo Impulse turbine is fitted with one or two needle valves and each one consists of three major component assemblies :

1.   Branchpipe

2.   Nozzle / needle tip

3.   Needle actuator

All components are sized and designed to cater for variation in head, flow and pipeline characteristics.

1.   Branchpipe

This is an angled pipe which supports the needle valve assembly. The internal diameter is chosen to keep the velocity of the water within a design limit. Velocity is a function of flow rate and head.

The wall thickness of the branchpipe is designed in common with other pressure vessels, to suit the internal pressure rating i.e. head of water at the site.

2.   Nozzle and Needle Tip

Designed to provide the optimum jet form of water at the runner, the spear tip and nozzle angles are the same for all duties. nozzle diameter is purely a function of head and flow as is the distance the needle travels from fully open to fully closed.

Branchpipes, nozzles and needle tips are all site specific but, as with runner diameter, there are a number of set branchpipe diameters with matching nozzles and needle tips - only the diameter to which the nozzle is machined and the spear travel is unique to each site.

3.   Needle Actuator

3a.   Electric   -   AC or DC, to suit customer requirements, with manual handwheel

281

override.

3b.   Hydraulic -   Taking the oil supply from power pack with AC or DC pumps (or both).

3c.   Manual -   Handwheel directly connected to the needle valve rod.

For electric and hydraulic actuators, the customer must advise the minimum time he can accept for totally closing the needle valve from full flow operation. The closure time will directly affect pressure surges in the pipeline which feeds the turbine.

The actuation of needle valves is an area where close co-operation between turbine manufacturer and customer/consultant is essential. Needle valves may be part of a governing system in which needles are required to economise water on changes of deflector position (see next section) or may be flow control devices due to downstream requirements (in irrigation or water supply schemes), or water availability at the pipeline intake.

For any of these applications, the actuator on the needle valve will receive a signal from an external device and by collaboration, the turbine manufacturer ensures the customer gets what he requires.

Deflectors Assemblies   (Fig. 7)

Deflectors are used to cut off or reduce the jet of water hitting the turbine runner.

They serve two main functions :

1.   To control the speed of the turbine by regulating the flow of water onto the runner. If the load on the generator is altered, the speed of the turbine changes and it is restored to the required running speed by a governing system moving the deflectors to a new position.

2.   To shutdown the turbine, usually in an emergency situation. Water is completely deflected away from the runner causing the unit to slow, and eventually stop.

The deflector assembly is mounted into the turbine case and the components are sized to suit the case and the maximum jet size which the runner will take. Deflectors are connected to vertical shafts which protrude through the top of the turbine case through bearings and seals and then are connected to a common operating arm and lever.

Actuation of the deflector is by either :

1.   A hydraulic cylinder with a supply from a hydraulic pumping unit with AC or DC pumps (or both).

    The governing controls porting of the oil to either side of the piston in the cylinder.

    or

2.   A weight which is normally held in the suspended position by a solenoid valve or electro magnet which, as required, de-energises causing the weight to drop pulling the deflectors fully into the jet.

    On resumption of power and the requirement to start the set again, the weight can be either manually reset or remotely by an electric linear actuator (AC or DC).

Governing (Fig. 8)

Three types of governing are used :

A.   Mechanical/hydraulic

B.   Electronic/hydraulic

C.   Electric load "dump" system

1.   Mechanical/Hydraulic

     1a.   With separate hydraulic pumping unit.

         This is typified by the Woodward UG system in which the speed of the turbine is sensed by rotating flyweights driven by an electric motor taking its supply from the control panel voltage transformers.

         The flyweights operate a pilot valve spool porting oil under pressure from a separate hydraulic console to the open or closing side of the cylinder operating the deflectors.

         The hydraulic pumpset has AC or DC driven pumps (or both) and accumulators to maintain sufficient oil under pressure to effect shutdown in case the pump(s) fail totally.

         The governor has adjustments for droop, load limit, speed setting and gate (deflector) limit. A feedback system is used to attain compensation of the speed control.

     1b.   Without separate hydraulic pumping unit.

         The Woodward PGG governor is used in this application and cannot be used over full governor capacity range of the UG unit.

         The main differences in design of the PGG governor are that the oil pump is integral with the speed sensing head and the drive for the pump and flyweights in common through a timing belt and pulley drive system from the turbine shaft.

2.   Electronic/Hydraulic

    This is a similar system to the mechanical/hydraulic unit with speed sensing and all mechanical links, speed setting etc. being replaced by electronics.

3.   Electronic Load Dump System

    Commonly called electronic load governors or controllers, this system is used for small outputs, typically up to 100kW.

    The turbine runs at full output available for the head and flow. If there is any load change, the electronic unit senses this change instantaneously by the effect on generated frequency. Excess load generated by the turbine/generator is immediately switched into a load bank system having the same capacity as the total generator output.

    This type of governing does not use any mechanical devices on the turbine.

CONCLUSION

We have now built up a complete turbine (Fig. 9)

It has been designed to suit a specific site and has taken into account the customer's requirements for operation and control.

We have made a "special" for a "standard" price.

David A Williams

GILBERT GILKES AND GORDON LTD

Fig 1 Specific speed and Turbine types

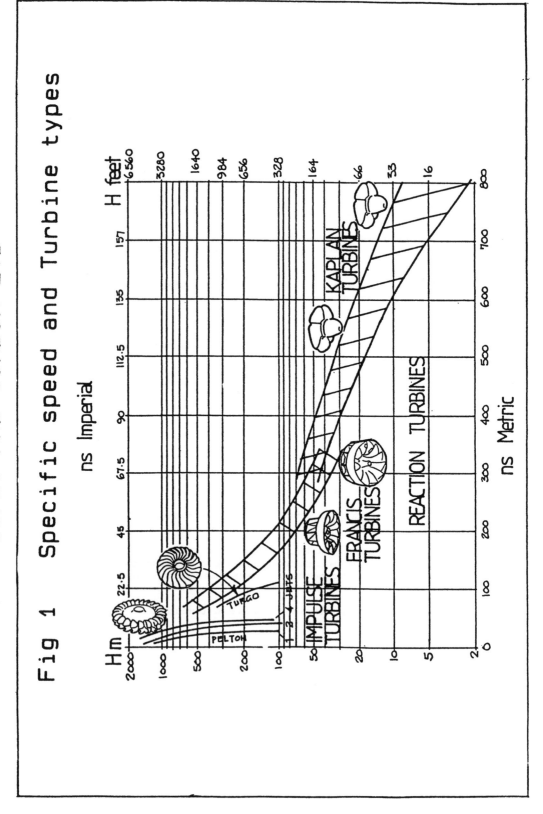

GILBERT GILKES AND GORDON LTD

Typical Model Curves

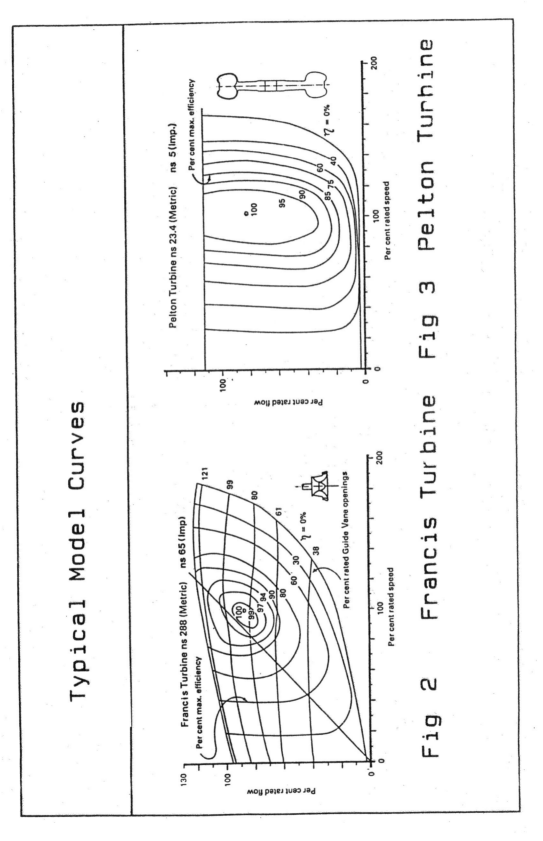

Francis Turbine ns 288 (Metric)   ns 65 (Imp)

Fig 2   Francis Turbine   Fig 3   Pelton Turbine

Pelton Turbine ns 23.4 (Metric)   ns 5 (Imp.)

GILBERT GILKES AND GORDON

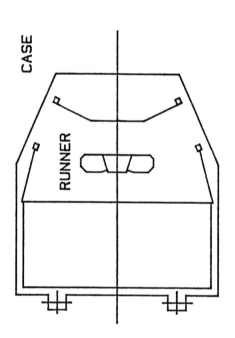

Figure 4   CASE AND RUNNER

GILBERT GILKES AND GORDON

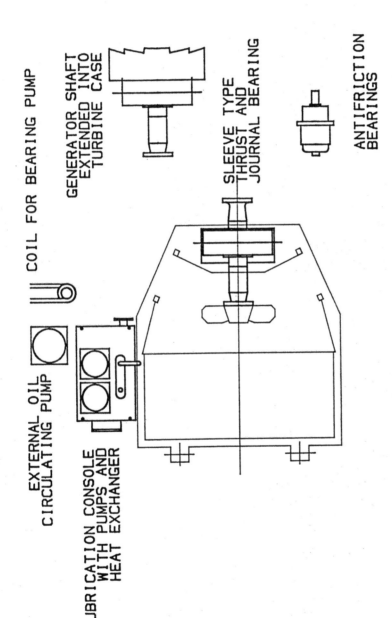

EXTERNAL OIL
CIRCULATING PUMP

COIL FOR BEARING PUMP

GENERATOR SHAFT
EXTENDED INTO
TURBINE CASE

SLEEVE TYPE
THRUST AND
JOURNAL BEARING

ANTIFRICTION
BEARINGS

LUBRICATION CONSOLE
WITH PUMPS AND
HEAT EXCHANGER

Figure 5   SHAFT AND BEARING

GILBERT GILKES AND GORDON

HYDRAULIC SERVO
CYLINDER

HANDWHEEL

ELECTRIC ACTUATOR

BRANCHPIPE

NEEDLETIP.

NOZZLE

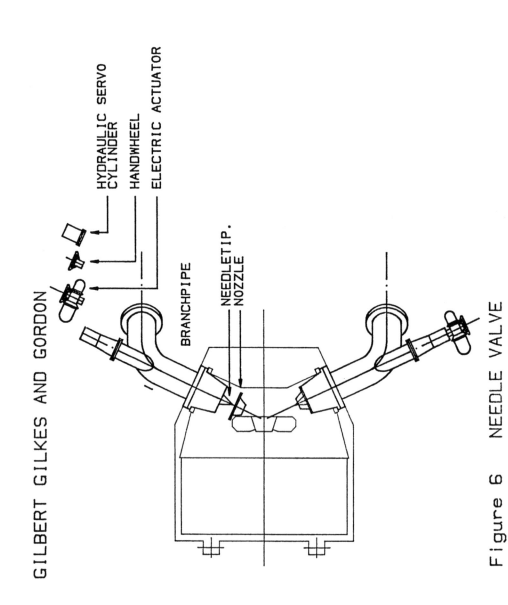

Figure 6   NEEDLE VALVE

GILBERT GILKES AND GORDON

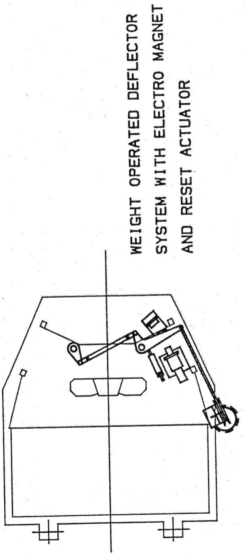

WEIGHT OPERATED DEFLECTOR
SYSTEM WITH ELECTRO MAGNET
AND RESET ACTUATOR

Figure 7   DEFLECTOR ASSEMBLIES

GILBERT GILKES AND GORDON

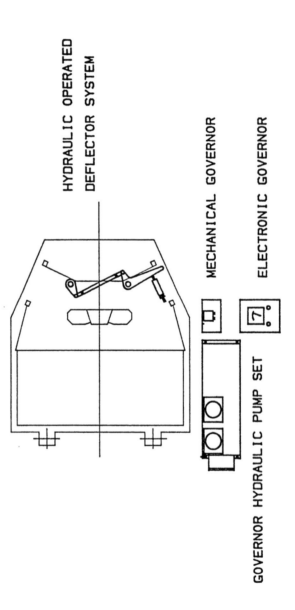

HYDRAULIC OPERATED
DEFLECTOR SYSTEM

MECHANICAL GOVERNOR

ELECTRONIC GOVERNOR

GOVERNOR HYDRAULIC PUMP SET

Figure 8   GOVERNOR

GILBERT GILKES AND GORDON

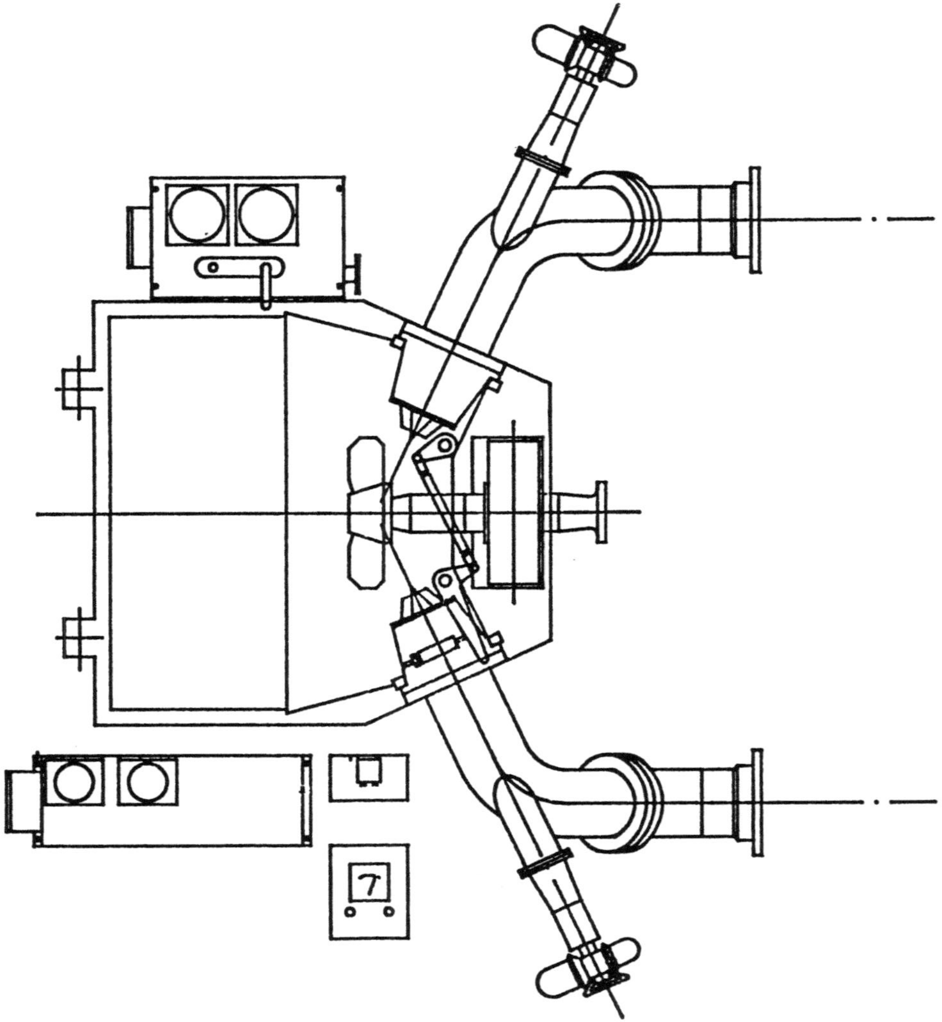

Figure 9   COMPLETE TURBINE ASSEMBLY

First International Conference on Renewable Energy–Small Hydro
3–7 February 1997, Hyderabad, India

GENERATING POWER FROM DRINKING WATER SYSTEMS

Martin Exner

Head of CINK Sales Office
Jerrytex, Konevova 31, 13000 Praha 3
Czech Republic

There is a lot of flowing water in pipes of drinking water systems feeding every city in the world. As one people working with hydro power equipment we follow the idea to utilise this flowing water for electrical power generation.

Existing state

There are drinking water systems with dams, water treatment plants water reservoirs and kilometres long pipings in between, where water flows down to the cities by gravity as shown in drawing enclosure No. 1. At the end of these pipelines are sometimes installed pressure absorbers (usually needle valves), because the pressure of water flowing kilometres by gravity must be kept within certain limits.

Technical problems

Installation of turbines on existing drinking water systems requires addressing following technical problems:

— Water contamination must be avoided.

— Hydraulic blows are not allowed.

— Water supply must not be interrupted.

— Turbines and the equipment must be installed in existing building in limited space.

— The pipings is designed more to favour of high hydraulic losses and the energy utilisation is not generally considered. The piping is sometimes heavily encrusted and in bad condition.

— Head fluctuation in drinking water dams.

Solution — CINK turbine

From the beginning, the CINK turbines were developed for use in drinking water systems and to have such performance data and quality on a repetitive basis alongwith ecologically harmless and without disturbance to the waterworks supply and technology.

CINK turbine is crossflow type turbine with original radial closing plate invented and patented by CINK with fully functioning draft tube. Because of this Cink turbine reaches much better efficiency than classical crossflow turbine.

Why is CINK turbine ideal solution for drinking water operation?

Ecological harmlessness

The control plate of CINK turbine is moved by electrical motor and no hydraulic regulation is required. So, no oil leakage into the drinking water is possible. CINK turbine has patented sealing system of the bearings and even uses biological grease for lubrication. In the year 1992 the Chief Hygienic Supervisor of Czech Republic, after detailed examination by the State Institute of Health in Prague, had granted permission to use CINK turbines in waterworks and a certificate issued to this effect.

Avoiding hydraulic blows

CINK turbine has low runaway speed (max 180%) by falling out of the electrical grid and there is almost no change in throughflow rate in runaway operation. On the graph enclosure No. 2 you can see the comparison of flow and speed changes of CINK, Francis and Kaplan turbines. The flow of other type of turbines is increased a lot and hydraulic blows occur. CINK's low runaway speed and no change in flow enable to keep the turboset running almost unlimitelly long time and to start its stopping gradually and in timing what will exclude any hydraulic impacts and thus hydraulic blows. CINK turbine doesn't need any hydraulic or mechanical brakes, any automatic by-passes, any surge tanks, any relief valves etc., what is generally required by other type of turbines or reversed pumps.

For instance, in front of water treatment plant Nova Ves near Frydlant in Czech Republic a pressure feeding piping DN 1200, 6,64 km long, which brings water from the dam Sance in quantity of 1300 to 2100 lit./sec. with net head fluctuating from 10 to 38 m according to the water level in the dam, two Cink turbosets are installed and during complex tests both were brought to runaway speed by a pretended grid defect during full intake of 2300 lit./sec and maximum water pressure. The measured pressure increase has shown the water column higher to 3 metres only.

Avoiding water supply interruption

Above mentioned quality of CINK turbine is a guarantee that there is no influence to continuous water supply during turbine operation.

Installation in the limited space in existing building

CINK turbine is much smaller then Francis turbine for the same output and can operate in various positions. Pelton turbine also can't be very often used because of building requirements. CINK can be adapted to considerably smaller room dimensions.

Typical example of installation into the water reservoir or treatment plant is shown on the drawing enclosure No. 4. The turboset with all auxiliaries and new pipe is build without water supply interruption upto final weld to the old piping. After everything is prepared, all the water tanks under installation are fully filled up, water supply to the installation is closed for several hours and two welders one on intake and one on outlet make the connection of old and new pipelines. In this time, usually at night, the water supply is covered by filled up reservoirs.

Coping with problematic hydraulic conditions

Prior to the turbine installation the water pressure on the intake must be measured, because as per our experience there is a high difference of theoretical hydraulic loss calculations compared to the exact measurement of pressure in relation to the discharge volume, because long feeding pipes with higher water velocity show big hydraulic loss which may even more increase in older pipings due to encrustation of the inside surface.

For example water treatment plant at Nova Ves showed that with flow 1300 lit./sec. the pressure loss 14,65m and with 2100 lit./sec. the pressure loss was 36,57 m. When we take into consideration frequent and high variation of reservoirs water levels (for instance in the dam Sance it is up to 11m) then a great difficulty of optimum design of turboset appears. It is not possible to achieve it without extra ordinary performance quality of turbines because the turboset has almost constant speed of 50 Hz frequency for public electric grid. Such turbines will be preferred which have a large regulating ability together with suitable efficiency curve over the whole regulating range. The efficiency should not drop too much with considerable deviation of effective head from the minimal projected value. Our turbines have the required qualities and that is why their application in water plants are successful.

Aeration of the water by CINK turbine, savings in chemicals for water clarification

Further important specification of our turbine design is intentional draw-in of air into the inside space of turbine in order to reduce the cavitation, to increase the efficiency and to improve the running behaviour. The turbine works as very good aerator and this has very favourable influence to technology processes in water treatment plants or in sewage plants. As per experience from treatment plant at Horka by mixing and aerating the water by CINK turbine saves a large quantity of chemicals used for flocculation by 20-50% according to water quality.

Economy, payback period

Installation of CINK turbines in drinking water systems is very beneficial and the payback period could be less than one year because of the following:

— The only costs are for turboset only, because the penstocks, dam, buildings and electrical connections already exist.

— There is no need for any expensive project, civil works permission, water right licensing and other bureaucratic delays.

— No flow measurement is required, because the flow data are known.

— CINK turbines are fully automatic, designed for unattended operation, controlled by PLC microcomputer. Occasional check can be done by waterworks personnel, hence no new staff is required.

— The generated power can supply the treatment plant itself and the electricity has not to be purchased from the grid at higher purchase rates. Excess power can be sold to the grid.

— As we are informed by waterworks personnel, the turbine can be used as a perfect control of water quantity after calibration of control plate positions.

CINK's experience

CINK turbines are installed up to date in 270 sites in 13 countries including Germany, Switzerland, Great Britain, France etc. More than 20 turbines operate in drinking water, some of them for many years as can be seen from the table enclosure No. 5.

We feel that there is a huge power potential in all drinking water pipes, dams, treatment plants and reservoirs in every place where the people live around the globe. It is our pleasure to offer clever solution to utilise this very cheap and environmentally friendly source of power.

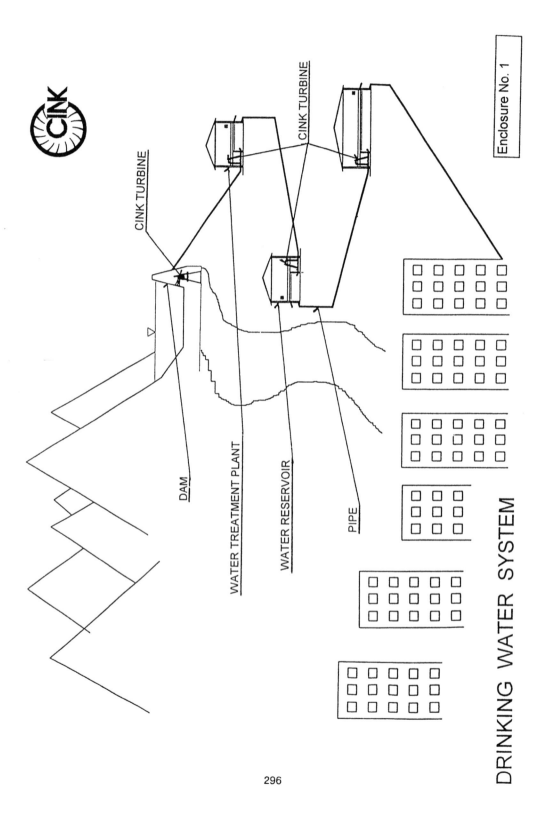

CINK TURBINE

CINK TURBINE

CINK TURBINE

DAM

WATER TREATMENT PLANT

WATER RESERVOIR

PIPE

DRINKING WATER SYSTEM

Enclosure No. 1

OTHERS

GUIDE VANES/FLAPS

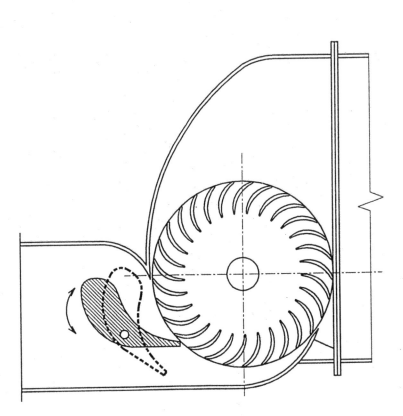

CINK

REGULATION PLATE

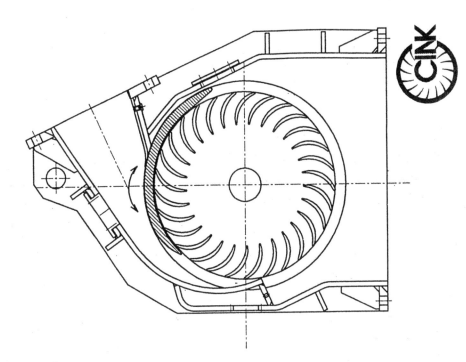

297

Enclosure No. 2

COMPARISON OF RUNAWAY SPEED AND FLOW

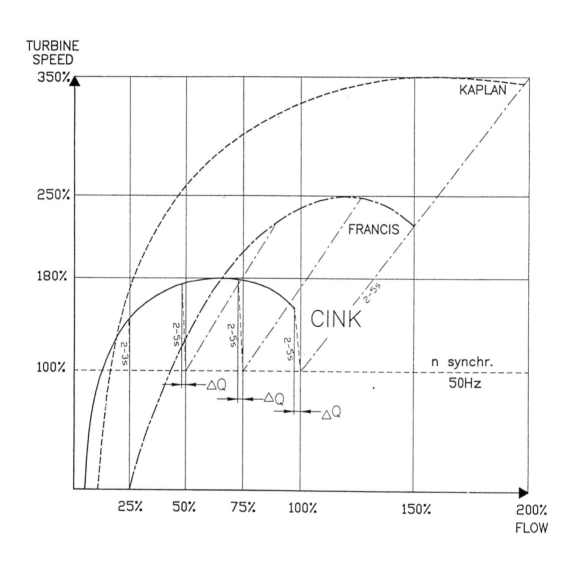

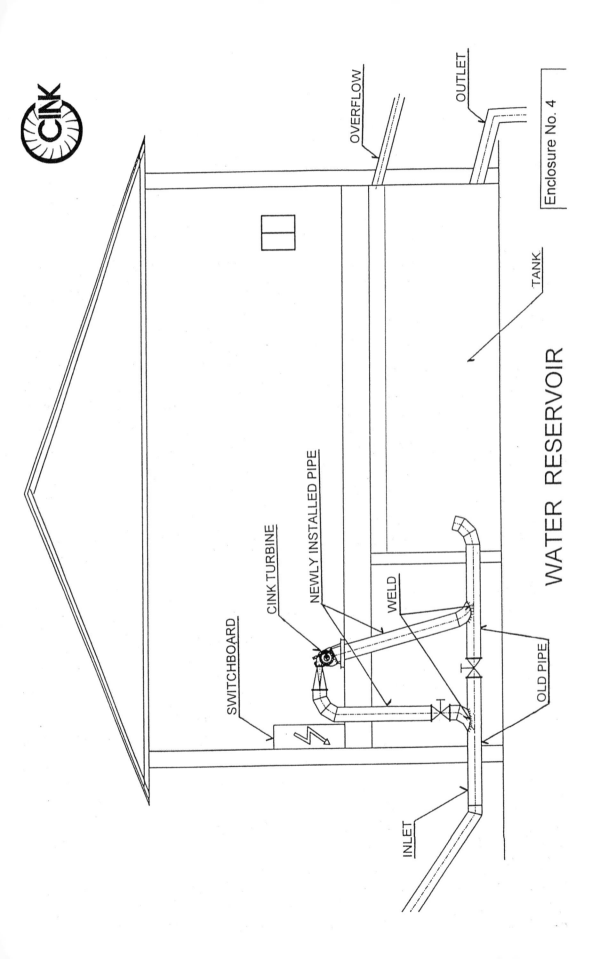

OVERFLOW

OUTLET

Enclosure No. 4

TANK

SWITCHBOARD

CINK TURBINE

NEWLY INSTALLED PIPE

WELD

OLD PIPE

INLET

WATER RESERVOIR

INSTALLATIONS OF CINK TURBINES IN DRINKING WATER

Locality	Operation in D/TP/R*	Total output [kW]	Number of turbines	Type of turbine	Year of instal.	Penstock length, diameter L[m]/ DN[mm]	Annually generated power [kWh]
Rimov	D	1000	2	6B×530	1985	20/800	4.000.000
Vrchlice	D	12	1	2,5B2×56	1989	50/500	60.000
Josefuv dul	D	220	2	2,5B2×198	1990	20/800	1.000.000
Marsov	R	40	1	2,5B3×248	1991	4060/400	250.000
Kruzberk	D	204	2	3,4B×360	1991	80/600	1.100.000
Josefov	R	45	1	2,5B4×100	1992	3500/500	240.000
Horka	TP	90	1	3,4B2×420	1992	750/600	400.000
Hrinova	D	352	2	3,4B2×290	1992	200/2000	1.200.000
Nova Ves	TP	368	2	4,5B2×800	1992	6640/1200	2.300.000
Sance	D	253	1	3,4B3×312	1993	280/800	1.500.000
Moravka	D	104	1	3,4B3×380	1994	180/800	400.000
Nova Bystrica	D	214	1	3,4B4×400	1994	40/400	1.500.000
Demanova	TP	37	1	2,5B5×105	1995	1200/500	250.000
Horka	D	152	1	3,4B4×314	1995	30/600	800.000
Frydek Mistek	R	37	1	2,5B7×90	1996	2000/500	240.000
Liberec	R	206	1	3,9B×175	1997	3000/700	1.400.000

*D = Dam, TP = Treatment Plant, R = Reservoir

First International Conference on Renewable Energy–Small Hydro
3–7 February 1997, Hyderabad, India

OPTIMAL PROTECTION OF HYDROTURBINE DRIVEN INDUCTION GENERATORS

Devadutta Das

Water Resources Development Training Centre
University of Roorkee, Roorkee 247 667, India

ABSTRACT

For harnessing small hydropower sources, very often use of induction generators is being suggested to effect economy. However it is often being observed that the protections provided against system faults and abnormal operating conditions are either inadequate or more than that required. In order that the protection provided meet the system requirements just adequately, the paper discusses the operating characteristics of the induction generator, of the power system and of the hydraulic system and proposes a set of protections which can provide it with an effective protective umbrella.

INTRODUCTION

In India, with a large network of irrigation canals, there lies immense possibility of harnessing the hydroenergy available in canal falls for power generation. The canal irrigated lands are generally electrified. However, the high cost of developing these canal falls for power generation has been impeding the development. Inorder to reduce the specific cost of installation, use of induction generators has been often suggested in preference to synchronous generator of equal output rating.

Electrical protections are provided to protect the generator against the adverse effects of different types of electrical faults and abnormal operating conditions. From experience it has been observed that the equipment is often provided with protections more than that the service conditions warrant. In view of this the paper discusses the operating characteristics and the system behaviour under dynamic conditions of induction generator and suggests an optimal protective umbrella which is adequate to meet all types of system contingencies.

INDUCTION GENERATOR CHARACTERISTICS

An induction generator is, simply speaking, an induction motor driven at a supersynchronous speed with the excitation power being supplied from an external source, normally the grid.

The real power delivered by an induction generator is a function of the slip between rotor speed and the rotational speed of the mmf (magnetomotive force) produced by the stator currents. The rotational speed of the magnetic field which rotates around the stator corresponds to the synchronous speed. The rotating magnetic field of the stator produces a magnetic flux in the airgap between the stator and the rotor. The stator air gap flux cuts the bars in the squirrel cage rotor thereby inducing a voltage across the rotor bars. As the rotor bars are short-circuited by end-rings, the induced voltage produce a current flow in the rotor bars. The rotor currents in turn produce a magnetic field which interacts with the stator field to produce a torque. If the rotor is driven by a prime-mover at a super-synchronous speed, its motion relative to the rotational direction of the stator magnetic field is in an opposite direction and energy is transferred from the rotor to the stator. However, reactive power for setting-up the stator magnetic field still flows from the power system to which the generator is connected.

The equivalent circuit of the induction generator is shown in Figure 1.

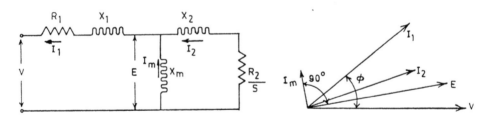

Stator   resistance : R_1 Stator   reactance : X_1
Rotor   resistance : R_2/s Rotor   reactance : X_2
Stator   current   : I_1 Rotor   current   : I_2
Magnetising reactance : X_m Magnetising current : I_m
Internal   voltage   : E Terminal voltage   : V

Figure 1 : Equivalent circuit and Vector Diagram of Induction Generator

The stator current leads the terminal voltage by the power factor angle, ϕ, and is a function of terminal voltage and slip as per relationship given below:

$$I_1 = - \frac{V}{Z}$$

302

and $Z = (R_1 + jX_1) + \dfrac{(\dfrac{R_2}{s} + jX_2)(jX_m)}{(\dfrac{R_2}{s}) + j(X_2+X_m)}$

The reactive component of the stator current is $I_1 \sin \phi$ and signifies the reactive power (kilovars) which is to be supplied by the system or the shunt capacitors if provided.

The power output of an induction generator has a definite limit and is given below [1]:

$$P_{omax} = V^2 \frac{[X_m+X_1-X_e-2R_1]}{2[R_1{}^2+(X_m+X_1)(X_e)]}$$

where, $X_e = X_1 + \dfrac{X_m \cdot X_2}{X_m+X_2}$

Thus, if the torque of the hydro-turbine exceed the pull-out torque of the induction generator, the latter cannot further accept the additional load. This would eventually lead to increase in the speed of the prime-mover, and also in increase of magnetising and output current. Therefore to avoid such an unstable condition, the maximum steady state torque of the prime-mover is kept much less than the pullout torque of the induction generator coupled to it.

The excitation power is supplied from the other synchronous generators connected to the power system or from a bank of capacitors connected to its terminals. The excitation power remains constant irrespective of the load on the generator. The reactive power required for setting up the stator magnetic field is approximately given by the equation:

Excitation VARs $\approx \sqrt{3}.V. I_0$

where, I_0 = no-load current of the induction machine in amperes

V = terminal voltage in volts

When the induction generator is switched onto the power system, there is inrush transient of magnetising current having a peak magnitude of about 6.5 times the normal full load current, and about 25 times the no-load current. The peak magnitude of the inrush magnetising current is independent of the speed of the machine at the which the generator is connected to the power supply system. Studies have indicated that this transient recovers in about 20 cycles [2]. During this period, there is a dip in the voltage of the power system due to flow magnetising current of such a large magnitude.

Fault Current Contribution

When a fault occurs in the system, the system voltage collapses to a very low value tending to zero. As the induction generator depends on the power system for its magnetising power, due to collapse of system voltage, the fault current contributed by it decays rapidly. For a fault at the generator terminals, the fault current is given by the equation:

$$I_f = (\frac{E'}{X'}) \ e^{\tau}$$

where,
E' = initial value of internal voltage before the fault
X' = transient reactance of the generator

$$= X_1 + [\frac{(X_m . X_2)}{(X_m + X_2)}]$$

T_s' = short-circuit time constant = $X'/2\pi f R_2$

t = time instant.

τ = $-t/T_s'$

The above values can be used for determining the making and breaking ratings of the circuit breaker and for protective relaying purposes.

Overspeeding :

The induction generators are generally coupled to small hydro-turbines with maximum output of about 5000 KW. As such, these generators have low inertia. Further, when used in conjunction with low head turbines of propeller type which has a high runaway speed to normal speed ratio (of the order of 2.7), the speed rise in the event of load throw off could be very rapid also. The closure time of the inlet valve is typically between 5 seconds and 30 seconds in small hydroplants. Normally, by suitably co-ordinating the inertia of the machine with the inlet valve closing time, the speed of the machine could be limited to 140% normal speed during full load rejection with full gate and maximum head conditions. However, if the induction generator is left connected to the power system after load throw off which has a high capacitance like capacitor bank, cable or long overhead line, the terminal voltage will rise due to the effect of the capacitance. This situation is further aggravated due to overspeeding of the machine.

Neglecting saturation of the magnetic circuits of the generator and the step-up transformer, and winding resistances, the approximate overvoltage attained during the above condition is given by the following expression [3].

$$\frac{V_t}{E'} = \frac{N. \ X_c}{X_c - N^2 (X' + X_t)}$$

where,

N = maximum overspeed

X_c = capacitive reactance left connected to the power system

X_t = transformer leakage reactance.

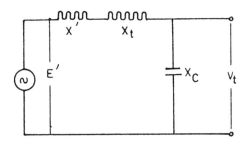

Fig. 2 : Equivalent circuit of induction generator with capacitance provided on generator terminals

Assuming that the generator is directly connected to the system (i.e., $X_t = 0$), $X' = 5.0\Omega$, $N = 140\%$, $X_c = 50\ \Omega$, the overvoltage that would occur in the event of load throw-off with capacitance remaining in circuit is approximately 174% of the rated voltage.

During overspeeding condition, the frequency rises which increases the current into the machine, as $I = (2\pi f). C.E.$, where C = capacitance in micro-farads, f = frequency. As the excitation current increases, the voltage induced in the machine also increases.

It is also likely that the capacitive reactance of the system and the inductive reactance of generator (and transformer), may also achieve resonant conditions leading to further overvoltage.

PROTECTION OF INDUCTION GENERATOR

In the foregoing section, the dynamic characteristics of the induction generator have been discussed which provides an insight into the protective requirements of the machine. However, the decision for providing suitable protections also depends upon the cost of providing such protections vis-a-vis cost of damage that would arise in the event of a fault. Induction generators are generally much smaller in output and hence much cheaper in initial cost. Therefore, the protection requirements must be also simpler. For larger rated induction generator of 500 KW and above it may be logical to provide slightly more exhaustive protections including protections against incipient faults. However, the decission for the same emanates from the considerations of the factors stated in the next page.

- machine down time and loss of revenue during the above period.

- cost of repairs to the machine.

- availability of spare generator/spareparts.

The protections to be provided are dependant upon the types of faults or abnormal conditions of operation that can arise in the induction generator. These are as follows :

(1) Stator - phase to earth faults

(2) Stator - phase to phase faults

(3) Overload

(4) Single phasing

(5) High/Low voltage

(6) Loss of load and overspeeding

(7) Reverse power flow

(8) Poor power factor.

Stator Phase-to earth and Phase to Phase Fault Protection

The induction generator has to be provided with adequate protection against the damaging effects of phase-to-earth and phase to phase faults. For small induction generators, a high set instantaneous overcurrent device is adequate. If the switching device contains a HRC fuse, no additional protective relay is required for this function. If circuit breakers are used, a high set instantaneous overcurrent relay can be used. These relays must be set high enough to permit the flow of initial magnetising inrush current on energisation. For faults deep inside the machine winding, the impedance could be high enough to reduce the magnitude of the fault current level below the pick-up value of the instantaneous overcurrent relay (Device 50) to operate. An inverse time-overcurrent relay can be provided in addition to instantaneous overcurrent relay to take care of the above contingency. This device would also provide protection against over loading. For small induction generators either thermal replica type relay or CT connected induction disc type relay can meet the requirements.

For medium and large induction generators (>500 Kw rating), differential relays (Device 87) could be considered as these provide positive protection for stator phase-to-phase and phase-to-earth faults. But the provision of this protection involves additional cost in the form of relays and additional current transformers.

A cheaper alternative which could be used for medium sized induction generators is the self balancing scheme. In this scheme, the phase lead and the neutral lead of each of the phases pass through a CT, the secondary of which is connected to an instantaneous overcurrent relay. This scheme is generally less costly than the differential scheme.

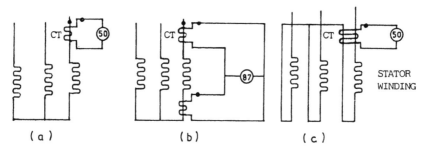

(a)   (b)   (c)

(a) with instantaneous overcurrent relay (b) with differential relay protection (c) with self balance differential relay protection using overcurrent relay.

(Protection scheme shown for one-phase only. Identical schemes for other two phases are adopted.)

Figure 3 : Protection of Induction Generator stator winding

Overload Protection

Overloading of an electrical machine is detrimental as it causes overheating of the winding thereby reducing the life and consequent damage to the insulation of the machine. As already stated, the total current of the machine increases as the slip increases. But since the prime-mover is provided with overspeed protection, it is possible to remove the machine from the power supply system much before the current exceeds damaging proportions.

In case of external faults, the voltage on faulted phase(s) collapses thereby preventing transfer of magnetising current to the induction generator. Consequently, the contribution from the induction generator to the fault current rapidly decreases. The rate of decay of the fault current is so fast, that operation of time-over current relay or even voltage controlled (or voltage restrained) relay is not possible. Hence provision of these protections do not protect induction generators against external faults and hence need not be provided.

In view of the above, if stator winding (high) temperature protection is provided, electrical protection against overload can be dispensed with.

Ground Fault Protection

The selection of ground fault protection scheme is dependant upon whether the generator neutral is grounded. As the induction generator does not contribute to sustaining the fault current, the neutral need not be grounded if not otherwise necessary. If the generator neutral is not connected to earth, to prevent system overvoltages, the powersystem connected to the generator need to be grounded. This is accomplished by having the generator side of the step-up transformer connected in star with star point grounded. If the generator side winding of the step-up transformer is connected in delta, a separate grounding transformer is required to be provided. The most widely used arrangement for this is to provide three (3) numbers of potential transformers (PT) with their primaries connected in star with star point earthed. The secondaries are connected in open delta formation with an overvoltage relay interposed.

Overvoltage Protection

Overvoltage in the system is caused by lightning and switching surges, and due to capacitors provided to improve powerfactor. When the load is thrownoff by an induction generator, it speeds up very rapidly. If the power factor improving capacitors are left in circuit with the induction generator terminals, very high terminal voltages can be produced.

For taking care of the overvoltages due to lightning and switching surges, surge arrestors with capacitors connected in parallel are provided as close as possible to the generator terminal.

To prevent generation of overvoltages in the system, the power factor improvement capacitors preferably should be provided on the high voltage side of the step-up transformer with provision of overvoltage relay to trip the generator circuit breaker. (Fig.4).

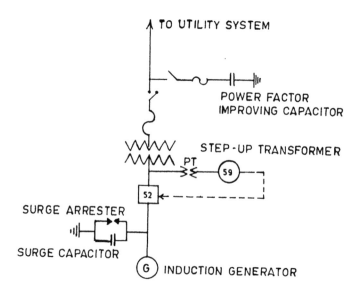

Fig. 4 : Overvoltage Protection of Induction Generator

Overspeed Protection

When the load is rejected, the hydro-turbines tend to overspeed thereby stressing the connected generators. The overspeed achieved in case of Kaplan (Propeller) turbines is as high as 2.7 times the rated speed. Francis turbines may reach speeds as high as 2.0 times the normal speed. Such small units with low inertia may reach such speeds which may be detrimental to the machine life. To prevent the machine from reaching higher speeds, overspeed device is provided to actuate closure of the wicket gates or the inlet valves. The overspeed device could be a tacho-generator with a centrifugal switch. The most reliable system is a non-contact type of tachogenerator with built-in system which produces an overspeed signal in the event of failure of the tachogenerator. The overspeed relay may be set to operate at 110% overspeed.

Reverse Power Protection

The need for the provision of a reverse power relay is to be determined on the basis of the adverse impact of reverse power flow on the generator and prime-mover. The hydro-turbine is neither affected under such conditions nor is the induction generator when it motors. However, when a hydraulic turbine driven generator motors, it draws power from the system to meet the magnetising power and to meet the windage and friction losses of the generator and turbine. This energy is a wastage and has a price. However, motoring is possible only in case of prime-mover (hydro-turbine) unable to deliver mechanical power output. This is only possible when water flow into turbine does not take place on account of low water level in the forebay. Such a situation can be better taken care of by providing low level switch in the forebay to trip the machine and close the inlet valve.

It is however possible to close the generator breaker without opening the inlet valve in which case the induction generator operates as a motor drawing power from the system. To prevent such occurrences, a speed relay contact from the tachnogenerator to close at about 95% speed may be provicded in the closing circuit of the cirucit breaker. This arrangement would inhibit closing of the brekaer until the generator speed reaches 95% rated speed thereby preventing motoring operation.

Protection against single phasing

In the event of single phasing, an induction generator cannot stall as it is driven by a prime-mover. The active power generation continues on the other two phases due to existence of the magnetic field set up by the magnetising power available from the system. However, the current in the healthy phases are unbalanced and of abnormal magnitude. Undervoltage relays provided on each phase can protect the generator against single phasing by removing the generator from service. The under-voltage relay will also trip the generator cirucit breaker in the event of failure of mains power supply.

Protection against stator overheating

The stator winding gets overheated when the generator is either overloaded or the ambient temperature for which the machine has been designed is exceeded or when the cooling system has failed. Though immediate failure of insulation is not likely due to overheating, irreversible deleterious change takes place impairing the life of the insulation. Temperature detectors (ETDs) either of resistance temperature detectors (RTDs) type or thermocouples are embedded in the stator winding slots equally distributed along the periphery. For medium and large sized induction generators six (6) to twelve (12) numbers of ETDs can be provided. The temperatures from the ETDs can be monitored and recorded. The information can be utilised to sound an alarm as the thermal-limit temperature is approached and can be made to trip the generator circuit breaker as the thermal-limit is exceeded.

For small induction generators, embedded thermistors can be used to trip the generator circuit breaker when the thermal limit is exceeded.

Bearing overtemperature protection

Overheating of the bearings are detrimental especially in case of sleeve type bearings. RTDs or thermocouples can be embedded in the bearings and below the bearing liner surface to monitor the temperature. Indicating type dial thermometers with alarm and trip contacts can be provided for protecting the bearings against high temperature.As the thermal-limit is approached, the alarm sounds and as it is exceeded, the generator circuit breaker is tripped.

Rotor overheating protection

Due to unbalanced loads and voltages in the powersystem, negative sequence currents are setup which causes double frequency current to flow in the rotor iron thereby overheating the rotor. Though negative sequence relay provides positive protection against rotor overheating, due to its high cost, it is not recommended. Operation of induction motor with voltage unbalance between 1 to 5 percent can be done with derated output and above 5 percent is not permitted. Due to similarity between induction motor and generator similar restrictions on induction generator can be enforced.

SWITCHING DEVICE

A switching device is required to switch on to and remove the generator under normal or fault condition from the power system.

For small induction generators ($\leq$ 200 KW), three phase coil operated contactors can be used. As the contactors have very low short-circuit interrupting rating as compared to the short-circuit current contribution from the power system, HRC fuses are to be provided in conjunction with the contactors. Contactors with HRC fuses for short-circuit protection are characterised by low initial cost. Further, contactors have a longer life as compared to breakers as these are designed for more load-switching operations than breakers.

The MCCBs are provided with thermal overload relay and magnetic release against short-circuits. MCCBs with contactors can also provide adequate protection to small induction generators.

For medium and large sized induction generators, circuit breaker is the preferred alternative. The circuit breakers are provided with shunt-trip device actuated by protective relays, and tripping operation is accomplished by electrical stored energy device.

CONCLUSIONS & RECOMMENDATIONS

The necessity and adequacy of the protections to be provided for hydroturbine driven induction generators is dependant upon its operational constraints and characteristics, and significantly differs from those applied for protecting synchronous generators.

Some of the protections like reverse power, over/under frequency, voltage restrained overcurrent relay are not justified due to their doubtful ability to respond under certain conditions of operations of hydro-turbine driven induction generators.

To sum up, the protections which are necessarily to be provided for induction generators are mentioned below in a matrix form.

Recommended Protections for Induction Generators		
Device No. Protection function	Small	Medium & Large
12 Overspeed	O	O
50 Instantaneous overcurrent	O	O
51 Time-overcurrent	X	-
27 Under voltage	O	O
59 Over voltage	O	O
38W Stator winding High temperature	X	O
38B Bearing High temperature	X	O
87 Differential	-	O
Legend : O - Recommended X - Optional		

Figure 5 shows the typical protection scheme for hydro-turbine driven induction generator of medium and large capacity. For small induction generators the protection and switching scheme are simplified keeping in view the basic requirements and economy of application.

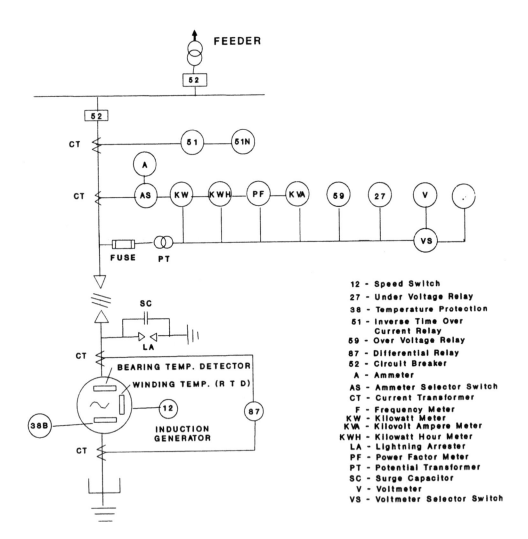

Figure 5: Typical Main Electrical Single Line Diagram

REFERENCES

Journals

1. Barkle,J.E. and Fergusion,R.W. 1954 . Induction Generator Theory and Application. AIEE Trans.**73**,Part III A : 12-19.

2. Allan,C.L.C. 1959 . Water turbine driven Induction Generators. IEE Paper No.3140 s : 529-550.

3. Pereiera,L. 1981 . Induction Generators for small hydro plants. WP & DC, November : 30-34.

4. Nailen,R.L. 1982 . Spooks on the power line? Induction Generator and the public utility. IEEE Trans. on Ind. Appls. **IA-18(6):** 608-615.

5. Parsons (JR),J.R. 1984 .Co-generation application of Induction Generators. IEEE Trans. on Ind. Appls. **IA-20(3):** 497-503.

6. Bailey,J.D. 1988 .Factors influencing the protection of small-to-medium size Induction Generators.IEEE Trans. on Ind. Appls. **IA-24(5):** 955-964.

Books

1. GEC Measurements. 1979 . Protective Relays Application Guide, Stafford,UK.

2. Electricity Council. 1981 .Power System Protection, Vols.I,II. Peter Peregrinus Ltd., Stevenage,UK and New York.

3. ASEA Publication. 1986 . Generator Protection - Application Guide

4. Blackburn.J. 1994 .Protective Relaying,Marcel Dekker,New york.

****

First International Conference on Renewable Energy—Small Hydro
3 — 7 February 1997, Hyderabad, India

BULB GENERATING SETS FOR LOW HEADS

S.R. Rathore

Hydro Turbine Engineering, Bharat Heavy Electricals Ltd.
Piplani, Bhopal 462 022, India

ABSTRACT

There has been a sharp rise of cost of energy with the decrease in water head. That is why during the past years, the high out put projects were preferably constructed on high and medium head sites. The availability of such sites being now reduced, the increased power demand requires to investigate possibilities of utilsing low head sites.

Bulb generating sets represent a successful step in the search for economical low head development. Axial machines are considered the best choice for increasing outputs and reducing excavation costs considerably due to their simplified forms and their smaller dimensions when compared to vertical kaplan machines of the same capacity.

The bulb turbine with an upstream generator is a direct result of the development for low head applications. It is now being substituted for the vertical Kaplan generating units under the lowest heads. The paper will briefly discuss the comparison in their designs and operating characteristics.

The name of "Bulb" given to this generating unit is due to the shape of the upstream water tight enclosure that accommodates the generator, which is submerged in water passage. This arrangement is quite suitable for low heads and large discharges and also where large variation in discharge and head. Bulb units have virtually replaced vertical Kaplan machines in this field. Refer Fig.1.

1. DEVELOPMENT OF BULB

The overall cost of a hydro power plant is highly dependent, for a given power output, on the relative discharge and water head values. The cost of a unit of electricity shows a sharp rise with a decrease in water head. That is why during the past years, the high output projects were preferably constructed on high and medium head sites. The availability of such sites are now less, and there is increased power demand. This requires more and more attention on low head sites.

Kaplan turbine with its controllable blade is quite suitable for low heads. However, the tendency to harness sites of even lower heads has led to the development of bulb units which are now substituted for the Kaplan turbines for large capacity under the lowest head.

2.0 BULB VERSUS KAPLAN

Following passages provide a comparison of vertical Kaplan and bulb units for low heads :

2.1 Hydraulic Design

Straight design of the water passage improves hydraulic characteristics of flow, which allows sizes to be reduced while keeping the same output. Further more pumping operation can be carried out (reversible machine).

Bulb units are the type of machines offering the best power/capacity ratio for low heads in the range of 3 to 25 meters, while presenting substantial savings in civil works.

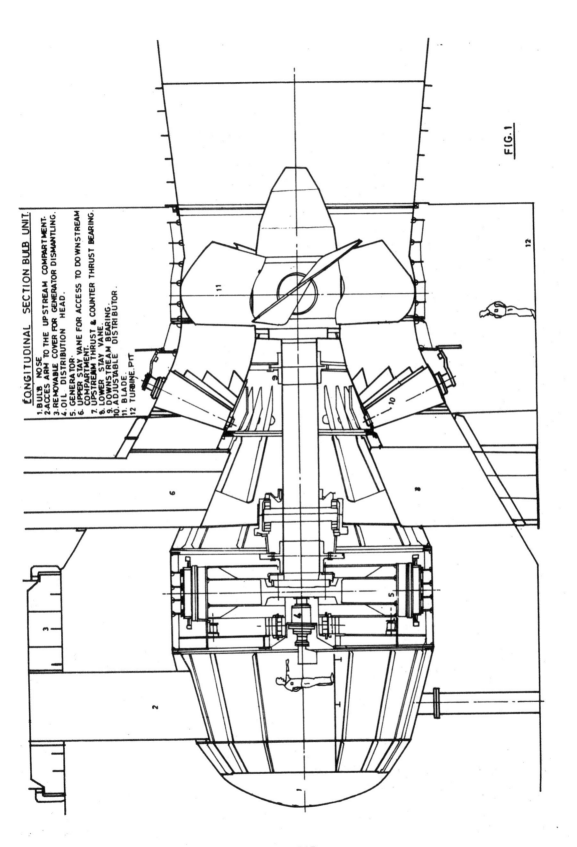

LONGITUDINAL SECTION BULB UNIT.

1. BULB NOSE
2. ACCES ARM TO THE UPSTREAM COMPARTMENT.
3. REMOVABLE COVER FOR GENERATOR DISMANTLING.
4. OIL DISTRIBUTION HEAD.
5. GENERATOR.
6. UPPER STAY VANE FOR ACCESS TO DOWNSTREAM COMPARTMENT.
7. UPSTREAM THRUST & COUNTER THRUST BEARING.
8. LOWER STAY VANE.
9. DOWNSTREAM BEARING.
10. ADJUSTABLE DISTRIBUTOR.
11. BLADE.
12 TURBINE PIT

FIG. 1

2.2 Proportion of Bulb for low heads :

M/s Neyrpic, France has carried out a statistical study on turbines designed since 1970 of 5 MW or more and for heads less than 20 meters. Out of 135 projects, the types of turbines used are :

Bulb Turbines	79
S-Type turbines	8
Propeller turbines	2
Kaplan turbines	46

Figure 2 shows output versus head diagram for bulb and Kaplan turbines. The proportions based on head for above turbines are as below.

Head range	Bulb and S-Type	Kaplan & Propellers
6.6 m	84%	16%
6.7 to 9.5 m	95%	5%
9.6 m to 12 m	77%	23%
12.1 to 15.5 m	57%	43%
15.6 to 20 m	32%	68%

The above table shows that upto 15 meters the trend is clearly towards bulb sets.

2.3 Reduction in diameter of runner while calculating the specific power,

$$P_{11} = \frac{P}{D^2 H^{1.5}}$$

Bulb units provide a reduction in runner diameter of 5% for a head of 15 m and about 10% for a head of 8m. This is illustrated on Fig.3.

2.4 Power house dimensions - civil work cost :

Compared to vertical Kaplan turbines, bulb turbines allow (i) less excavation, (ii) smaller upper power house structure and (iii) about 35% reduction in power plant length (See Fig.4)

This finally provides around 20% of saving on civil work costs.

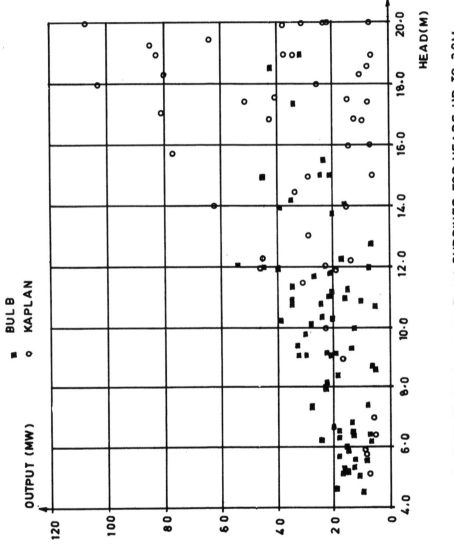

FIG. 2 –RECENT BULB AND KAPLAN TURBINES FOR HEADS UP TO 20M

319

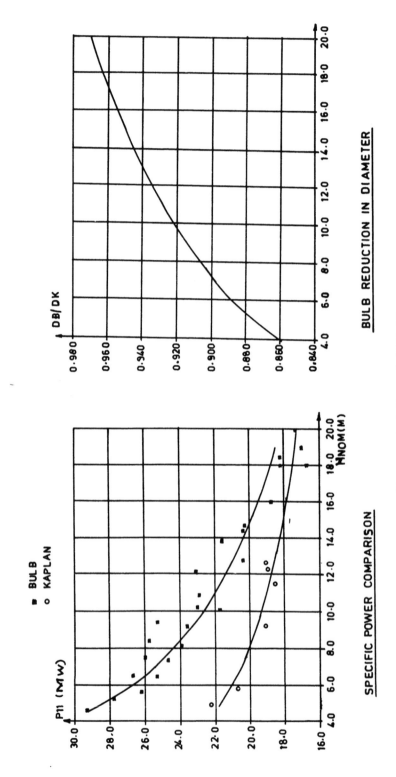

DB/DK

BULB REDUCTION IN DIAMETER

■ BULB
○ KAPLAN

P11 (MW)

SPECIFIC POWER COMPARISON

FIG.3 BULB AND KAPLAN TURBINES

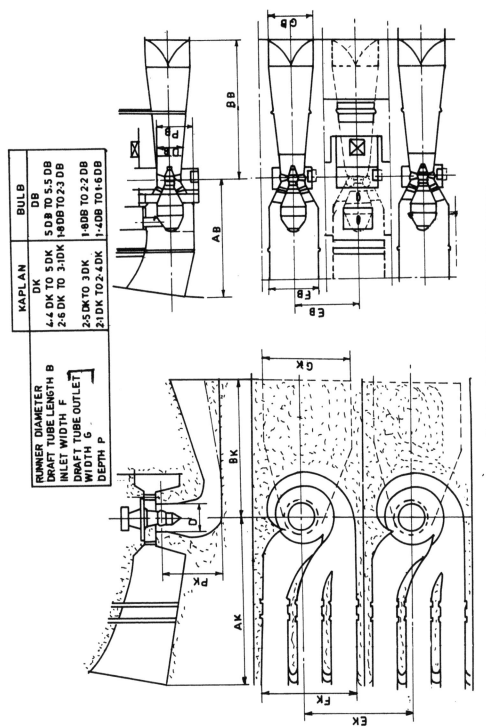

RUNNER DIAMETER	KAPLAN	BULB
DRAFT TUBE LENGTH B	DK	DB
INLET WIDTH F	4·4 DK TO 5 DK	5 DB TO 5.5 DB
	2·6 DK TO 3·1 DK	1·8 DB TO 2·3 DB
DRAFT TUBE OUTLET		
WIDTH G	2·5 DK TO 3 DK	1·8 DB TO 2·2 DB
DEPTH P	2·1 DK TO 2·4 DK	1·4 DB TO 1·6 DB

FIG 4 BULB AND KAPLAN TURBINES POWER HOUSE DIMENSIONS

2.5 Mechanical Design: Significant Aspects ;

i) Supporting Structure -
All torques, axial loads and nearly all radial forces are taken up by the stay ring, fitted with six stay vanes. The stay vanes are hydraulically shaped for less obstruction to water.

Two stay vanes are hollow to allow easy access to inside of the bulb for maintenance and for pipings.

ii) Shaft, bearing and runner:

The most efficient design is with two guide bearings having generator rotor and runner overhung at each end. It is selected generally upto 15m head and 50 MW power capacity mainly because of its simplified erection process and its cost effectiveness.

The runner servomotor is housed in the hub and fed with governing oil through shaft. The thrust bearing and the counter thrust bearing are located in the plane of the stay ring so as to minimise extra torques.

iii) Governing and Lubricating System:

The oil supply system is common to governing and lubrication. This arrangement reduces the number of auxilliary equipment, simplifies automation system and decreases power consumption.

iv) Use of Speed increasers :

Speed increasers are employed for improving economics of low head projects. They enable generators of more conventional design, better efficiency and lesser cost.

The design is well suitable for very low head projects where speed is low and direct driven generators would become two much expensive.

2.6 Erection:

Bulb sets provide around 30% reduction in erection time mainly due to -

i) There is no interruption in the erection process because of no spiral casing.

ii) Erection may proceeds with generator and turbine separately, once the shaft is in position, at three places simultaneously.

a) in the turbine pit :

- installation of the lower half runner throat ring, runner and shaft seal.
- installation of the upper half runner throat ring and concreting of draft tube, anchor ring / foundations.

b) Installation of the generator-rotor , stator and bulb-nose.

installation of access arm and closure of the hatch cover.

c. Installation of the governing and lubricating equipment.

2.7 Maintenance

Main features of maintenance in bulb sets are :

i) removal of the blades individually without removing the runner hub.

ii) maintenance of the shaft seal without disturbing the turbine guide bearing due to a removable one, allowing access to the shaft seal from the conduit after dewatering.

iii) removal of all guide and thrust bearing elements through the hollow stay vane.

2.8 Operation

Two types of regulations or controlling are available:

i)   Double Regulated Bulb turbines -

Both guide vanes and runner blades are adjustable by means of hydraulic servomotors and a cam ensures correct combination of guide vane angle and runner blade angle for any condition of discharge and head.

ii) Single Regulated Bulb Turbines:

To economise in the cost for low head projects, two types of simplified controlling are being adopted :

 a) Adjustable distributor - Fixed blades :AD-FB

 b) Fixed distributor - Adjustable blades:FD-AB.

In the first one, runner is a propeller type and the guide vanes are adjustable.

In the second one, the runner is an adjustable blade Kaplan type and the distributor is fixed type, an integral part of the stay ring. The stay vanes are hydraullically shaped to act as the guide vanes. For this type of regulation, turbines have to be equipped with a down stream gate for starting and stopping of the unit. But even with additional cost of the downstream gate, this variant is still cheaper than the AD-FB variant.

3.0 CONCLUSION

In view of the above comparison, the bulb units are better alternatives to vertical kaplan, particularly for low heads in every respect; hydraulic,mechanical, civil, erection, maintenance and operational.

In spite of their lower rotating inertia, bulb units behave quite satisfactorily with respect to frequency control or load rejection. Downstream gate control for sluice operation is an important technical feature of the bulb units.

REFERENCES:

1. Strohl,T., "Bulb Turbine Development" presented at Bangkok Symnpozium on 10th and 11th March 1987.

2. Catalogue of "Bulb Turbine" by M/s Neyrpic, France.

HIMALAYAN HYDRO (UNDP-GEF PROJECT)

First International Conference on Renewable Energy–Small Hydro
3–7 February 1997, Hyderabad, India

CHALLENGES IN IMPLEMENTATION OF SMALL HYDRO SCHEMES ON HILLY STREAMS

Y.K. Murthy[1] *and B.S.K. Naidu*[2]

[1] Consultant, World Bank, D-44 Gulmohar Park, New Delhi 110 049, India
[2] General Manager, Indian Renewable Energy Development Agency Ltd.
 India Habitat Centre, Lodhi Road, New Delhi 110 003, India

Synopsis

Small Hydro, in hilly and remote areas cut off from the electricity network, is of great relevance to a developing country like India having enormous potential in Himalayan ranges. Higher engineering challenges are involved in implementation of hilly hydels. This has been realised by Govt. of India whose incentives are on a higher scale for such schemes which are also supported by GEF, UNDP, World Bank & IREDA etc.

The paper attempts to address the various technical issues that pose challenge while implementing small hydro in hilly regions.

INTRODUCTION

In the recent years, greater attention is being paid all over the world on the implementation of small hydro electric schemes, in the light of their minimal impact on the environment, less initial capital investment and more rapid implementation. Their relevance is being realised particularly in the developing countries, where large rural areas are cut off from the national electricity grid. In India, it has been estimated that a potential of 10,000 MW exists in small hydro. Also over 2000 potential sites with an aggregate capacity of 5000 MW have been identified in various parts of the country. Uptil now, small hydro schemes with a total installed capacity of nearly 500 MW are under operation and schemes with another 500 MW are under various stages of implementation.

Though small hydro schemes have many attractive attributes mentioned above, and vast potential is available throughout the world, exploitation of the potential has been limited. Only about 20,000 MW small hydro capacity is reported to have been developed world-wide so far (excluding China) of which about 50% is developed in Europe alone. In China, where the definition of a small hydro plant comprises station capacity upto 25 MW , the power generated in small hydro plants was 15,055 MW upto the end of 1993 with an annual energy generation of 47.0 billion kWh, and the country has a target of 26,000 MW installed capacity by 2000 A.D.

Of late, Ministry of Non-Conventional Energy Sources(MNES), Government of India has been encouraging the exploitation of this potential by providing 100% cost of survey and investigation and 50% of the cost of DPR preparation as a grant subject to certain bench mark ceilings. Further more, private sector entrepreneurs have been given an incentive of interest subsidy for developing small hydro schemes upto 3 MW station capacity through financial institutions and capital subsidy upto 100 kW station capacity. The subsidies are on a higher scale for Hilly Hydels because of the higher challenges involved in implementing them.

In addition, Government of India has received a sanction from Global Environment Facility(GEF), a consortium of World Bank, UN Development Programme (UNDP) and UN Environment Programme to implement 20 commercially viable small hydro demonstration schemes in hilly regions under a $ 15 million project including upgrading of water mills besides preparing a master plan of development of hilly hydels in 13 Himalayan States of India. Thus Government of India has launched a vigorous programme of exploitation of the potential of small hydro schemes to benefit the rural and remote areas. Having financed more than 100 MW Small Hydro, World Bank is considering extension of second line of credit to IREDA for development of another 200 MW of which 100 MW is likely to be from Hilly Hydels.

In India, small hydro schemes are defined as those with an installed capacity of the power plant not exceeding 15000 kW with individual units rated from 1000 KW to 5000 KW whereas Mini Hydro Plants with an installed capacity not exceeding 2000 KW with individual units rated from 100 KW to 1000 KW and the micro power plant with an installed capacity of not more than 100 KW with units rated not more than 100 KW.

In India, there are two distinctive categories of small hydro schemes. First are those that are to be located in the irrigation canal system using the canal drops or to be located at the toe of the small irrigation dams and second category are those located in the hilly streams. In the power plants of the former category, hydraulic and structural parameters are more defined whereas in the case of the second category, each hydro scheme is unique and needs to be planned and designed to suit the topographical, geological and geomorphological characteristics of the hilly terrain to obtain the most economical and structurally safe alternative.

In this paper, an attempt is made to highlight the various technical issues, which pose challenging situations to evolve commercially viable schemes and build them expeditiously and at the same time ensure that the components of the schemes are structurally safe and hydraulic performance is satisfactory.

CHALLENGES IN IMPLEMENTATION

General

Case histories have shown that a number of small hydro schemes have failed and many of them have also yielded poor results because they were poorly designed, built and operated. Unfortunately there appears to be no clear statistics about small hydro failures, which could give an idea of the number of plants affected by lost power production and the corresponding economic loss.

As per authors' experience about small hydro projects on hilly streams, at least 3 schemes, 2 of them located in Himachal Pradesh and one in Manipur, failed during unprecedented rainfall in the area. The plants were in operation for about 4 years in Himachal Pradesh and more than 10 years in Manipur. The locations of these schemes were visited and main cause was the failure of the diversion weir and/or power house due to extensive erosion of the bed of the stream, which consisted of bouldery strata caused by unprecedented high flood discharge in the stream.

However, on the basis of the meager available data on the failure or poor performance of the small hydro power plants, it has been concluded that the basic deficiencies in the planning of the layout and design and construction of the various civil engineering components of the schemes and also sometimes faulty operation of the plant have been responsible for the failure of the plants. Some of the salient technical issues which pose challenging situations specific to the site conditions of a scheme particularly located on hilly streams are discussed below under various heads.

Planning and Layout of Small Hydro Schemes on Hilly Streams

Selection of an appropriate site for power development along a hilly stream is the most difficult task and needs a team of specialists competent in river hydraulics, foundation engineering and a hydro scheme designer with a background of the construction during implementation. One or two visits to the prospective sites to reconsider and critically inspect the areas of the proposed scheme to get first hand assessment of the site conditions at the proposed site for the scheme are essential. During these visits, a layout most suitable to the site conditions has to be prepared for each of the major components namely diversion arrangement across stream to lead the discharge into the water conductor system, layout of the water conductor and the power house complex including the tail race after a study of the available geological and hydrological data. During the visit of the team, layout should be marked on the site to confirm that each component of the scheme is appropriate to the site conditions. The layout of all the major components should be evolved considering various feasible alternatives so that a final layout plan of the scheme is prepared. Any field investigations which are considered essential should also be undertaken concurrently and completed within a fixed time frame and the result of the investigations and other relevant studies should be taken into account while finalising the layout of the scheme under the guidance of the team of specialists.

After the layout of the scheme has been finalised design of each component of the schemes should be undertaken on the basis of the recognised standards. Some of the important design aspects of various major components of the scheme are discussed below.

Design Aspects

Diversion Weir across stream

Diversion weir and the power intake complex should be designed taking into account the existing foundation conditions based on a geological appraisal by an expert engineering geologist. It is also important to estimate the design flood for which the spillway should be provided with the help of a specialist in hydrology. Hydraulic and structural design should be computed for ensuring stability of the diversion weir complex. In some cases such as the schemes where trench weirs are designed across the stream, it is essential to carry out hydraulic model tests to determine the vital hydraulic parameters of the weir as well as the power intake complex. Structural failures have occurred in many schemes since the complex had been designed with poor appreciation of uplift pressures, foundation weakness zones, slope stability problems of the adjoining abutment hill slopes.

The flash floods that occur in the hilly streams of small catchment areas not only cause extensive erosion in the stream bed but also convey large amount of coarse as well as fine sediments into the power intake. There are several instances where the diversion weir designed with some pondage behind the weir to be used to produce peak power, got filled up with sediments and hardly any pondage was available and the plant had to be operated as a run-of-the river scheme. It becomes obligatory to provide suitable silt excluders for flushing out the coarse sediments before the flow enters into the power intake. If the boulders are rolling during the floods, some boulder catches located upstream may be necessary. In addition, it will be necessary to provide a desilting basin for excluding hard particle size larger than 0.2mm to prevent erosion of the turbine blades. There are several case histories in our country, where the silt erosion has been experienced in the turbines requiring large scale repairs to the runners resulting in closing down the units and less than satisfactory performance of the plant.

If the diversion weir is provided with gates, it would be necessary to formulate a gate operation schedule not only for the spillway but also for the power intake to flush out the silt through the spillway to a maximum extent and also simultaneously plan to close the power intake gates when the silt charge in the hill stream exceeds permissible limits. Case histories of the hill stream out flanking the weir during the floods is also not uncommon.

Water Conductor System

The scheme may have a power channel taking off from the power intake terminating in a forebay or a penstock pipe line taking off directly from the power intake located in the abutment feeding the turbines. Hydraulics of the power channels and the forebay and structural stability of the related structures should be assessed according to the site conditions. Adequate protection should be contemplated to avoid land slips and settlement along the power channel. Special care is necessary for the upper reaches of the power channel to protect against the onslaught and damages during high floods and also to ensure smooth flow conditions into the penstock pipe line without any air vortex formation at the power intake. Geological conditions along the power channel and the foundation conditions of the penstock intake structure should be closely examined and designs should be made to ensure structural stability.

There are instances where some reaches of the power channel aligned along the side-long slope of the hill have developed cracks and slide down if the hill stream at the toe of the hill slope is eroded or undercut during floods. Protection at the toe of the hill slope may be necessary to prevent undercutting of the hill slope.

If the penstock pipe lines are laid over the ground, extreme care should be taken to ensure that there is sufficient surface drainage to prevent erosion from high velocity flow developing along the penstock slope during heavy rains. It is also necessary to prevent any damage to the foundations of the anchor blocks and saddle supports provided for the penstock pipe line. Surface drainage should be planned and provided along the penstock slope with stone pitched drains.

In some of the small schemes, there are instances when the penstocks are badly damaged by rolling boulders from the unprotected excavated side slopes and they should be examined at site and loose boulders lying on the side slopes should be removed before the onset of rains.

There are also instances where the penstock slope develops differential settlement due to uneven loading coming from the penstock particularly when the penstock pipe line does not sit properly on the saddle supports and the entire load is transferred to the anchor blocks.

Power House and Tail Race Channel

It is most important to locate the power house at a site free from problems of landslide from the adjoining side slopes and also away from the effects of high tail water during floods. Tail race channels upto its confluence with the main stream should be oriented in such a way that sediments from the stream should not enter into the tail race channel thereby creating higher tail water level and reduction in the operating head of the plant.

The foundation of the power house to be located on the bank of the stream should be carefully examined by a competent engineering geologist to ensure that the foundation is not subjected to erosive forces of the flow of the stream during floods. Hill slopes or excavated slopes adjoining the power house structure should also be examined to make sure that there are no possibilities of land slides of the slopes endangering the power house structure. The designer of the power house complex should visit the site before finally issuing the construction drawings and confirm that all aspects of the site conditions are reflected in the design. There are also instances of the power house structure being washed away during high floods.

CONSTRUCTION ASPECTS

All the components of the small hydro schemes as designed should be constructed under strict technical supervision. Generally these schemes are located in remote areas and are not easily accessible except by foot and therefore timely site inspections are not undertaken by the supervising engineers. Many of the damages to various components of small schemes are also attributed to construction defects on account of poor supervision. Poor qualilty of work during execution of the small structures involved in the scheme due to lack of timely supervision has also been responsible for extensive damages or failure. Some system of quality control during construction should be enforced and monitored. It should be made mandatory that the team of specialists which was associated during the initial formulation of the scheme should not only review broadly all the key drawings of the scheme before final construction drawings are issued to the field, but also should undertake visit to the scheme during construction twice or thrice particularly during the critical stages of construction of various components.

HILLY SMALL HYDRO IN PRIVATE SECTOR

In the context of the recently adopted liberalisation policy in the power generation sector in the country and the incentives being given by the Government , several private developers are eagerly coming forward to undertake these small hydro schemes in the hilly regions. Because of little or no experience of undertaking similar works, the private developers approach a consulting agency or sometimes hydro-electric equipment manufacturers who are often less conversant with the river hydraulics or foundation engineering. The schemes are evolved, designed and cost estimated by such organisations. With least financing at this stage of scheme formulation by the developers, computer models are used to obtain the projected annual power production and the returns to the developer on his investment are worked out by the consultants or the equipment manufacturers mentioned above. At this stage, everything seems to be in order and the developer confidently moves forward and secures the financing arrangements and awards the work to a civil contractor on the basis of least tendered cost. As the work on various components of the scheme commences, the developer discovers that the features of the scheme envisaged during the formulation of the feasibility report of the scheme are quite different from the exposed site conditions such as:

i.   Excessive quantity of excavation required for obtaining the competent foundation conditions of the component structures.

ii.   The alignment of the water conductor system needs to be shifted away from the hill stream course to prevent erosion and damage during the freshly computed flood discharge.

iii. The location of the proposed forebay needs to be shifted towards the hill to ensure competent foundation for the power intake structure involving excessive hill slope excavation.

iv.   The alignment of the penstock pipe line taking off from the forebay upto the proposed power house needs re-orientation to ensure a smooth gradient. Consequent to the change in the penstock alignment, location of the power house may need shifting also.

These modifications, which become obligatory to ensure safe operation of the scheme after completion, will obviously result in higher cost and longer period for construction which will adversely affect the projection of his returns. Furthermore, when the construction of the scheme is over, the investor may consider that the worst period is now behind him and he is ready to earn the income on his investment for selling his kilowatt hours. But even during operation of the plant, the same characteristics of the nightmare may appear because :

i.   The power output might have been over-estimated because of lack of sufficient hydrologic data analysis during the formulation of the scheme.

ii.   Trend in the increase in the flood discharge during rainy season causing hydraulic problems as a result of poor evaluation of the behavior of the hill stream and threatening the safety of the component structures.

iii. Some of the component structures are beginning to show signs of distress and recurring damages due to poor assessment of the site conditions and poor quality of work during the construction.

iv.   Failure of the generating equipment and loss of efficiency due to rapid wear of the runner caused by high concentration of sediment in the river discharge which was poorly evaluated.

The foregoing important issues discussed above amply demonstrate and forewarn that the developer, enthusiastic to undertake the new venture of small hydro scheme either on BOT or BOO basis should engage a competent and experienced consulting agency which is able to mobilise a team of specialists to be able to advise and guide the necessary site specific investigations and optimisation studies to formulate the feasibility report of the scheme and cost estimate with a fair and acceptable percentage of variation provided under physical contingencies to cover the surprises during implementation, which are inevitable in a hydro scheme whether small or large. Efforts should also be made to evolve innovative designs for the civil structures of various components of the scheme, which could be suitably and economically adapted to the site conditions instead of the conventional designs.

CONCLUSION

In conclusion, it is important and essential that the site selection for a small hydro scheme whether it is for a private developer or a public utility, should be conducted in a systematic way with a competent team of specialists in river hydraulics, hydrology of small streams, engineering geologists, hydraulic and structural engineers with a background of design and construction. There should be a design organisation who should be responsible to prepare design and construction drawings with frequent visits to the site to fully assess the site conditions as and when required. Advice and guidance from the team of specialists should be obtained during any critical stage of construction to ensure that the structures are appropriately constructed to meet the challenging situations that arise during the implementation of small hydro schemes in hilly regions.

First International Conference on Renewable Energy—Small Hydro
3—7 February 1997, Hyderabad, India

THE DEVELOPMENT OF TRADITIONAL HIMALAYAN WATERMILLS FOR SUSTAINABLE VILLAGE-SCALE MICRO-HYDROPOWER

O. Paish,[1] *R. Armstrong-Evans,*[2] *R. Saini,*[3] *D. Singh*[4] *and D. Kedia*[5]

[1] I.T. Power Ltd., The Warren, Bramshill Road, Eversley, Hants RG27 OPR, UK
[2] Evans Engineering Ltd.,Trecarrel Mill, Trebulett, Launceston, Cornwall, UK
[3] Alternate Hydro Energy Centre, Roorkee University, Roorkee 247 667, India
[4] T.E.R.I., Darbari Seth Block, India Habitat Centre, Lodi Road
   New Delhi 110 003, India
[5] Industrial Consultants, GS Road, Bhanga Garh, Guwahati 781 005, India

ABSTRACT

The Watermills Block of the Hilly Hydro Project has set out to demonstrate technologies for upgrading and replacing some of the many thousands of traditional watermills, or gharats, in the Himalayan regions of India.

The paper covers:

(i)   the conclusions of site surveys, identifying the range of local circumstances across the regions.
(ii) the major design issues and different technologies proposed to meet the needs of the mill-owners
(iii) elements of the strategy planned to ensure the projects undertaken are successful and sustainable.

1. INTRODUCTION

1.1 The Hilly Hydro Project

The UNDP-GEF Hilly Hydro Project is an ongoing initiative supported by the World Bank *Global Environment Facility* and the Government of India to demonstrate and promote the use of small-scale hydropower in the 13 Himalayan states of India.

The principal existing use of hydropower in the Himalayas is through the use of traditional wooden vertical-axis watermills (*gharats*) for grinding grain. These operate off 2-6m head, developing typically 0.5kW. There are believed to be up to 200,000 mills in the Indian Himalayas, a further 25,000 in Nepal, and many more in Pakistan, China, Afghanistan, Myanmar and parts of Turkey. This indigenous technology is built and maintained by the miller himself using local materials. However in recent years, gharats have started to fall into disuse: owners have descended to the plains to seek more lucrative employment, more effective diesel powered mills in nearby towns have reduced their market, and major deforestation has caused some water supplies to disappear. Yet, if this abundant and renewable waterpower resource could be exploited more effectively with appropriate and modernised equipment, it could play a key role in driving sustainable economic development in the hilly regions. This is the task which the India Hilly Hydro Project is now endeavouring to address.

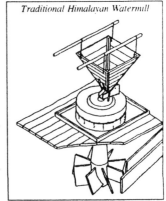

Traditional Himalayan Watermill

The Watermills component of the project is seeking to upgrade and develop 100 watermills in different regions with improved technology which will serve as prototypes for upgrading the remaining watermills in the region. The upgrades will cover two different types of development:

- upgrading existing mills for improved productivity
- replacing existing mills, or developing new sites, with modernised multi-purpose units, designed to operate a range of end-use equipment

1.2 Background

1.2.1 The Himalayas

The populations of the Himalayas, from Afghanistan to Myanmar, still live predominantly in agricultural economies, often at subsistence level. The market for milling is reasonably well served by traditional watermills spread throughout the region, but most other essential agro-processing services (rice-hulling, oil-expelling, juice-extraction, etc.) are non-existent and have to be done manually (ie. by women), or by many hours' walk to electric or diesel machines in town. In Manipur, an alternative waterpower technology has been identified: the *Pani Dhenki*, a drop-hammer used for de-husking rice.

1.2.2 Beneficiaries

Improving traditional watermills is intended to benefit :

- **millers and their families**, whose economic situation will be improved by greater productivity and income generation.
- **local communities**, who would stand to benefit from faster and more efficient agro-processing services, a wider range of mechanised crop-processing machinery, plus electrical services via an add-on generator, such as evening lighting, or water-heating, battery-charging, crop-drying, or irrigation pumping at night.
- **women**, who will have a reduced burden of either manual crop-processing, or of hours or days of walking to electric or diesel machines
- **manufacturers** who will have the possibility of producing and selling large batches of simple, standardised equipment.
- **local workshops**, who will be needed to supply components and spares, or provide repair services, or who may choose to develop/copy their own watermill upgrade.
- **the environment**, by displacing harmful emissions and reducing the reliance on wood-fuel

1.2.3 Traditional technology

It is important to note the advantages inherent in the indigenous watermill technology, in particular it is:

- simple technology
- locally designed and built
- involving mainly local materials
- low cost

At the other extreme, a state-of-art mini-hydro plant may be able to increase the useful energy output at a watermill site by several times, but will inevitably involve:

- imported components
- manufacture and installation by trained engineers
- training of local operators
- specialised spare parts
- transport of heavy equipment to remote areas
- high capital costs

Such modern installations can prove to be unsustainable in the longer-term without accompanying advances in local and regional infrastructure, local skills and engineering facilities, and major technology transfer initiatives. For example, once the crossflow (Banki) turbine had been proven and demonstrated by Swiss engineers in Nepal, it took 7 further years of concentrated promotional activities to establish a workable level of dissemination [1].

A sustainable approach to developing and transferring appropriate technology therefore implies a compromise strategy, where modern know-how (eg. on hydraulics and turbine technology) is applied in the local context. This approach, adopted from the start by the Watermills Block [2], encourages local manufacture and support of the technology, and designs that are simple both in concept and in construction.

1.2.4 Nepal experience
Nepal is the only country where significant progress has been made to upgrade watermill designs, where developments occurred in 2 phases. Initially a packaged steel assembly was developed by a local workshop which incorporated a vertical-axis impulse turbine made with fabricated steel buckets and a penstock pipe [3]. This became known as the MPPU (Multi-Purpose Power Unit) and in the early 1980s this system outstripped the sales of crossflow turbines being promoted through foreign aid programmes because of its simplicity, its similarity in principal to the traditional watermill, and its low cost [1]. When the manufacturer ceased production of MPPUs to take up more profitable business, GTZ of Germany supported a programme to develop a cheap 'construction kit' which involved supplying the MPPU runner and a few components to enable millers to install upgrades themselves [4]. This 'improved ghatta' programme ran from 1984-1988, with further support from 1991-1993, and there is now a steady market for these kits of over 100 per year, providing improved milling plus a rice-hulling option. GTZ however recognise that there is scope for developing and improving the technology, which still has a number of limitations, in particular:
- low efficiency (around 40%)
- low speed (around 150 rpm - too slow for electricity generation)
- primitive bearings - which are unsafe to run unattended for long periods and require changing every 3 months
- labour-intensive manufacture

1.2.5 Indian experience
There has been more than one attempt in India to copy and disseminate the Nepali design of watermill upgrade [5] [6] but these initiatives have failed to make a significant impact, through either institutional reasons (highly subsidised, no training or technical support, owners not helped to develop their businesses), or technical reasons (designs were copied, not understood and transferred, so systems were inappropriately specified and installed).

2. SITE & EQUIPMENT SURVEYS

2.1 Overview
In the first phase of the Watermills project, the team from IT Power, UK, the Alternate Hydro Energy Centre at Roorkee University, the Tata Energy Research Institute in Delhi, and Industrial Consultants, Guwahati, has completed numerous site surveys in the Northern and North-East states and studied the needs, capabilities and aspirations of mill-owners through both one-to-one interviews and an organised forum of mill-owners in March 1996. A standard questionnaire was developed for completing each survey. A census is also ongoing which is attempting to quantify the total number of gharats in the region and therefore the potential for implementing new technology.

A broad conclusion from the surveys has been that nearly all sites are technically feasible for upgrade machines to be installed, but the more critical questions in selecting a viable site are:

1. is the owner willing and motivated to invest in, operate, and maintain an upgraded system ?
2. will there be sufficient new business to justify the cost of the upgrade ?
ie. the priority is less on finding the right sites, but more on identifying the right owners.

2.1.1 Case study - Gadora Bridge

The Gharat Owners Association in Chamoli District (UP) have nominated a site at Gadora Bridge for demonstrating a multi-purpose system to replace the existing, traditional gharat. The site is an attractive choice for a demonstration scheme since it is directly by the roadside on the tourist route to Bhadrinath, and the bridge is also a crossroads for local movements. The miller is young and enthusiastic, and keen to adopt changes if it is worth his while.

The existing gharat is one of 5 at Gadora Bridge, and one of 21 on this stretch of the river. It was therefore agreed by the Association that this gharat should be converted for uses other than milling, so as not to take business away from the other millers.

An available head of 8m and design flow of 150l/s implies a gross potential of 12kW, therefore a working capacity of around 6kW.

Discussions on possible end-uses concluded that the following activities might be feasible and profitable in the context of the businesses operating locally:

- Daytime: oil-expelling, rice-hulling, chilli-grinding, and juice extraction, plus water-heating and cooking for a road-side café
- Evening: lighting and heating for the café
- Night: drying of chillies

The existing mill-house would have to be demolished and replaced with a new building to accommodate the proposed end-use equipment plus café, and a small amount of work would have to be done on the channel. It was agreed that these civil works could be carried out by the Association.

2.1.2 Case Study - Dehra Dun

Another typical example is a 1.5m head site near Dehra Dun in Uttar Pradesh. The owner mills with a traditional watermill by day, but also wants to generate electricity to be able to run a jam-making business from the produce of his farm. He also wants power for lighting, TV, and fan in the evenings and night. He has a flow of 150 litres/sec and power potential of 2kW. He would be keen to invest in improved hydropower technology, but has found nothing available to meet his need. In this case the miller's requirements would be served by a modernised gharat-type system which could still be used for milling, but also adapted to allow an add-on generator to provide electrical power.

2.2 Establishing Local Contacts

A further important issue during the 1st phase was the need to identify local organisations to support the technology at the local level and to take responsibility for managing and supporting new projects.

The Himalayan Environmental Studies and Conservation Organisation (HESCO), an Indian NGO based in Chamoli District, UP, was responsible for setting up the Gharat Owners' Association in Chamoli and has been working with them to design and implement simple upgrades for traditional gharats. The improvements have involved lining the open chute with galvanised steel sheet, using a Teflon bush for the top-bearing (at the centre of the bed-stone), and using a fabricated ball-bearing assembly as the footstep bearing at the base of the rotor.

About 12 gharats have been upgraded to date in Chamoli District. The component and material costs are about 1500Rp and the upgrade work is undertaken by the miller himself. Increased output of 2-3 times has been experienced, with stone speed increasing typically from 70 to 120rpm.

2.2.1 Gharat-Owners' Meeting

On 17th March 1996 a meeting of the Association's 30 members was staged by HESCO. The meeting enabled the owners to express their outlook on their current livelihoods, as follows:

- Competition is increasing from diesel and electric mills, despite the fact that these mills charge in cash (0.5-0.6 Rp/kg). the quality of the flour is inferior, and they won't accept small quantities. Traditional gharats charge 10% in kind, but are much slower.
- They would like to be able to diversify into other services and products, but need help to be shown how they can modify their mills to achieve this.
- They spoke enthusiastically for adopting more advanced and multi-purpose technologies. in particular expressing a preference for systems with as little maintenance as possible, even if this cost more.

2.3 End-use equipment survey

A brief survey was also carried out to establish the current costs and power ratings of end-use equipment relevant to the setting up of multi-purpose watermill upgrades. The following table summarises the general conclusions from a survey in Saharanpur (U.P.):

Equipment	Rating	Price
Good quality single-phase generators	2-5kW	4000-9000Rp
Grindco Mills	1-4kW	1000-2500Rp
Rice huller	1.5-3kW	1500-2000Rp
Oil expeller	3-4kW	5500Rp
Spice-grinder	0.75-1.5kW	1300-1800Rp

2.4 Technology needs

The general engineering strategy for upgrading the mills should be to establish the very best systems that can be coped with by the mill-owners. The needs and capabilities of the person who is going to install and operate the equipment should have as much bearing on the overall design as the hydraulic and engineering criteria.

It is proposed that there are two general approaches that can be applied successfully in the local context:

1. a simple upgrade that can be maintained locally. either by the miller himself, or a local craftsman.
2. a design that is still simple in concept, but is well-engineered and sophisticated enough to be maintenance-free for at least 5 years.

The danger lies in adopting a half-way solution which is both unreliable and too sophisticated for local expertise to repair. The need for sustainability would indicate that the designs for upgraded gharats should be maintainable locally (approach 1). and the multi-purpose units should be designed to be robust and maintenance-free (approach 2).

3. DESIGN ISSUES

Because of the great diversity of the sites and the economic and personal preferences of the millers. the array of possible detailed design choices is large and there are virtually no areas which are totally clear regarding either the general approach or the precise selection of equipment.

Outline designs for the first demonstration units are proposed in Section 4. Some of the main issues which have been considered are as follows.

3.1 Cost vs Quality

Most areas are characterised by the phrase 'Pay now or Pay Later' i.e. either a lot of time and effort is expended at the outset to achieve little or no maintenance, or minimum capital is spent, accepting that there will be regular on-going maintenance efforts and costs. For example, an intake structure can be an expensive but permanent concrete structure, or it can be a seasonal low-cost diversion weir which needs to be re-built every year.

3.2 Manufacturing Strategy

- For the upgraded units there is a basic choice between (i) *complete packaged units* or (ii) *kit units*. The choice amounts to spending time and money in the workshop under 'controlled' conditions, or in field supervision. A kit of many parts will also take many times longer to assemble and install.

- The designs should not be outside the capabilities of modest workshops, so regional or local workshops may be employed to build the mechanical components. However pilot batches from larger, centralised workshops will have the benefits of economies of scale and of establishing 'quality benchmarks' for others to follow. For example, manufacture via casting, readily available in many areas of India, might enable low-cost mass production of large batches.

3.3 Maintenance Strategy

- The level of maintenance that is required by design rather than default is a critical issue. You can design for 'no maintenance' and only require maintenance in unforeseen circumstances or when the design life (of say 15 years) has been achieved. Such infrequent maintenance can be carried out by trained field technicians.

- The alternative is to stick to very simple technology which is a relatively small step from existing equipment and would be maintainable by the miller. However, there is a danger that this may not be a sufficiently large step in the technology to keep him in business. With increasing competition from diesel and electric mills, he may have to exploit his greatest asset of being able to operate 24 hours per day at virtually no extra cost. It is doubtful whether a locally-made design using low-cost materials could be made to survive 24-hour, unattended operation.

3.4 Rotor Design

- The runner design should be chosen on the basis of *strength, cost* and *performance*.

- The turgo-type from Nepal (also copied in India) is perfectly satisfactory, giving 40-50% efficiency. However it has to be fabricated by hand. A design that can be sand-cast in one piece would be much cheaper for batch production, and would guarantee quality.

- High 'specific speed' rotors, eg. propeller turbines, lead to a lighter turbine but may also cause some customer resistance since the concept of a propeller turbine, for example, is very different from a traditional Gharat.

3.5 Bearings

- The design of bearing system is likely to differ significantly between the simple gharat upgrade and the construction of a multi-purpose scheme for 24 hours operation.

- For long life and low maintenance, recommended for the multi-purpose units, the key issues for bearings are: good alignment, good sealing, and adequate rating.

3.6 Drive train

- The drive options for a multi-purpose unit include flat belts, Vee-belts belts and link belts. Flat belts are widely used but transmit limited power and are difficult to keep on the pulleys. Vee-belts give a smooth drive and stay in place on vertical and horizontal drives but availability of the correct sizes may cause a problem. Link belts of various materials are available and can be stocked in reel form and made up to length as required. They are more expensive than conventional Vee belts but some types are claimed to last twice as long.

3.7 Flow Regulation

- Flow variation is catered for automatically in the traditional Gharat by corresponding changes in the depth of water in the open chute. Using a penstock would reduce turbulence losses, and mechanical flow-control (eg. with basic spear valves) can be relatively simple with a single-Jet impulse turbine.
- On low head sites fed from long, level channels, consideration may be given to running a fixed flow turbine (ie. unregulated) on a 'stop/go' regime during low flow periods (ie. waiting for the forebay and channel to fill up, then operating for a short time at full flow). This may be the best solution for driving equipment such as oil expellers which require a certain minimum power to operate effectively.
- Overspeeds are unlikely to damage most machinery in the short term but sparks from mill-stones could cause fire unless an automatic shutdown is fitted. This could be a simple lever which trips a jet deflector, for example, when the grain hopper is empty or the shaft speed increases.

3.8 Load Governing

- Mechanical Load Governing is not commonly used in micro-hydro, but may be the cheapest way of protecting very small systems from overspeed. It involves loading the turbine or generator shaft directly to dissipate excess power in proportion to the shaft speed. For example a simple water-immersed centrifugal governor is an option.
- Electronic Load Governing in various forms is now widely used, but is not that cheap (although this needs to be checked with Indian suppliers). It is most suitable for plants in the 5 to 50 kW range.

3.9 End-Use Issues

Perhaps even more critical than the design of hydropower unit is how the power will actually be used.

3.9.1 Mechanical vs electrical
Although the local viewpoint may often be in favour of electricity generation as part of a new scheme, this may not be the best use of the hydropower resource to meet local needs and guarantee the success of a project. In most cases using the majority of power for mechanical end-uses has proven to be more cost-effective and sustainable; almost all of the 1000 micro-hydro schemes in Nepal operate one or more agricultural machines. The key issues can be summarised as follows:

- Electrical power is very versatile and can be used for electrical, mechanical and thermal end-uses; but it requires the necessary appliances to be locally available and affordable.
- Electrical energy is often used non-productively ie. only for lighting. Providing mechanical power effectively forces the use of income-generating end-uses.
- An electrical installation is much more expensive because of the additional, sophisticated equipment required (generator, speed governor, controller).
- Mechanical power is easily understood, it is more likely to be replicated elsewhere in the region, and machinery can usually be repaired with local know-how. Electrical equipment can not generally be repaired locally and the consequences of generator failure can be severe.
- Very little power is wasted with direct mechanical end-uses. In contrast, small generators and motors are typically less than 80% efficient, so a generator operating motorised equipment is less than 65% efficient.
- Mechanical power is restricted to on-site consumption, whereas electricity can be transmitted any distance. Gharats are often hindered by an accessibility problem: people may have to carry their produce down the valley to the gharat and back again, whereas diesel mills may be available in the village. Hence an electric mill operating in the village using hydroelectric power transmitted up from a modernised gharat site may find a bigger market.

- Starting loads, which can be a problem with electrically driven equipment, are easier to handle using mechanical drives. Processing machinery can become clogged if run too slowly and will then take a higher torque to restart.

3.9.2 Charging and income-generation

- Maximum utilisation of the available power is vital for profitability. The long-term target should be, for example, to run agro-processing machinery by day, lighting by evening, and a further useful service at night, for example water-heating, crop-drying, battery-charging or irrigation pumping.
- Agro-processing is a widespread need and charging for the service is straightforward, whether in kind or in cash.
- Water-pumping is also a major need in hilly areas, but defining the service received and setting up an appropriate charging system is extremely difficult.
- An electrical supply can be defined more clearly, but people in the villages have so little cash that securing monthly payments for each domestic supply can be time-consuming and unprofitable. Overnight battery-charging may be a better method of providing a clear-cut electrical service with immediate payment.

3.9.3 Electrical generation

- There are pros and cons for choosing any of a variety of generator types, depending on the application and local availability: Synchronous, Induction, Homo-Polar, or Permanent Magnet.
- Constant voltage and variable frequency is a simple option for lighting loads. This requires a synchronous generator with a field control (voltage regulator) and no other components.
- Underspeeds caused by inadequate power or heavy loads can damage electrical generators. Reduced frequency can cause the transformer (if fitted in the excitation system) to overheat. Induction generators do not like running underspeed which is a condition that often exists where the power is being used for domestic purposes.

3.10 Areas for Development

The technology for village-scale hydropower is generally under-researched and could benefit from R&D in a number of fields, for example:

- Hydraulic testing of alternative runner designs
- Hydraulic testing of open chutes relative to penstocks
- Design of low-cost nozzle valve or simple spear valve.
- Design and test a simple mechanical overspeed trip.
- Design and testing simple mechanical governor.
- Design and testing robust generator.
- Design and testing improved footstep bearing and adjustable top bearing for gharat upgrades.

4. PROPOSED SYSTEM DESIGNS

4.1 Overview

The preliminary designs proposed for the demonstration units, are as follows:

- a vertical-axis *new gharat* design with a simplified metal runner suitable for casting or fabrication, and designed for upgrading existing mill installations, but with an option for running a further device off the main shaft.
- a horizontal-axis *open crossflow* design for multi-purpose operation, either to replace traditional mills or for new sites.

4.2 'New Gharat' Watermill Upgrade

Layout and Upgradability
A principal aim behind the proposed design is that millers should be offered a simple concept, but one that can be upgraded. Therefore the layout of the first demonstration units will continue to be vertical-shaft systems, replacing the existing wooden construction, and used primarily for milling in the traditional manner; the millstones, mill-house and open chute can remain the same during the upgrade process. However options for driving other machinery (either an electrical generator or a rice-huller, spice-grinder, etc.) will also be immediately possible by removing the top mill-stone and replacing it with a single pulley and belt-drive. Further sophistication can be achieved by replacing the open chute with a penstock pipe and control valve. In the longer-term, greater options can be opened up by using the same rotor but with a horizontal axis and belt-driving two or more machines off the main shaft in an enlarged mill-house.

Rotor design
The traditional wooden rotor is less than 20% efficient. The steel runner design used in Nepal is limited hydraulically to about 50% efficiency, and contains two-dimensional curvature which requires laborious fabricated manufacture. Efforts have therefore been made to design a runner which can exceed 50% efficiency but have a geometry suitable either for casting, or low-cost welded fabrication. This would have major implications on its suitability for both low-cost mass production, and for being replicated at local-level. Casting also guarantees the quality of the runner construction. Furthermore, this design of runner is suitable for converting to a horizontal axis layout at a later date.

Speed
Traditional watermills run at less than 100rpm. The Nepali design increased the speed to around 150rpm, which was insufficient to be able to attach a standard 1500rpm generator with a single belt-drive, since 6:1 speed-increase is generally regarded as the maximum. The proposed rotor design is intended to operate in the region of 250-300rpm. This is an important advance in enabling electricity generation, plus other faster-running agro-processing machinery, to operate with a single belt-drive from the main shaft.

Figure 1 shows a schematic of a simple 'new gharat' upgrade, and Figure 2 illustrates the same runner used with a horizontal axis in multi-purpose mode.

4.3 'Open Crossflow' Multi-Purpose Schemes

The 'Multi-Purpose Unit' will be appropriate where there is adequate head and flow to produce at least 5kW of power at the shaft. These projects will require considerably greater engineering and financial input than the simple mill upgrades. The key features of this unit must be robustness and very low maintenance. The competitive advantage of these 'Multi-Purpose Mills' must be so marked that the enterprising millers that have set up diesel and electric mills, will want to get back into watermilling because it is cheaper and better than diesel.

The open crossflow design is recommended for a number of reasons: the crossflow turbine is well-disseminated through publications by SKAT and others; it can be replicated in modestly equipped workshops; and the 'open' arrangement allows the technology to be more transparent and it is also significantly cheaper (and more efficient) without the casing. Furthermore the unit needs to have a horizontal axis because of the horizontal alignment of most end-use equipment and the desirability of avoiding twist belts wherever possible.

Figure 3 and Figure 4 illustrate the main elements and proposed layout of the multi-purpose design.

It is not proposed that the components for such a system should be manufactured in the remote areas, but that 'benchmark designs' should be built at a centralised workshop, at least for the first batches. If village workshops wish to assemble or modify the units in the future, this should be encouraged. Small diesel engines are now assembled by hundreds of small workshops using components and spare parts supplied by larger companies in the regional centres. The original engines from which the designs were copied were long lasting and the same approach could be adopted with the manufacture of turbine components.

The 'overhung' layout allows the wet components to be kept outside the mill building, with the bearings inside. It is important that the bearings and shaft are of adequate specification and that the overhung loads are kept as close to the bearings as possible. Running the end-use equipment off the main shaft is intended to keep the system compact, cheaper and safer to operate.

5. CONCLUSIONS AND IMPLEMENTATION STRATEGY

The Watermills Block is addressing the immediate requirements of rural people by aiming to demonstrate the best technical solutions to meet their technical capabilities and economic needs. The technology must be both <u>affordable</u> and <u>locally acceptable</u> - technically and culturally. The history of small hydro shows that ignoring these factors has sometimes led to the transfer of efficient and powerful technologies which have had no chance of being replicated, maintained, or owned by local people, and have therefore been unsustainable unless propped up by foreign aid programmes.

The upgrading of traditional watermills is an effective and sustainable way of meeting the energy needs of a major section of the rural poor. Rather than attempting overly ambitious leaps in technology, the need for sustainability means starting at the current point of development and moving forward in steps that can be understood, afforded, and bring immediate and worthwhile benefit. There is a danger in adopting 'half-way' solutions which are too simple to be reliable, but too sophisticated for local expertise to repair.

The danger with any Short Term Development Programme' is that a lot of effort is put into establishing as many projects as possible within the time and budget allocated, without making provision for the long term training and backup. The Watermills programme is endeavouring to ensure that there will be groups of trained local engineers who are well-equipped to attend to problems as they occur. Gharat-owners will also need training (eg. via a local Association) to maintain their own mills.

If unwisely managed, one negative impact of watermills development can be anticipated (from experiences elsewhere) as occurring through market forces. If one watermill increases its business, other millers may suffer and come to resent the new technology. The Hilly Hydro Project is already taking active steps to prevent this problem by supporting the setting up of Watermill Associations, adding to the existing one in Chamoli. Groupings of local watermill owners in a valley or region will agree by consensus which mills to upgrade and what services should be provided so as to minimise any local conflicts of interest. Past experience [7] has shown that conflicts can be avoided by involving local people in the decision-making process and ensuring private purchase and ownership of the new technology.

There is also a danger of making the pilot projects too sophisticated, so that they are then too difficult to replicate. The project aims to keep the technology simple, robust and transparent, while providing appropriate levels of training and backup.

References:

1 *Micro-hydropower in Nepal: development effects and future prospects with special reference to the heat generator*, D.Jantzen, K.Koirala, FAKT, 1989

2 *Engineering small waterpower schemes from traditional mills*, R.Armstrong Evans, O.Paish, India Hilly Hydro Project - Brainstorming Conference, New Delhi, November 1995

3 *Multi-purpose power unit with horizontal water turbine: operation and maintenance manual*, A.Nakarmi, A.Bachmann, UNICEF/Nepal, 1984

4 *Improved ghatta construction manual*, Manfred Bach, GATE Publications, 1985

5 *Performance testing of modified gharats in Kumaon region*, R.Saini, N.Ahmed, Non-Conventional Energy Development Agency, Lucknow, U.P., 1989

6 *Rural Technology Manual - improved water mill*, A.Ahmad, CDRT, IERT, Allahabad, India, 1992

7 *Key Factors for the Success of Village Hydro-Electric Programmes*, N.P.A. Smith, World Renewable Energy Congress, 1994

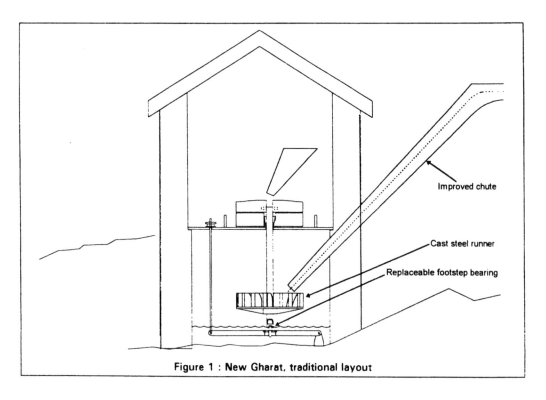

Figure 1 : New Gharat, traditional layout

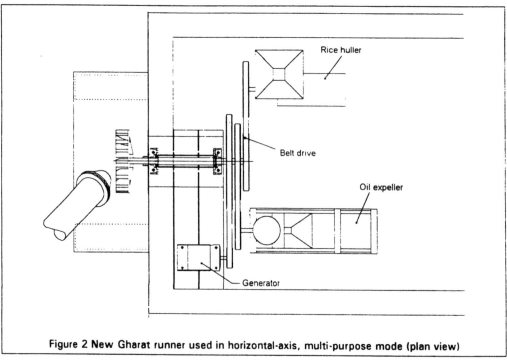

Figure 2 New Gharat runner used in horizontal-axis, multi-purpose mode (plan view)

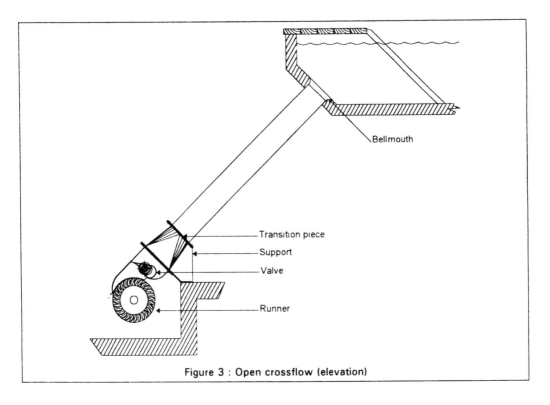

Figure 3 : Open crossflow (elevation)

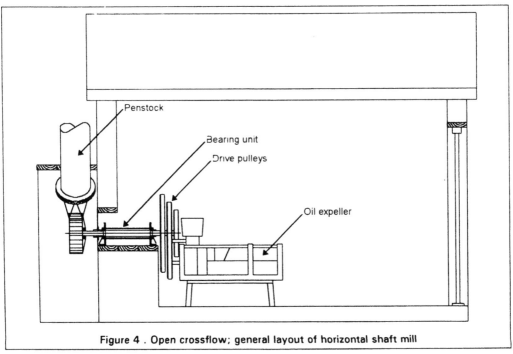

Figure 4 . Open crossflow; general layout of horizontal shaft mill

First International Conference on Renewable Energy—Small Hydro
3 – 7 February 1997, Hyderabad, India

LOAD DEVELOPMENT

Andy Brown, Bill Langley and Ian Tansley

Micro Hydro Power Group, Dulas Ltd.
The Old School, Eglwysfach Machynlleth, Powys SY20 8SX, UK

ABSTRACT

In the Himalayan foot hills, there is a significant amount of electricity generated by isolated run-of-river hydro schemes. Load factors are often as low as 18%, and yet evening peaks are sometimes difficult to meet.

Against this background of "spare" energy, there is enormous pressure on fuel wood, and suppressed demand for alternative fuels such as LPG. This paper looks at ways of improving load factors, to reduce CO_2 emissions, and improve the economics of present and future mini hydro schemes.

In an example from Sri Lanka, the rural grids in some areas were suffering "brown-outs" during the peak evening load time, and a massive upgrading of the lines looked inevitable. In an innovative move, several rural industrial consumers (mainly tea and rubber factories) were offered a reduced tariff if they accepted dual tariff meters which charged high rates between the hours of 6pm and 9pm("peak" times). The factories found no problem in modifying their production schedules, they made savings to their bills, and the "brown-out" problem disappeared. All parties gain in this situation, at very little added cost or complexity.

Storage type appliances (storage heaters, insulated hot water geysers and storage cookers) can be combined with dual tariffs to significantly increase domestic load factors. Low cost timer units are available which can tolerate power cuts, and switch appliances on and off automatically at pre-set times. At least 6 countries including UK use this method. The paper looks at pricing off peak electricity to compete with domestic charcoal, LPG, coal and fuelwood.

In India, small connections are sometimes given on a fixed monthly payment basis. This has advantages as no meter or meter reading is needed, but can lead to overloading problems if there is no precise control on the peak current drawn by the consumer.

In Zimbabwe, Nepal, Eritrea and elsewhere, current limited connections are common, where the customer buys 100W, or 250W, or even 1kW, on a fixed tariff basis. Tamper proof auto-resetting current limiting devices then control the maximum current going to the dwelling. Load limiting connections can be 50% cheaper than metered connections to install. The paper looks at the potential for these techniques in India.

The paper also looks at international experience in innovative tariff collection systems, including community involvement and pre-payment meters, and how it may help in the Indian situation.

Electrification is one of the largest industries in the world. New consumer connection technology is being developed in India and elsewhere. Indian industry is well placed to benefit from innovations in this sector.

This section of the Hilly Hydro Project is very challenging and complex, but the relative lack of innovation in demand side management, combined with an ample resouce of off-peak electricity, provides significant opportunities for change. Whilst reductions in fuelwood use for poor rural consumers will be difficult, there are many other users who can switch from fuelwood and other CO2 producing fuels. The national potential is very significant.

BACKGROUND

In many regions with extensive hydropower, such as countries in the Himalayan foot hills there is a significant amount of electricity generated by isolated run-of-river hydro schemes. Load factors are often as low as 18%, and yet evening peaks are sometimes difficult to meet.

Against this background of "spare" energy, there is often pressure on fuel wood, and suppressed demand for alternative fuels such as LPG. This paper looks at ways of improving load factors, to improve the economics of the present and future mini hydro schemes.

The information is drawn from work in many countries, but a focus has been drawn by the UNDP/GEF Hilly Hydro project in India, which includes a section of work on load development with a particular emphasis on reducing CO_2 emissions.

TARIFFS LEVELS

The World Bank recommends a minimum of US$0.08/kWh for domestic consumers and it has been observed that electricity costs of US$0.2/kWh are still more attractive than any other form of energy for lighting and powering radio/TV.

India has a rather simple tariff structure which, in the main, does not allow for demand side management of the domestic load. The rate is also quite low, with many domestic users paying less than US$0.03/kWh. Nepal and Sri Lanka, both with extensive hydro, have slightly higher rates, but many developing countries have subsidised rates which can make rural electrification unattractive, if viewed in terms of short term profitability. "Social" tariffs often discriminate in favour of the small consumer, increasing prices as consumption exceeds a certain level.

At present, the average price for a standard single rate tariff domestic consumer in the EC is around US$0.08/kWh.

DUAL TARIFF SYSTEMS

Most countries use a range of tariffs to apply some demand side management to their grid systems. In an example from Sri Lanka in 1985, the rural grids in some areas were suffering "brown-outs" during the peak evening load time, and a massive upgrading of the lines looked inevitable. In an innovative move, however, several rural industrial consumers (mainly tea and rubber factories) were offered a slightly reduced tariff if they accepted dual tariff meters which charged high rates between the hours of 6pm and 9pm. The factories found no problem in modifying their production schedules, they made savings to their bills, and the "brown-out" problem disappeared. Everyone gains in this situation, at very little cost or complexity. Importantly, no coercion or enforcement is required as the consumers have a free choice. Three phase dual reading meters with built in clocks (see Fig 1) are available for $120 or less in most countries, and have hold over batteries to protect timers against power cuts.

Although almost all industrialised countries offer a dual or triple "off-peak" or "time of day" tariff, most developing countries either have no such option, or offer this only to industry.

INTERRUPTABLE TARIFFS

This option is not so widespread, but is offered in perhaps 30% of industrialised countries. UK, USA, Spain and Sweden all operate this system with small industrial consumers. The consumer agrees to have appliances connected to this line switched off automatically, by the utility, for a stated number of peak hours per year. The tariff is typically set at 50% of the standard tariff, and the maximum number of hours might be 700 per year. In a corollary to this system, small hydro generators are sometimes offered a very high rate if they agree to keep enough water to provide an immediate grid input on demand.

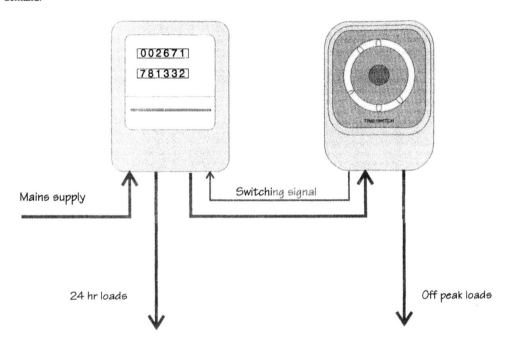

Figure 1 - Dual tariff meter and timer (Dulas ltd.)

Interruptable tariffs can be a useful peak reduction technique, but the automation and control required means that large consumers are the most obvious first target market. Small, plug in load management units are now available for around $200, which disconnect individual appliances or groups of appliances when frequency or voltage drops. These are then reconnected when normal generating conditions resume.

APPLIANCES

In order for significant domestic up take of off-peak electricity to take place, new appliances are needed. Countries with well established off peak tariff systems, such as the UK, Sweden, Japan and New Zealand, have a range of low cost timers, for switching on washing machines and hot water immersion heaters during off peak times (usually between midnight and 8am), and a range of appliances incorporating storage such as storage heaters (fig 2) and storage cookers (fig 3).

Such devices can be combined with dual tariffs to significantly increase domestic load factors. In UK,

20 million storage heaters, each of around 3kW, operate on timers or carrier signal operated devices to increase off peak loads. The off-peak tariff is 25% of on-peak. In this way, the load factor of an individual dwelling can be brought to over 50%. The UK off-peak system "economy 7" is supplied for only 7 night time hours. Such a system in rural India could be available for up to 15 hours per day, as industrial loads are not yet developed. This makes the storage devices smaller and cheaper.

Conventional electric cooking rings are cheap to buy, but have low efficiencies and cause very poor load factors. Low wattage cooking devices, such as rice cookers, can be very efficient and make for good load factors.

Domestic hot water ("geysers"), if operated on off peak power, can improve load factors significantly, as witnessed by the large number of countries (almost all western European countries for example) which heat water in this way.

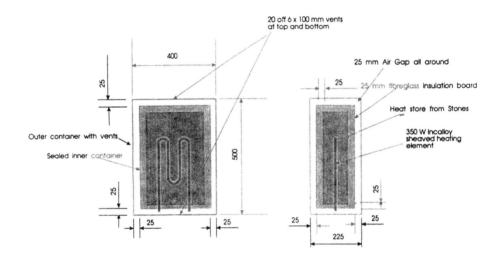

Figure 2 - Storage Heater (Dulas ltd.)

Many small isolated hydro schemes in temperate industrialised countries such as UK and USA, and some in the highlands of more tropical countries such as Sri Lanka and Nepal, achieve 100% load factors for most of the year by automatically diverting excess power to water or space heating loads, such as storage heaters or water heating. This is most simply achieved by the use of electronic load control (ELC) governors, still rare in India but very widely used for schemes of up to a few hundred kW in most countries where such schemes are common.

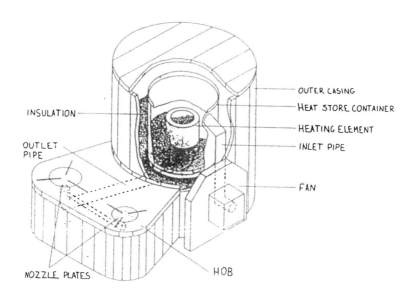

OUTER CASING
HEAT STORE CONTAINER
HEATING ELEMENT
INLET PIPE

INSULATION

OUTLET PIPE

FAN

NOZZLE PLATES

HOB

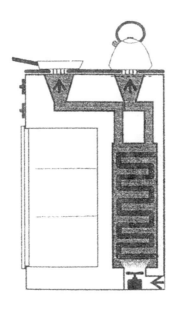

Figure 3 - Heat storage cookers, Nepali and concept (Dulas ltd.)

FIXED CURRENT CONNECTIONS

In India, small connections are sometimes given on a fixed monthly payment basis. This has advantages as no meter or meter reading is needed, but can lead to overloading problems if their is no precise control on the peak current drawn by the consumer.

In Zimbabwe, Nepal and elsewhere, current limited connections are common, where the customer buys 100W, or 250W, or even 1kW, on a fixed tariff basis. Tamper proof auto-resetting current limiting devices then control the maximum current going to the dwelling (see figures 4 and 5). Load limiting connections can be 50% cheaper than metered connections to install. Eritrea is one country presently testing this option, motivated by potential savings in meter reading and revenue collection in remote villages.

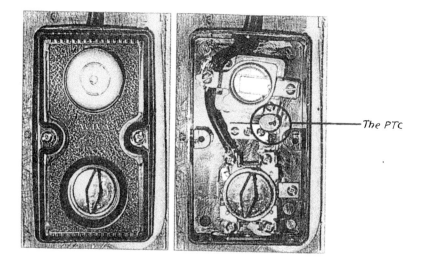

Figure 4 - Current limitter - Positive Temperature Coefficient Thermistor (PTC) (from reference 1 above, I.T. Consultants)

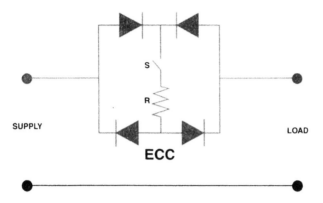

Figure 5 - Current limitter - Electronic Current Cut out (ECC) (from reference 1 above, I.T. Consultants)

Although a very tempting option for utilities suffering from extensive illegal connections, it is generally thought to be a dangerous option for regulating existing connections. In one example in Malawi, consumers rejected the system strongly, spoiling the market for such connections for some years to come. It has been successful where the customer has chosen the option (e.g. Zimbabwe) or where an isolated village scheme has been designed from the start around fixed current connections (e.g. Nepal)

Load limited connections, whilst increasing load factors, have been observed not to waste electricity in warm climates, probably because the cost of wear on the appliances (e.g. bulbs) is enough to discourage wastage. Typical load factors for individual dwellings without heating are 24% to 37%.

Electrification is one of the largest industries in the world, and is growing faster than the car industry or the electricity generation industry, yet relatively little international research and communication takes place. New consumer connection technology is being developed in India and elsewhere. Local industry is well placed to benefit from innovations in this sector.

INNOVATIVE REVENUE COLLECTION SYSTEMS

We might also look at international experience in tariff collection systems. In Thailand, 10% of electricity is collected by a member of the community, who typically receives 5% of the payments as an incentive. This reduces the cost of collecting from poor consumers. In Bangladesh, informal sub connections mean that on average 2.5 houses per meter are connected. This also reduces collection costs but can reduce levels of safety. Pre payment, card operated meters are widely used, in South Africa and the UK for example, but require an infrastructure. They typically reduce demand by up to 50%, which may be useful in some cases.

In a pilot project currently taking place in Eritrea, the combination of fixed current connections, and village based collection are expected to reduce the cost of revenue collection from $0.65 to $0.25 per consumer per month - significant when typical monthly payments are $2/month.

FUELWOOD AND CO_2 ISSUES

The Hilly Hydro project sets out to reduce CO_2 emissions, in part by displacing fossil fuel use in cooking, heating etc., and in part by displacing thermally generated electricity with hydro electricity.

Fuelwood is a complex issue, and some of the factors brought out by projects around the world are:

Labour - many families spend 25% of their available time on fuel wood collection, and see reductions of this figure as a benefit. On the other hand, many people are employed in the collection of fuelwood, and see a reduction in use as a threat.

Sustainability - although fuel would can be a major cause of deforestation, in many cases, fuel wood use is carried out in a sustainable way, and only the clearance of land for cultivation, timber resources or the effects of livestock, cause unsustainable depletion. Sometimes electrification can boost agricultural activity, causing a negative CO2 effect.

Health - smoke related diseases and kitchen hygiene are greatly effected by electrification, and by moves away from open wood burning cookers.

Social changes - In cold climates, electrification can increase fuelwood use as people stay up later and use more heat. In warm climates this effect is not so significant. The social focus of an open fire is a significant factor in many countries, so much so that many electric heaters, even in industrialised countries such as UK, imitate wood fires.

Non-cooking use - In many areas, fuelwood use for the preparation of animal feeds and alcohol is equal to that used in cooking. Alcohol production is often illegal, however, which can lead to distortions in data collection.

Target groups - Consumers who pay for fuelwood, such as bakeries, schools, hospitals and middle income rural households, may be the first consumers to change from fuelwood to off-peak electricity devices.

CONCLUSIONS

Where load factors are low, and energy from run of river hydro is available, there are a number of well proven, demand side management steps which can be considered by those countries (mainly developing countries) who have not explored more complex tariff systems.

Although quantitative information is difficult to obtain and to transfer from one country to another, there is evidence that complex tariff systems work in that so many countries have adopted, and stayed with, these systems.

For utilities in areas such as the Himalayan foothills, the adoption of new tariff structures, if implemented with programmes to promote new appliances, could have a greater effect on the economic viability and social contribution that the hydro industry can contribute than most other initiatives. Such positive effects help hydro to compete with thermal stations, in turn reducing the CO_2 output of the regional grid.

Appliances designed to operate in a dual tariff environment, such as storage heaters and cookers, have the potential to displace wood, coal and paraffin, and thus reduce CO_2 emissions.

REFERENCES

1. Dr Nigel Smith, June 1995. Low Cost Electrification. Study Report by Intermediate Technology Consultants.
2. Electricity Association Services ltd, 1994. International Electricity. Annual prices report. Registered Office: 30 Millbank, London SW1P 4RD.
3. GEF Hilly Hydro Project. April, 1996. Block 2 interim report.
4. Dulas ltd. January 1996. Eritrea low cost grid connection. Project report.

First International Conference on Renewable Energy–Small Hydro
3–7 February 1997, Hyderabad, India

THE VILLAGE BENEFICIARY AS A DECISION MAKER

R. de S. Ariyabandu

Irrigation Water Management and Agrarian Relations Division
Hector Kobbekaduwa Agrarian Research and Training Institute
P.O. Box 1522, Colombo, Sri Lanka

ABSTRACT

Micro hydro has the potential of providing an alternative source of power supply to the rural population in Sri Lanka. However, this option is limited mainly to the central and southern hill country where the annual average rainfall is above 2000 mm.

Though Sri Lanka had immense experience in estate micro hydro, it was not thought as an option for rural electrification. Sri Lanka has a comprehensive rural electrification programme, which intends to supply much of the rural households with electricity. According to the Ceylon Electricity Board (CEB) work plan, 80% of rural households would be provided with grid power while the balance 20% to be supplied with renewable sources such as micro hydro, solar etc.

Innovative rural folks in many parts of central and southern hills had experimented with water wheels from natural stream flows or Irrigation canals to produce electricity.

However, it is only in 1990 an NGO called Intermediate Technology Development Group (ITDG) embarked on the task of developing village hydro as an option for rural electrification.

The key factor in the ITDG development approach is making the village beneficiaries responsible for the decision making process, actual implementation and post construction, operations and maintenance. The beneficiaries organized themselves into a Electricity Consumer Society (ECS) which acts as the prime decision making body. The approach adopted by ITDG entails that the civil and transmission cost of project implementation to be borne by the beneficiaries. In this respect the ECS decides on the initial monetary contributions by beneficiary households and the mode of participation in civil works. Initial contributions have varied between Rs. 500 - 3000/= per household among different projects. Participation in civil works is mainly by way of labour. Among the 16 implemented projects the labour contribution varies between 10%-42%. In many projects supply of electricity is dependant on the initial payment and active participation in civil works. On completion of the project the ECS decides on the amount of electricity usage. Usually the limit had been set at 100 watts per households but it varies between 40-120 watts depending on the capacity and number of beneficiary households. Deciding household tariff is the sole responsibility of the ECS. The present tariff rates vary between Rs. 0.5 - 1.0/= per watt per month. The decision on tariff is an economic dependant factor. The low tariff rates decided by ECS is an indication of the economic situation of the villagers. Unfortunately when tariff rates are decided the ECS does not take into account the cost of replacement of machinery at the end of its useful life period. When tariff was calculated according to an accepted formula, the actual tariff collection was less than the realistic tariff in at least 83% of the projects in operation.

Hence the decision taken by the ECS is purely based on the welfare of the community than sustainability of hardware. The present tariff collection incidently is adequate only for routine operation and maintenance. It is a common scenario in many projects that the ECS prefer to extend the supply to additional beneficiaries by using ECS funds for expansion than invest in productive end-uses to make projects more viable.

This paper attempts to analyze the degree of decision making by the Electricity Consumer Societies with respect to the rationality of the decision in terms of social welfare, institutional sustainable and economic viability. The information for the paper is based on annual monitoring of 16 micro-hydro projects currently under operation.

INTRODUCTION

Sri Lanka, an island with an annual average rainfall of 2000 mm. However, the central and southern hill country receives an average annual rainfall of 3000 mm. Traditionally, Sri Lanka had been dependant on hydro power for most of it electricity needs. At present 84% of the countries power needs are met through large-scale hydro power schemes. The over dependency on large scale hydro power has led to the neglect of the small-scale micro-hydro sector. Other reasons for omitting this sector from the national generation expansion planning process are, higher production cost per KW, inadequately proven technology and low overall demand compared to high growth rates expected (2). At present only 40% of Sri Lankan population enjoy the benefit of electricity. The balance 60% which are mostly in rural areas have to depend on fossil fuel (kerosine) for household illumination.

This paper reviews the experiences of an NGO, the Intermediate Technology Development Group (ITDG), in its efforts to electrify remote rural villages through the "village hydro programme". The paper critically looks at beneficiary participation in village hydro project development and attempts to assess the village beneficiary as a decision maker and its impact on project sustainability.

VILLAGE HYDRO PROGRAMME

In 1991, ITDG's Micro-hydro Programme began working in rural areas - particularly in the hilly areas of the Matara District. The Programme used its estate hydro experience and expertise to work with local innovators and manufacturers to implement the village hydro schemes.

The first four schemes, implemented in 1992 and 1993, were considered pilot projects. Even though the projects were, by nature, somewhat experimental, part funding was received from government and other donors. At present, 26 village hydro power schemes have been implemented.

The Village Hydro Project attempts to improve local technical capabilities related to equipment manufacture, repair, operation and maintenance, and to learn about and develop institutional arrangements which would be suitable for village power generation and supply.

The Programme works with a Technical Advisory Committee (TAC) which includes engineers and social scientists from the government, the private sector and universities (3).

DISTRIBUTION OF VILLAGE HYDRO SCHEMES

Though the initial ITDG was limited to one district (Matara) in the southern hills, during the subsequent three years the programme expanded to four other districts of Galle, Kandy, Badulla and Ratnapura. The total number of projects as at June 1996 is 26 with 16 projects having more than one year's operational experience. ITDG did not advertise its programme but the rapid expansion in the programme can only be attributed to its success of the initial projects and the demonstrative effects it had in adjoining villages and districts. At times people as far as 180km from the initial project sites have visited the projects to educate themselves and learn about the technology. It is only in late 1995 that the National TV carried documentaries on success stories for village hydro programmes. Since receiving public attention, there had been many inquiries from interested people where hydro potential exist. At present there are over 200 such applications received by ITDG and other catalyst (individuals capable of guiding people to set-up their own village hydro projects). ITDG finds it difficult to handle the present influx of inquiries and as a solution to this problem, they are directing such requests to village catalyst in their respective districts.

SELECTION PROCEDURE

Selecting projects for implementation needs careful consideration of many factors. Technical feasibility, social and economic feasibility, financial support and project monitoring and supervision.

At the initial years of programme implementation all requests that came to ITDG were assessed for its technical feasibility by the "Energy Team" of ITDG. Technical feasibility takes into account the height of the water head (usually kept at 15m minimum head due to use of peltons) and continuous power for 12 months of the year or designed for minimum flow rate. With the emergence of catalyst in different districts, ITDG now sends such requests to catalyst in these districts. They are capable of carrying out technical feasiblities by themselves. Once its technically proven to be feasible, ITDG conducts a socio-economic survey to assess the potential of the beneficiary capacity to contribute to the project (in cash and kind), pay monthly tariff and operate and maintain the plant by themselves. Even if a project is technically feasible, it can be rejected on the outcome of the socio-economic feasibility. If a project passes both these stages, the ITDG would attempt to bring potential donors (Rotary Club, Janasaviya Trust Fund, Department of Energy Conservation) and the committee members of Electricity Consumer Societies (ECSs) to negotiate independently. Though negotiation take place independently, many times the donors wanted the involvement of ITDG as a surety. This indicates the lack of confidence on the part of donors to deal directly with village organizations. However, building confidence with ECS could only be a matter of time. Once the project is commissioned, ITDG monitors the progress at least in the initial years since commissioning the project. As a matter of fact, it is the responsibility of ECS to manage and monitor the progress of the project after commissioning.

REQUESTING FOR FUNDS AND FUND DISBURSEMENT

Number of NGOs, state trust funds, state organizations etc. have shown interest in funding village hydro projects as long as it serves the needs of the villagers. However, as stated earlier, funders are more comfortable in channelling funds through ITDG than directly to village ECSs. Incidently some of the village ECSs too think that funds should be handled purely by ITDG even if the donors are willing to fund the ECS directly. This is a system by which ECSs evade responsibility and accountability to beneficiaries. Justification on the part of ECSs is that accusations of fund abuse by village level committees is a common phenomenon in rural Sri Lanka.

Total fund allocation is dependent on the "project feasibility report" submitted to donors. Donors would negotiate with ITDG on the strength of the "project feasibility report" and agree for funding on a mutual understanding. In the project budget, the cost of civil works and transmission is not solicited from donors as these are considered to be the total responsibility of the beneficiaries. Fund disbursement is the responsibility of ITDG. Usually hard cash is not given to ECSs. All material required for project construction is either delivered to the site from Colombo or bought from local town centres (if available). Equipments like turbines are manufactured to order by local manufacturers. Usually ITDG keeps a marginal consultancy cost which invariably exceeds as the project progresses.

TECHNICAL INNOVATION AT LOCAL LEVEL

In most of the present project sites, there were local technological innovations by individuals to generate electricity. The most common of these technologies is the "water wheel" in which one turns a large wheel made out of small buckets, through water to charge a dynamo that gives sufficient power for a household. These type of technologies have been evident from medium run-of-the river type of irrigation canals. It is these enthusiastic individuals that later take-up the leadership in ECSs to expand their idea to electrify the village.

VILLAGE HYDRO TECHNOLOGY

Village hydro is an indigenous, renewable source of energy, the development of which is generally not associated with adverse environment consequences (3). Village hydro can provide energy through direct mechanical power or by linking a turbine to a generator to produce electrical power. A typical village hydro unit would consist of a weir to impound water from a run-of-the river type, a forebay tank to collect water (size vary with generation capacity), a PVC penstock to convey water to the turbine (usually the turbine is covered with some local material to prevent U.V. rays damaging the PVC) and an induction generator controller coupled to a hot plate as ballast. Besides these main equipment, the power house also has a locally manufactured switch board with earth-fault relays, miniature circuit breakers and separate kWh meters connencted to each distribution line provided to monitor the power consumed by houses connected to each line. A butterfly valve is used to control the flow of water (3).

VILLAGE HYDRO DEVELOPMENT APPROACH

Village hydro projects are a community property. Hence, it is mandatory to establish a Electricity Consumer Society (ECS). This concept was first introduced at Kalugaldeniya in one of the initial projects. The necessity of these societies were so well-established, that in all future projects, beneficiaries formed their own societies even before they made the first request to ITDG. This approach became easier for ITDG to implement projects, because all functions with respect to project implmentations were subsequently channelled through ECSs.

Electricity Consumer Societies (ECSs) are village organizations established at village level for the purpose of managing village hydro projects. The ECSs are usually formed at project inception on their own initiative. ECS, as any other village society has a President, Secretary, Treasurer and a committee of members.

Once the funds are approved by the donor, ECS is informed to collect a initial contribution from each beneficiary household. Also the ECS is expected to get beneficiary participation in civil works, such as construction of the weir, power house, forebay tank. Subsequently in the implementation process, ECS is also expected to get beneficiary participation in installing transmission and to a lesser extent in the laying of the penstock. The initial monetary contributions from beneficiaries are expected to cover the cost of material required for civil works and transmission.

The ITDG's concept is that they would not support (financially) beneficiaries for civil and transmission costs. It is the beneficiary equity for the project. This concept though it may not be the best for poor village communities, has made the beneficiaries work hard to make their project a success. There had been instances (in Dolapalledola project) where the beneficiaries had to wait for nearly a year before they could convince the local member of Parliament to give them funds to purchase transmission wire.

INVOLVEMENT OF BENEFICIARIES IN PLANNING AND CONSTRUCTION

One of the reasons why village hydro beneficiaries find it difficult to bear the full financial cost of civil works and transmission is inadequate financial planning at the commencement of the project. At project initiation beneficiaries are allowed to decide on the amount of beneficiary initial financial contributions. The amount is usually decided on economic status of beneficiaries disregarding the actual cost of civil and transmission costs. Table 1 shows the relationship between total village contribution to total project cost.

Table 1 :   Relationship between Initial Village Contribution and Total Project Cost

Name of the Project	Generated Capacity (KW)	Benefi. House-holds	Total Project Cost (Rs.)	Initial Village Cost	% Total Initial Cost to Project Cost
Katepola	25	88	2,102,838	23,350	11
Athuela	35	97	1,971,971	39,705	02
Illukpitiya	5	51	557,676	55,000	10
Niriella	5	38	534,774	37,500	07
Yagirala	1.5	14	346,110	29,250	08
Nakiyadeniya	2.5	26	137,589	62,395	45
Kalugaldeniya	3.5	36	245,063	32,500	14
Beralapanathara	4.5	65	467,549	41,800	09
Udagedara	1.0	08	146,177	19,200	13
Umangedara	1.7	17	197,391	48,200	24

Source : Monitoring Survey, 1995 (1).

365

According to Table 1, except one project Niriellawatte in all other projects, initial beneficiary contribution had been less than 25%. In three projects it is as low as 10%. The large variation in beneficiary contribution between the projects (02% - 45%) indicates that individual beneficiary contributions is a village decision disregarding the cost of the project. This means the beneficiaries need to be supported in both civil and transmission works or in either one. Usually the beneficiary contribution could cover the initial civil works but cost of transmission had been a problem in many projects. Table 2 indicates the donor beneficiary contribution in civil and transmission components.

Table 2 : Donor Beneficiary Contribution in Civil and Transmission Costs (in Rupees)

Project	Civil Cost				Transmission Cost			
	Donor Contri.	%	Benef. Contri.	%	Donor Contri.	%	Benef. Contri	%
Katepola	0	0	163,570	100	350,000	72	140,000	28
Dolapalledola	0	0	52,723	100	88,200	100	0	0
Illukpitiya	50,000	38	80,000	62	115,000	79	31,000	21
Niriella	0	0	23,115	100	60,450	58	44,184	42
Yagirala	0	0	23,927	100	0	0	14,900	100
Nakiyadeniya	0	0	21,319	100	0	0	21,740	100
Kalugaldeniya	0	0	24,998	100	0	0	17,051	100
Berelapanathara	17,810	25	521,515	75	24,587	70	10,543	30
Udagedara	0	0	32,585	100	11,579	90	1,340	10
Umangedara	0	0	19,733	0	0	0	30,964	100

Source : Monitoring Survey, 1995 (1).

Table 2 illustrates that when projects are small (low generated power capacity) and the houses are not scattered widely, the beneficiary contribution is adequate to meet the costs of both civil and transmission work. However, this assumption does not necessarily hold true in all cases. In Dolapalledola, a small project of 2.1 KW, beneficiaries could not find funds for transmission. Though the project serves only 18 households, the houses are scattered and they required 300 meters of wire. The ECS did not have sufficient funds and ITDG also held on to its concept of not funding for components which are the total beneficiary responsibility.

This made the ECS look for other funding sources, which they (ECS) did on their own. The process took almost one year, during which time the beneficiaries did not have electricity though the project (electro-mechnical) was almost completed. Finally, the ECS managed to convince the local Member of Parliament to allocate funds from his decentralized budget. The importance of this processes is the perseverance of the ECS which finally brought results. However, the reason for the delay, as shown in the above example is bad financial planning at project initiation.

In Illukpitiya (5 KW project), though the project was relatively large, with 51 beneficiaries, the ECS was unable to raise adequate funds to cover both civil and transmission cost. In Beralapanathara (4.5 KW project) one of the earliest projects (1992), both civil and transmission had to be supported. In these two examples, initial per household contribution had been only Rs. 1,100/= and Rs. 650/= respectively. However, most of these beneficiaries in both projects are small tea-land holders and their monthly income vary from Rs. 5,000/- - 12,000/=. Hence, they could have contributed more at project initiation. Another deciding factor for initial contribution is how much project beneficiaries in other already implemented projects pay. This again shows that monetary contributions have followed a tradition without considering the cost of the project.

ELECTRICITY TARIFF

All beneficiaries in village hydro-projects have accepted the payment of tariff for using electricity. However, the tariff rate is usually a decision of the ECS. The actual tariff paid by beneficiaries varies among projects. Most projects appears to follow a tariff rate of Rs. 0.50 per watt of power used (Table 3).

Table 3 :   Tariff Rates Paid by Beneficiaries Different Projects

Project	Monthly Average Tariff Rate (Rs.)	New Tariff Rate (Rs.)	Difference (Rs.)
Katepola	100	148	-48
Dolapalledola	35	118	-83
Illukpitiya	100	67	+33
Niriella	100	87	+13
Yagirala	75	294	-219
Nakiyadeniya	60	67	-07
Kalugaldeniya	36	58	-22
Beralapanathara	31	56	-25
Udagedara	50	275	-225
Umangedara	50	85	-35

Source : Monitoring Survey, 1995 (1).

Except for the Beralapanathara project all other beneficiaries are allowed a maximum use of 100W per household. The tariff rates given in Table 3 are decided upon welfare and to be uniform with "other" projects. In this situation no consideration is given to tariff collection, subsequent operation and maintenance and replacement.

Another reason for collecting tariff, is because it was given by the project implementors.

At project implementation, the electro-mechanical component has an estimated life period of 5 years. Hence, the tariff collected over five years should be in a position to replace the electro-mechanical components if required and still have adequate funds for routine operation and maintenance.

TARIFF FORMULA

The principal behind the tariff formula developed by ITDG is to "arrive at realistic tariff rate and to be able to pay back the cost of hardware and software component within a specified period of time" Ariyabandu, 1995. The tariff formula adopted is as follows:

$$\text{Tariff Rate} = \underbrace{\frac{(C-S+D-V)F}{12 \times H}}_{\text{"P"}} + \underbrace{\frac{R}{12 \times H \times 4}}_{\text{"Q"}} + \underbrace{\frac{O-B-E}{12 \times H}}_{\text{"X"}} + \underbrace{\frac{W+I)i}{12 \times H}}_{\text{"Z"}}$$

(Please refer Annex I for details)

This formula considers replacement of the electro-mechanical component "R" in five years. The tariff thus calculated therefore is given in column number III in Table 3. The difference between actual average monthly tariff and tariff calculated according to the formula is given in column IV. Accordingly, two out of the 10 projects collect underrated tariff, which means the ECS will not be able to handle major maintenance or replacements at the end of the 5 year period.

Projects like Yagirala and Udagedara where the tariff difference is over Rs. 200/= are potentially more vulnerable than other projects when faced such situations in future. When number of beneficiaries are small vulnerability to face a crisis situation is grater due to two reason. Low monthly tariff collection as in Yagirala and Udagedara and inadequate membership to raise sufficient funds through spot collections. Under this situation the only hope for ECS's is to salvage a project is to look for outside donors. However, it is unlikely in future, that outside donors will be willing to fund maintenance as it is the responsibility of the beneficiaries.

Fortunately such a situation has so far not arisen in any of the projects. However, the oldest projects are now showing signs of fatigue. Recently the project in Kalugaldeniya had to replace the switch board which cost the ECS Rs. 10,000/=. Other older projects in Udagedara, Dolapalledola and Dolakadawatte are also anticipating major repairs soon. With low total membership of less than 20 (less than 10 in Udagedara and Dolakadawatte), the ECS will have a difficult time without outside donor support.

IMPACT OF TARIFF RATE ON ECSS

As stated earlier monthly tariff collection is expected to meet the routine O/M cost and any replacement in hardware. In one of the largest projects (25KW) inadequate ECS fund (collected through tariff) had led to a contravacy between two groups in the ECS. When one group wanted to spend ECS funds for replacing old electricity post (125 in number) at the cost of Rs. 125,000/=, the other group wants to save the money for a bigger contingency in the power house. The debate had been so intensive the present ECS committee made representation to ITDG Director to solve the problem for them. At the representation the ECS office bearers had even threatened to lock the power house and leave the key with ITDG. This issue has created so much of confusion among the ECS members, it had become impossible to conduct a ECS meeting without prolong arguments among the membership on the issue of electrical posts. When one analyses the reason behind this situation it is inadequate funds collected through tariff which had caused the problem. If a realistic tariff rate had been adopted, such problems which eventually lead to instability of the ECS and the project as a whole would not have occurred.

RATIONALE AND JUSTIFICATION OF VILLAGE DECISIONS

In most pre-feasibility socio-economic studies conducted on village hydro projects, majority of the beneficiaries say that they would agree with any tariff payment decided by the ECSs. The ECSs in turn, mostly attempt to adopt tariff rates used in other existing projects. However, beneficiaries have got the notion that standard should be Rs. 1/= per watt. Hence, Rs. 100/= per 100 watts is an accepted rate. The origin of this rate can be traced backed to small projects (less than 5 KW) initiated at the inception of the programme.

Given the rural economy in these villages, Rs. 100/= was thought to be an accepted amount a beneficiary can afford without having to be a burden on the household budget. The Rs. 100/= rate, over time came to be an accepted standard and ECSs in future projects adopted this rate as the maximum. Later, in larger projects of over 5 KW generated capacity, the ECS adopted the same rate as monthly tariff. The beneficiary idea of collecting only Rs. 100/= was to meet the routine O/M, since there had been no need for hardware replacement. However, as stated earlier signs of major replacements are becoming obvious in older projects. In most new projects and some of the old projects ECSs justify the present tariff as it had taken care of all the operation and maintenance needs. Another reason for ECSs to be satisfied with the present tariff rate is they (ECS) think that in the event of a major repair some external agency would support. More often than not ECS feel ITDG as the initiater of most village hydro projects should also help them in the event of a major repair. However, with the development of the programme, most ECSs are beginning to realize that post-project O/M and replacements are the responsibility of the Electricity Consumer Societies.

REFERENCES

1. ITDG, Intermediate Technology Development Group. 1995. Monitoring of Micro-Hydro Projects Synthesis Report. Colombo, Sri Lanka.

2. Lilly, D. 1996. Mini Hydro's Role in Sri Lanka Power Generation. ITDG, Colombo, Sri Lanka.

3. ITDG. Intermediate Technology Development Group. 1993. Setting-up and Sustaining Village Hydro Projects: Some Community Experience in Matara District. ITDG, Colombo, Sri Lanka.

TARIFF FORMULA

$$\text{Tariff Rate} = \underbrace{\frac{(C-S+D-V)F}{12 \times H}}_{"P"} + \underbrace{\frac{R}{12 \times H \times 4}}_{"Q"} + \underbrace{\frac{O-B-E}{12 \times H}}_{"X"} + \underbrace{\frac{(W+I)i}{12 \times H}}_{"Z"}$$

Tariff $= P + Q + X + Z$

Where :

P = Loan Repayment
Q = Replacement Fund
X = Operation and Maintenance Minus Revenue
Z = Welfare and Development Fund

C = Cost of Installation
S = Subsidy
L = Connection Charge
D = Cost of Distribution
V = Village Total Contribution (L x H)
F = Interest Factor
H = No. of Households/Consumers
R = Replacement Cost
Y = Years of Operation
O = Annual Operation, Maintenance and Management Costs
B = Net Annual Earnings from Commercial End Use of which no tariff is charged
W = Annual Allowance for Welfare
I = Annual Allowance for Income Creation

First International Conference on Renewable Energy—Small Hydro
3–7 February 1997, Hyderabad, India

MODERN CONTROL AND GOVERNING SYSTEMS FOR SMALL HYDRO PLANTS

Bernhard Oettli

Swiss Centre for Development Cooperative in Technology & Management (SKAT)
Vadianstr, U2 CH-9000, St. Gallen, Switzerland

ABSTRACT

Sophisticated electronic control- and governing equipment has become a world-wide standard for medium and large scale hydro plants. Although a wide range of options is now also available for the field of small hydro, the latest generation of such devices as digital turbine governors, electronic load controllers, PLC-type plant controllers, remote control systems and SCADA-type control- and monitoring systems has apparently not found it's way yet into Indian small hydro schemes.

The Hilly Hydro Project aims at promoting the introduction and spreading of such modern but proven, not necessarily more expensive control technology at a wider scale. This paper stresses the need for reliable, high performance control and governing equipment at competitive prices, with special emphasis given to a new class of so-called integrated digital controllers, providing features for both turbine governing as well as complete plant control.

1   Introduction - Categories of Control Systems

Control and governing systems are needed in order to keep certain variables of a system within defined limits and stable under varying operation conditions. For electrical power generation these are mainly, depending on the operation mode, the frequency, the voltage, the headpond water level and the power factor. In addtion, these variables are monitored and normally indicated through adequate instruments. If the control system is not able to maintain these crucial system variables within safe limits, a protection system has to properly disconnect the generating unit from the load and shut down the plant.

In other words, there are in fact a number of different functions to be performed to control a Small Hydro Plant (SHP). To make it even more complicated, some terms are not very well defined but can have different meanings. An overview of the main sub-systems and their dedicated functions is given below:

Usual Designation of Main System Function	Description / Main Purpose	Typical Examples
Control	can have three meanings: a) mostly used as general term for the overall control of the system, incl. protection, instrumentation, etc. b) often linked to the control of a certain level or certain procedures c) sometimes limited to the effective control function of specific variables	plant automation, automatic start-up & shut-down- or synchronisation-procedure water-level control voltage control
Governing	two meanings: a) would actually include all the governing functions within a SHP b) but mostly used in relation with turbine control/governing	water-level, voltage frequency, power factor turbine governor
Protection	for the protection of human beings and the equipment, nowadays mostly through protective relays	over-voltage, over-current earth-fault
Instrumentation	for the display of main variables, normally through indicating instruments	V-meter, A-meter, f-meter, kW- and kWh-meter
Data Acquisition / Monitoring	for the recording and/ or remote monitoring of crucial variables	water-level, kW, kWh min-max-values
Communication	for the communication (data-/ signal- transfer) between host/ control centre and remote (unattended) schemes	remote control & monitoring remote data acquisition

Although this paper includes at least some remarks on most of the above shown specialised plant- subsystems, the main focus, specially in the following chapter, will be given to the topic of Control and Governing in the sense b) (see explanations in the table above).

2   Dedicated Controllers

In view of the variety of control functions summarised in the table above it becomes evident that a range of dedicated controllers will be required to operate a SHP-scheme. The number and performance of the controllers is of course depending on the complexity of the set-up.

2.1   Basic Control Functions: Turbine- and Generator- Control

In the simple case of an attended SHP-scheme with a single unit, operating in isolated mode, at least

- the speed of the turbine/generator set (in this case the system frequency) and

- the voltage of the generator (the system voltage)

have to be maintained..

It might be a bit confusing that, depending on the mode of operation (isolated mode or grid connection), the turbine governor is not always used to control the speed but might also be used to control a water level or the share in kW in a multi-unit scheme. Similarly, the excitation control of the generator does not always affect the voltage only but can also serve to control the power factor or the share of reactive power if the unit is grid connected.

It is not the aim of this paper to go more into theoretical details at this point. Nevertheless, the diagram below gives an overview of the various principles to control turbine and generator and also indicates how such control functions can be implemented.

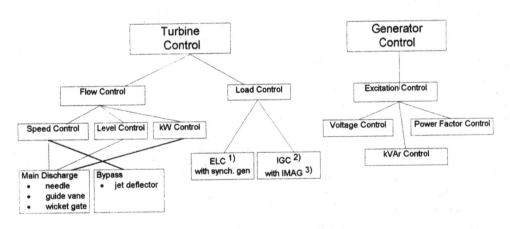

Explanations:
1) IMAG   Induction Motors used as Asynchronous Generators
2) IGC   Induction Generator Controller
3) ELC   Electronic Load Controller

This paper will not elaborate further on voltage-/ excitation- control, but it will provide a closer look at the technologies available for turbine- (speed-) governing in the following paragraphs.

2.1.1 Flow Controllers

As their name indicates, flow controllers are governors that control the discharge through a turbine by, depending on the turbine type, adjusting the wicket gate, guide vane or needle.

In their traditional form, flow control turbine governors are mechanical hydraulic devices. Though still quite popular in India, these conventional turbine governors are loosing ground rapidly. The main reasons are the complexity, the high precision skills required at the time of manufacture, the very demanding maintenance and the relative high investment cost. There is a clear tendency, therefore, to replace such governors by electro-mechanical or electro-hydraulic devices.

The application range of flow control governors is virtually unlimited but is nowadays, for cost reasons, in practise restricted to unit capacities above 100 kW.

2.1.2 Load Controllers

A completely different approach in order to control the speed of a turbine-/ generator- set is to adjust the total load (diverting excess power from the consumer to a dummy- or ballast load) to the turbine input power instead of matching the latter with the consumer load.

This is normally done by electronic devices, i.e. a complex system like a mechanical hydraulic flow governor, with probably more than 100 moving parts, hence subject to wear and tear, is replaced by a solid state, virtually maintenance-free device. Other major benefits are the much lower investment cost and the quick reaction time. The main disadvantages are the "waste of water" (hardly acceptable in case of storage schemes) and the higher load to the equipment, always running at "full load". Electronic load controllers therefore have a limited application range which is normally considered to be the micro range (up to 100 kW) but which can, under certain conditions be easily extended to 200 kW or even 400 kW.

Within this range, electronic load controllers have meanwhile been world-wide and well accepted. But, despite its habitual use in thousands of off-grid schemes in at least 30 countries, this technology appears to have hardly spread in India. Still, a number of Indian manufacturers have meanwhile developed indigenous prototypes or have even launched the regular production of electronic load controllers recently.

2.2 Enhanced Control: Plant Automation / Station Control

As soon as there are 2 or more units in a SHP-plant or in case the scheme is unattended, there is some obvious need for additional control functions.

2.2.1 Upper Level Control Functions

The main functions in order to enhance a control system can be designated as:

- plant automation
- data acquisition and -retrieval

Moreover, plants might be

- fully remote controlled

from a central control room, a bigger head scheme or from a special load dispatch centre. In such case additional subsystems for communication, remote control and - data acquisition and the like are required.

A schematical, though incomplete overview of some dedicated subsystems for more sophisticated plant control is given below.

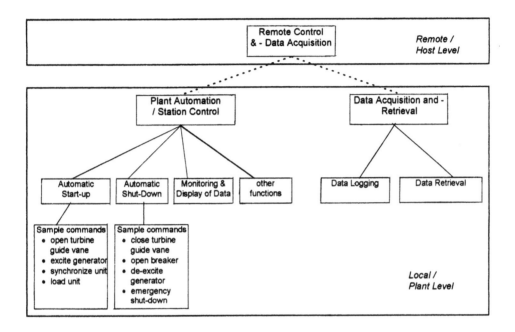

2.2.2 Common Technologies

For the upper level of control functions, there are typically three types of controllers:

PLC- (Programmable Logic Controller-) type Plant Controllers

These universal, software programmable controllers, normally of compact, modular, racked design, are most commonly used to incorporate plant control functions such as automatic start-up and shut-down and monitoring- and display-functions such as the monitoring and indication of bearing- or alternator winding temperatures which the PLC can directly convert into a command initiating e.g. an emergency shut-down. To enable communication with other plants or remote control stations, most of the PLC's meanwhile also have at least basic communication features incorporated.

However, PLCs are normally used for open-loop control functions and do not provide any closed-loop governing functions. Hence, PLC based plant controllers do not replace the basic turbine and generator control system but have to be considered as a supplementary system. Moreover, they can hardly serve for data acquisition, as they are normally not designed for mass data storage.

PC- based SCADA- Systems

Full-fledged **S**ystem **C**ontrol **A**nd **D**ata **A**cquisition units, nowadays usually based on a high performance PC, are to be positioned at the high end of plant automation, i.e. they incorporate most of the control, monitoring, data acquisition and communication functions.

On the other hand, they mostly are high priced too and there application domain thus lies typically above 1 MW.

Like PLC- based controllers, SCADA- systems do normally neither include the turbine governing- nor the voltage control function. This again has to be provided by the separate basic control system.

Data Loggers

At the lower end, specially in smaller off-grid schemes and for data acquisition functions only, a simple dedicated battery powered data logger, using replaceable chips or telemetric links, might be an appropriate technology choice. When combined with a basic turbine governor or load controller and a simple, purely hardware-type protection- and instrumentation system, such a system can fulfil all the needs e.g. of a stand alone attended scheme of a few hundred kW.

3   Optimised Independent Subsystems or an Optimal Cost-effective Integrated Control System ?

A common face of all the control systems that the preceding chapter dealt with is the fact that they are optimised to control a specific, technically fairly independent sub-system. From the point of view of the operator however, there is in many cases a strong dependency on a sole supplier (single source) as, most typical in the case of the turbine- and generator- controller, these are normally supplied by the manufacturer, i.e. the turbine governor, as a mechanical hydraulic device, remains in the domain of the turbine manufacturer, the automatic voltage regulator (AVR), as an electronic device, is integrated in the generator.

The fact that each such specific control device, developed and in operation since decades, has certainly reached a high level of perfection does however not necessarily mean that the complete control system is an optimal solution. This applies in particular with regard to

- the interaction / communication between the sub-systems
- the benefit / cost ratio of such a system, specially when considering control systems for schemes in the mini hydro range.

While, in case of medium and large, the cost of the control system, compared with the total investment, is not so significant, the share of the control system is much higher with small hydro plants. Moreover, it is often overlooked that, in order to effectively integrate optimal sub-systems into a good overall control system, competent plant- engineering is required wich in turn demands for qualified Know-How and a lot of experience. The risk is high that in such case, even though a technically fine solution might be found, the result is questionable as the costs are far too high.

The keyword is clearly optimisation: improve the performance of small hydro through the application of cost-competitive control-systems.

In view of these requirements, several engineering companies have recently started to develop a new class of control systems, which incorporate most of the features of the traditionally technically independent but probably single-sourced (e.g. turbine manufacturer), specialised sub-systems, and that are competitively priced. Such systems are often called **Integrated Controllers** and are normally based on latest PLC- or PC-technologies, using software as the main component to implement all the functions that are required for the control, governing, monitoring, etc. of a small hydro scheme.

Based on the latest experience with such an Integrated Digital Controller developed in Switzerland, the new approach and some specific features of this new class of control systems is outlined in the following chapter.

4 Integrated Digital Controllers (IDC) - a perspective

Since the early days of computers, control systems for hydro power plants were transferred to this new technology. Completely different approaches to system integration became possible and as soon as computer hardware was reliable and the performance sufficient this technology was state of the art. For SHP applications however, the adoption of integrated digital controllers proofed slower though it appears that time has now come for a breakthrough.

4.1 Main Characteristics

The main characteristics of an IDC are quickly named: all control functions are transferred to a software programme which is running on a computer. All tasks handled by conventional control devices can be performed but a completely new range of system optimisation can be tackled now, as all controllers can interact easily on a software base.

Block diagram of an IDC

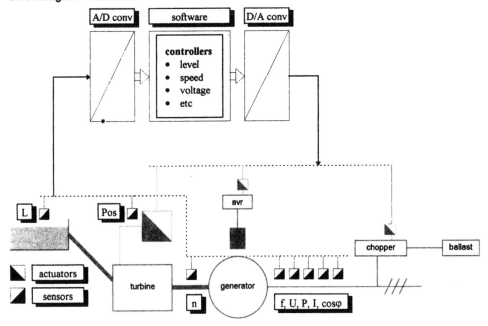

To transfer the former dedicated controllers functions into a computer, all signals from and to the SHP equipment, required for the closed loop control, have to be transformed into numbers. To achieve this a number of interfaces (in terms of hard- and software) between the equipment and the computer is required.

This is normally done in two steps:

- sensors read a physical parameter and convert it into an electrical signal (often standard signals 0-10V, 4-20mA or pulses) which is transferred to the computer
- Analogue to Digital (A/D) converters translate this electrical signals into numbers (digital values)

The opposite way

- Digital to Analogue (D/A) converters translate numbers into electrical signals
- actuators convert these electrical signals into physical actions.

Besides the pure control functions, all other systems (e.g. the protection system) can easily be implemented and integrated in the IDC. An overview on the standard main features and the additional potential of an IDC is given in the following two chapters.

4.2 Main Features

An IDC will normally come with a standard package of application software, with full featured control functions that can be modified, re-defined and expanded to cater the customer specific needs. **Standard features** will typically be:

Automatic Starting	either fully automatic, mainly with grid-connected schemes, or semi-automatic (i.e. automatic start after the manual operation of the main inlet-valve) in an isolated scheme
Speed Control	achieved through a mixture of load control (for fast reaction to minor changes in load and quick synchronisation) and flow control (electronic control signals fed to hydraulic servo or servo motor)
Synchronisation	for the automatic paralleling of two units or of a unit with the grid
Water Level Control	when operating on a grid
Normal Shut-Down/ Emergency Shut-D.	To run down a plant, be it in case of normal operation, an abnormal speed increase or a malfunctioning of a component or sub-system

Typical **additional features** are:

Load Balancing / kW-Control	to control of share of active power of two or more units operating in isolated mode
Reactive Power- (Cos φ) Control	to control the share of reactive power of a unit in a multiple unit- or grid operated scheme
Data Acquisition	for local- or remote- data acquisition and -retrieval
Display Functions	to display main operating parameters/ alarms on CRT-monitors or alpha-numeric displays
Remote Control	providing all the necessary communication-functions to control and monitor a plant from a central control room or a head scheme

4.3 Major Advantages of IDCs

The most important advantages like reduction of system integration cost and better overall control system optimisation have already been mentioned before. In addition, there are some technical and organisational benefits that should be named:

Communication	One of the main advantages of IDC is the possibility, at virtually no additional costs, to communicate all data, plant status, emergency conditions, etc. to other systems or the operator. The distance is almost irrelevant; via a modem & telephone line for instance, the IDC can communicate around the world. The operator can manipulate the plant with visualised commands and can be assisted with on line help and recommendations.
System Integration	the barrier between the different systems of an SHP (control, protection, communication) disappears as soon as all functions are integrated within a IDC
Data Acquisition / Reporting	operating & performance data of an SHP plant can easily be collected and for documentation be stored or printed.
Updating	as control algorithms are improved and new control techniques (for instance fuzzy logic) are available these can be downloaded to the IDC.
Adjustment / Adaptive Controllers	often the adjustment of the control parameters for a specific site needs professionals. Is the process on a IDC already much easier and consistent, the use of adaptive controllers, which are self adjusting, opens the possibility of self tuning and optimising SHP plants.

4.4 Concerns

The typical SHP plant today is kept as simple as possible. There are many reasons to avoid complex systems, the main arguments being:

- economic non-viability
- non-sustainability, non-transferable skills and Know-How
- maintenance- problems: black box units with no interference possibilities etc.

It should however been noticed that the same applies to most of the conventional control systems, too. The skills to maintain or even repair a mechanic- hydraulic turbine-governor or an electronic automatic voltage-regulator (AVR) properly has probably in few cases been successfully transferred from the manufacturer to the operator of a SHP-plant, but in most cases the parts will simply be replaced (AVR) or the supplier is quickly called in case there are serious problems.

Nevertheless, there are other concerns like:

- longevity of components
- rate of evolution of computer technology (today's generation outlived after 3 - 5 years)
- sensitivity to EMP (lightning)

4.5   Buala: A Demonstration Project

To highlight the usefulness of an IDC the relatively complex task which formed the basis of a project to be executed by MHPG-partners over the last two years is described below:

> The province capital on a South Pacific Island is provided with electricity from a 150 kW Diesel genset. Its operation is costly and fuel supply is precarious due to a difficult transport logistic.
>
> A possible site for an MHP scheme exists nearby, however, as the potential is not sufficient to simply replace the Diesel due to low water during dry spells. The MHP scheme has a design output of 150 kW (90 l/s, 240 m, availability ca. 75%). It includes a storage pond to supply water for evening peak loads or to accumulate water during the dry season in order to operate the MHP unit intermittently.
>
> The task at hand is to make best use of the available water resource to consequently minimise fuel consumption and CO_2 emission. As a solution to this management problem an IDC is proposed which, besides simply controlling the two generating units, also optimises the use of the water resource based on some predictions on the daily load curve and minimises the fuel consumption and operation time of the Diesel genset.
>
> The IDC acquires all necessary data of the scheme, calculates the remaining sto·ed energy, analysis trends of the inflow and the power consumption and deduces the optimal operation for the to generating units taking into account their best performance.

The scheme has been commissioned successfully in December 1996. All the above mentioned functions are effectively handled by the IDC supplied and programmed by a Swiss engineering company.

5   Conclusions for the Hilly Hydro Project

During the Technology Selection procedure, a separate activity block of the Hilly Hydro Project (HHP), optimal and cost-effective solutions have been discussed concerning all major components of a SHP. Apart from the civil works, where major attention was given to properly designed intakes and desilting arrangements, and the penstock, where the discussion focused on new material, the generating equipment was of course a main field of our studies and negotiations.

With regard to the control systems to be selected for and to be installed in the demonstration schemes, the local and international consultants agreed to vote in favour of an innovative approach, thereby promoting some latest designs and technologies and not to rely simply on standard, well known but obviously not very reliable control equipment (like the hydraulic turbine governors). As a result the following general recommendations were made:

Which Type of Controller?

Size of Plant	Mode of Operation	
	Isolated Mode	Grid Connection
up to 200 kW	ELC	ELC (IDC)
200 - 500 kW	ELC (IDC)	IDC
500 - 3000 kW	IDC	IDC (SCADA)

Explanations: ELC   Electronic Load Controller
                 IDC   Integrated Digital Controller
                 SCADA   Full Fledged Control and Data Acquisition System (PC- or PLC-based)

The final Technology Selection Matrix applied to the demonstration schemes, recently published in Issue 4 of SPLASH, the quarterly newsletter of the HHP, reflects these recommendations and contains in fact just these tree types of controllers.

INDEX OF CONTRIBUTORS